Genetic Engineering:
A Primer

Genetic Engineering: A Primer

Contributors

Takeshi Yamagami, Sonoko Ishino et al.

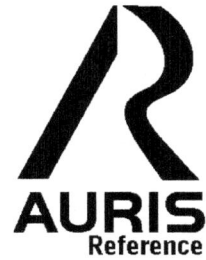

AURIS
Reference

www.aurisreference.com

Genetic Engineering: A Primer

Contributors: Takeshi Yamagami, Sonoko Ishino et al.

Published by Auris Reference Limited

www.aurisreference.com

United Kingdom

Genetic Engineering: A Primer

ISBN: 978-1-78154-953-7

British Library Cataloguing in Publication Data
A CIP record for this book is available from the British Library

Printed in the United Kingdom

Exclusively distributed by CBS Publishers & Distributors Pvt. Ltd.

Sales & Distribution Rights only for India, Pakistan, Bangladesh, Sri Lanka, Nepal and Bhutan.This book is not to be sold outside these territories.

Contents

List of Abbreviations

ADCC	Ab-dependent cell-mediated cytotoxicity
ABI	Association of British Insurers
CEX	cation-exchange chromatography
CAR	chimeric antigen receptor
CHO	Chinese hamster ovary
CFU	colony forming units
CSR	compartmentalized self-replication
CDC	complement-dependent cytotoxicity
EMSA	electrophoretic mobility-shift assay
EPC	Endothelial progenitor cells
EPCs	Endothelial progenitor cells
GFP	green fluorescent protein
HLA	human leukocyte antigen
IFO	Institute for fermentation, Osaka
LTR	long terminal repeat
NLSs	nuclear localization signals
PCR	polymerase chain reaction
PERV	porcine endogenous retrovirus
Q-PCR	quantitative-polymerase chain reaction
RNAi	RNA interference
SIN	self-inactivating
SNP	single nucleotide polymorphism
SB	Sleeping Beauty
SCNT	somatic cell nuclear transfer
TRE	tetracycline-responsive element
TF	tissue factor
Ti	tumor inducing
TIP	tumor inducing principle
ZFN	zinc finger nucleases

List of Contributors

Takeshi Yamagami
Protein Chemistry and Engineering, Department of Bioscience and Biotechnology, Graduate School of Bioresource and Bioenvironmental Sciences, Kyushu University, Fukuoka, Japan

Sonoko Ishino
Protein Chemistry and Engineering, Department of Bioscience and Biotechnology, Graduate School of Bioresource and Bioenvironmental Sciences, Kyushu University, Fukuoka, Japan

Yutaka Kawarabayasi
Protein Chemistry and Engineering, Department of Bioscience and Biotechnology, Graduate School of Bioresource and Bioenvironmental Sciences, Kyushu University, Fukuoka, Japan
Health Research Institute, National Institute of Advanced Industrial Science and Technology, Amagasaki, Japan

Yoshizumi Ishino
Protein Chemistry and Engineering, Department of Bioscience and Biotechnology, Graduate School of Bioresource and Bioenvironmental Sciences, Kyushu University, Fukuoka, Japan

Sonja Billerbeck
ETH Zürich, Department for Biosystems Science and Engineering (D-BSSE), Mattenstrasse 26, 4058 Basel, Switzerland

Sven Panke
ETH Zürich, Department for Biosystems Science and Engineering (D-BSSE), Mattenstrasse 26, 4058 Basel, Switzerland

Julio Perez-Marquez
Department of Biomedicine and Biotechnology, University of Alcalá de Henares, Spain

GuanghuaYang
Division of Gene Therapy and Hepatology, Center for Applied Medical Research (CIMA), University of Navarra, Pamplona, Spain

Research Center for Translational Medicine, East Hospital, School of [1]Medicine, Tongji University, Shanghai 200120, China

M. Gabriela Kramer
Division of Gene Therapy and Hepatology, Center for Applied Medical Research (CIMA), University of Navarra, Pamplona, Spain

Veronica Fernandez-Ruiz
Division of Gene Therapy and Hepatology, Center for Applied Medical Research (CIMA), University of Navarra, Pamplona, Spain

Milosz P. Kawa
Division of Gene Therapy and Hepatology, Center for Applied Medical Research (CIMA), University of Navarra, Pamplona, Spain

Xin Huang
Present address: Department of Biotechnology, Instituto de Higiene, Facultad de Medicina, Universidad de la República, Av. A. Navarro 3051, 11600 Montevideo, Uruguay

Zhongmin Liu
Research Center for Translational Medicine, East Hospital, School of [1]Medicine, Tongji University, Shanghai 200120, China

Jesus Prieto
Division of Gene Therapy and Hepatology, Center for Applied Medical Research (CIMA), University of Navarra, Pamplona, Spain

Cheng Qian
Magee-Womens Research Institute, University of Pittsburgh School of Medicine, Pittsburgh, PA 15213, USA

Mihaela Škulj
Sandoz Biopharmaceuticals, Mengeš, Lek Pharmacetucals d.d

Dejan Pezdirec
Sandoz Biopharmaceuticals, Mengeš, Lek Pharmacetucals d.d

Dominik Gaser
Sandoz Biopharmaceuticals, Mengeš, Lek Pharmacetucals d.d

Marko Kreft
Celica Biomedical Centre
Laboratory of Neuroendocrinology – Molecular Cell Physiology, Institute of
Pathophysiology, Faculty of Medicine, University of Ljubljana
Biotechnical Faculty, University of Ljubljana

Robert Zorec
Celica Biomedical Centre
Laboratory of Neuroendocrinology – Molecular Cell Physiology, Institute of
Pathophysiology, Faculty of Medicine, University of Ljubljana

Z Jin
Division of Pediatrics, Children's Cancer Hospital, The University of Texas
Graduate School of iomedical Sciences, The University of Texas MD Anderson Cancer Center, Houston, TX, USA
S Maiti
Division of Pediatrics, Children's Cancer Hospital, The University of Texas
Graduate School of iomedical Sciences, The University of Texas MD Anderson Cancer Center, Houston, TX, USA
H Huls
Division of Pediatrics, Children's Cancer Hospital, The University of Texas
Graduate School of iomedical Sciences, The University of Texas MD Anderson Cancer Center, Houston, TX, USA
H Singh
Division of Pediatrics, Children's Cancer Hospital, The University of Texas
Graduate School of iomedical Sciences, The University of Texas MD Anderson Cancer Center, Houston, TX, USA
S Olivares
Division of Pediatrics, Children's Cancer Hospital, The University of Texas
Graduate School of iomedical Sciences, The University of Texas MD Anderson Cancer Center, Houston, TX, USA
L Mátés
Max Delbrück Center for Molecular Medicine, Berlin, Germany
Z Izsvák
Max Delbrück Center for Molecular Medicine, Berlin, Germany
University of Debrecen, Debrecen, Hungary

Z Ivics
Max Delbrück Center for Molecular Medicine, Berlin, Germany
University of Debrecen, Debrecen, Hungary

D A Lee
Division of Pediatrics, Children's Cancer Hospital, The University of Texas

Graduate School of iomedical Sciences, The University of Texas MD Anderson Cancer Center, Houston, TX, USA

R E Champlin
Stem Cell Transplantation and Cellular Therapy, University of Texas MD Anderson Cancer Center, Houston, TX, USA

L J N Cooper
Division of Pediatrics, Children's Cancer Hospital, The University of Texas Graduate School of iomedical Sciences, The University of Texas MD Anderson Cancer Center, Houston, TX, USA

Béatrice Godard
INSERM SC11, Paris, France
Current address: Faculty of Law and Department of Social and Preventive Medicine, University of Montreal, Montreal, Canada.

Sandy Raeburn
Department of Medical Genetics, Nottingham, UK

Marcus Pembrey
Institute of Child Health, London, UK

Martin Bobrow
Department of Medical Genetics, University of Cambridge, UK

Peter Farndon
Department of Medical Genetics, Birmingham, UK

Ségolène Aymé
INSERM SC11, Paris, France

S Le Bas-Bernardet
INSERM, U643, Nantes, France; CHU Nantes, Institut de Transplantation et de Recherche en Transplantation, ITERT, Nantes, France; Université de Nantes, Faculté de Médecine, Nantes, France

I Anegon
INSERM, U643, Nantes, France; CHU Nantes, Institut de Transplantation et de Recherche en Transplantation, ITERT, Nantes, France; Université de Nantes, Faculté de Médecine, Nantes, France

G Blancho
Service de Néphrologie, Immunologie Clinique et Transplantation, CHU Nantes, Nantes, France

Eugene W. Nester
Department of Microbiology, University of Washington, Seattle, WA, USA

Elizabeth Kelly
Molecular Medicine Program, Mayo Clinic College of Medicine, Rochester, Minnesota, USA

Stephen J Russell
Molecular Medicine Program, Mayo Clinic College of Medicine, Rochester, Minnesota, USA

Joanna I Katashkina
Closed Joint-Stock Company "Ajinomoto-Genetika Research Institute"

Yoshihiko Hara
Fermentation and Biotechnology Laboratories, Ajinomoto Co, Inc

Lyubov I Golubeva
Closed Joint-Stock Company "Ajinomoto-Genetika Research Institute"
Closed Joint-Stock Company "Ajinomoto-Genetika Research Institute"

Irina G Andreeva
Closed Joint-Stock Company "Ajinomoto-Genetika Research Institute"

Tatiana M Kuvaeva
Closed Joint-Stock Company "Ajinomoto-Genetika Research Institute"

Sergey V Mashko
Closed Joint-Stock Company "Ajinomoto-Genetika Research Institute"

David W Ussery
Center for Biological Sequence Analysis, Department of Systems Biology, The Technical University of Denmark, Kgs. Lyngby

Preface

Genetic engineering is the direct manipulation of an organism's genome using biotechnology. It is a set of technologies used to change the genetic makeup of cells, including the transfer of genes within and across species boundaries to produce improved or novel organisms. The text *Genetic Engineering: A Primer* provides an excellent introduction to the area of genetic engineering of plants and animals. The goal of first chapter is to create PCR enzymes with superior performance, as compared to that of WT Taq polymerase. A genetic replacement system for selection-based engineering of essential proteins has been presented in second chapter. The utility of designing homologous primers for the genetic analysis based on the PCR has been discussed in third chapter. Fourth chapter focuses on development of endothelialspecific single inducible lentiviral vectors for genetic engineering of endothelial progenitor cells. In fifth chapter, the expression of PAM, and consequently the C-terminal amidation of recombinant mAbs, has been reduced by two approaches: gene manipulation using RNA interference (RNAi) and zinc finger nucleases (ZFN). In sixth chapter, we report a comparison of SB100X, a newly developed hyperactive SB transposase, to a previous generation SB11 transposase to achieve stable expression of a CD19-specific chimeric antigen receptor (CAR3) in primary human T cells. Seventh chapter examines the professional and scientific views on the social, ethical and legal issues that impact on genetic information and testing in insurance and employment in Europe. In eighth chapter, we report on the genetic modification of pigs to reduce porcine endogenous retrovirus infection risk in the xenogeneic context. In ninth chapter, we discuss on agrobacterium, an agent causing the plant tumor. The history of oncolytic viruses has been revealed in tenth chapter. The goal of eleventh chapter is to widen use of the λ red-recombineering technology to *P. ananatis*, a bacterium of interest in the field of metabolic engineering. Last chapter focuses on natural genetic engineering.

Chapter 1

MUTANT TAQ DNA POLYMERASES WITH IMPROVED ELONGATION ABILITY AS A USEFUL REAGENT FOR GENETIC ENGINEERING

Takeshi Yamagami[1], Sonoko Ishino[1], Yutaka Kawarabayasi[1,2] and Yoshizumi Ishino[1]

[1]Protein Chemistry and Engineering, Department of Bioscience and Biotechnology, Graduate School of Bioresource and Bioenvironmental Sciences, Kyushu University, Fukuoka, Japan

[2]Health Research Institute, National Institute of Advanced Industrial Science and Technology, Amagasaki, Japan

ABSTRACT

DNA polymerases are widely used for DNA manipulation in vitro, including DNA cloning, sequencing, DNA labeling, mutagenesis, and other experiments. Thermostable DNA polymerases are especially useful and became quite valuable after the development of PCR technology. A DNA polymerase from Thermus aquaticus (Taq polymerase) is the most famous DNA polymerase as a PCR enzyme, and has been widely used all over the world. In this study, the gene fragments of the family A DNA polymerases were amplified by PCR from the DNAs from microorganisms within environmental soil samples, using a primer set for the two conserved regions. The corresponding region of the pol gene for Taq polymerase was substituted with the amplified gene fragments, and various chimeric DNA polymerases were prepared. Based on the properties of these chimeric enzymes and their sequences, two residues, E742 and A743, in Taq polymerase were found to be critical for its elongation ability. Taq polymerases with mutations at 742 and 743 actually showed higher DNA affinity and faster primer extension ability. These factors also affected the PCR performance of the DNA polymerase, and improved PCR results were observed with the mutant Taq polymerase.

INTRODUCTION

In addition to their fundamental roles in maintaining genome integrity during replication and repair, DNA polymerases are widely used for genetic engineering techniques, including DNA cloning, dideoxy-sequencing, DNA labeling, mutagenesis, and other in vitro DNA manipulations. Among them, thermostable DNA polymerases are particularly useful for PCR and cycle-sequencing (Perler et al., 1996; Ishino and Ishino, 2013; Terpe, 2013).

The fundamental ability to synthesize a deoxyribonucleotide chain is conserved in relation to the structural conservation of the DNA polymerases. However, the more specific properties for this catalysis, including processivity, synthesis accuracy, and substrate nucleotide selectivity, differ among the enzymes. These factors should be considered when evaluating a DNA polymerase as an enzyme for genetic engineering (Ishino and Ishino, 2014). An enzyme possessing faster extension with better accuracy and higher efficiency is more preferable. In addition to these catalytic properties, thermostability is necessary for practical PCR. DNA polymerases are now classified into seven families, based on the amino acid sequences (Braithwaite and Ito, 1993; Ishino and Cann, 1998; Cann and Ishino, 1999; Ohmori et al., 2001; Lipps et al., 2003). The enzymes within the same family have basically similar properties. Commercial genetic engineering reagents have originated only from families A and B to date. The family A enzymes are used for dideoxy-sequencing, and the family A and B enzymes are used for PCR. None of the DNA polymerases from the other families are suitable for general use in genetic engineering experiments.

The 3'-5' exonuclease activity, which contributes to the proofreading of DNA strand synthesis, is generally associated with the family B enzymes, but not with the family A enzymes, although some family A enzymes have a weak 3'-5' exonuclease activity (Joyce and Steitz, 1994; Villbrandt et al., 2000). Based on these differences, family A is advantageous for the efficient amplification of a long DNA region, and family B is generally more suitable for the precise amplification of a shorter region by PCR (Eckert and Kunkel, 1991). Researchers in this field have been making continuous efforts toward longer extension and better accuracy in PCR, and have succeeded in developing practical and reliable PCR methods. One notable example is the development of LA (long and accurate)-PCR, which is performed with a mixture of two DNA polymerases, one each from family A and family B (Barns, 1994). Further improvements of PCR have included the identification of a processive enzyme (Takagi et al., 1997) and the modifications within family B DNA polymerases that confer higher accuracy (Wang et al., 2004; Ishino et al., 2012).

Protein engineering techniques, using site-specific or random mutagenesis, are powerful ways to create mutant enzymes from the known DNA polymerases. Several useful enzymes were successfully produced by these procedures. A cold-sensitive mutant of DNA polymerase fromThermus aquaticus (Taq polymerase) was developed with markedly reduced activity at 37°C, as compared with the wild type (WT) enzyme (Kermekchiev et al., 2003). This mutant may be applicable to hot start PCR. Another example is a mutant Taq polymerase with enhanced resistance to various inhibitors of PCR reactions, including whole blood, plasma, hemoglobin, lactoferrin, serum IgG, soil extracts, and humic acid (Kermekchiev et al., 2009).

The molecular breeding ofThermus DNA polymerases by a direct evolution technique (Brakmann, 2005; Henry and Romesberg, 2005; Holmberg et al., 2005; Ong et al., 2006), compartmentalized self-replication (CSR) (Ghadessy et al., 2001), also generated a PCR enzyme with striking resistance to a wide range of inhibitors (Baar et al., 2011). Furthermore, enzymes with a broad substrate specificity spectrum were also obtained by the CSR technique (Ghadessy et al., 2004; d›Abbadie et al., 2007), and are thus useful for the amplification of ancient DNA containing numerous lesions. Mutational studies in the O-helix of Taq polymerase produced enzymes with reduced fidelity (Suzuki et al., 1997,2000; Tosaka et al., 2001), which may be useful for error-prone PCR. These studies have contributed to the elucidation of the detailed structure-function relationships of DNA polymerases, as well as to the creation of novel enzymes with different substrate specificities, stabilities, and activities from those of their naturally evolved counterparts.

In addition to the engineering of characterized enzymes to convert PCR performance, the screening for a suitable DNA polymerase activity from known organisms is the most conventional way to discover useful enzymes. However, the culturable organisms are limited, and large-scale cultivation is needed to purify an enzyme to homogeneity for precise characterization. In this study, we analyzed the DNAs from microorganisms within various soil samples obtained from a hot spring area, and compared the sequences of a region within the pol genes included in the environmental DNAs.

We then predicted the amino acid residues that are critical for the primer extension reaction of Taq polymerase, by constructing numerous chimeric Taq polymerases including the pol gene fragments from the various environmental DNAs. A mutant Taq polymerase with the E742 and A743 substitutions possessed more efficient DNA strand synthesis ability and better PCR performance. This polymerase will contribute to the development of high-speed PCR with the standard PCR conditions optimized for Taq polymerase.

MATERIALS AND METHODS

Enzymes and Substrates

Enzymes for in vitro DNA manipulation and oligonucleotides were purchased from New England Biolabs (Ipswich, MA, USA) and Sigma Aldrich (St. Louis, MO, USA), respectively. The [methyl-^{3}H]TTP was purchased from Amersham (Buckinghamshire, UK) and the [γ^{32}-P]ATP was purchased from NEN Life Science Products (Boston, MA, USA).

DNA Extraction

Extraction of DNA from the environmental specimens was performed with an UltraClean Soil DNA Isolation Kit (MO BIO, San Diego, CA, USA), according to the manufacturer's instructions. The extracted DNAs were assessed by agarose gel electrophoresis and were quantified by spectrophotometrical measurement.

Construction of the Expression Plasmid for Chimeric Taq Polymerases

The expression plasmid for Taq polymerase, pTV-Taq, which contains the entire region of the structural gene encoding Taq polymerase in the pTV118N vector (Takara Bio, Shiga, Japan), was constructed exactly as described (Ishino et al., 1994). The gene in the pTV118N vector was expressed under control of the lac promoter and SD sequence. The pTV-Taq plasmid was subjected to site-specific mutagenesis, using QuikChange™ kit (Agilent Technologies, Santa Clara, CA, USA) to introduce BlpI (GCTNAGC) and BglII (AGATCT) restriction sites into the positions corresponding to the 5'- and 3'-termini, respectively, of the substitution region within the Taq polgene. The insertion of a BglII site leads to the mutations of amino acids, Leu787Ile and Val788Leu. The resultant plasmid was named pTV-Taq›, and it was used for the expression of chimeric Taq polymerases, produced by the direct substitution of the BlpI–BglII fragment with the pol gene fragments from the environmental DNAs, as described in detail below.

Amplification of the pol Gene Fragments from Metagenomic DNA

A region of the pol genes was amplified directly from the environmental DNA by PCR, using a primer set with the sequences 5'-dCGCAGGC TAAGCAGCTCC**GAYCCHAACYTSCARAAYATHCC**-3' and 5'-dGAG**YAAGATCT**CRTCGTGNACYTG-3', which correspond to the degenerate codons for DPNLQNIP (forward) and QVHDEIL (reverse), respectively (Y indicates C and T; H indicates A, C, and T; S indicates C and G; R indicates A and G; N indicates A, G, C, and T). The above two regions, which are conserved in the family A DNA polymerases, were successfully used to make a mixed primer set for PCR (Uemori et al., 1993). The nucleotide sequences corresponding to the conserved regions are shown in boldface, and the restriction endonuclease recognition sequences are underlined in the above primers. PCR was performed in a 50 µl reaction, containing 10 ng DNA, 25 pmol of each primer, 0.2 mM dNTP, and 1 unit of PfuUltra DNA polymerase (Stratagene). After an incubation of the mixture without the enzyme for 3 min at 95°C, thirty-cycles of PCR, with a temperature profile of 30 s at 95°C, 30 s at 55°C, and 1 min at 72°C, were performed. The reaction mixtures were electrophoresed in a 1% agarose gel, and the amplified fragments were visualized by ethidium bromide staining.

Analysis of the Amplified Gene Fragments

DNA fragments with a 600 bp size, amplified from the environmental DNA by PCR, were excised from the agarose gel, digested with the BlpI and BglII restriction endonucleases, and ligated into the pTV-Taq' plasmid predigested with the BlpI and BglII restriction endonucleases. The ligation mixtures were introduced into E. coli JM109 cells (TaKaRa Bio.), and 20 clones were picked independently from the transformants for each amplification. Plasmid DNAs were extracted from these clones, and the nucleotide sequences of the DNA inserts were determined by CEQ2000XL DNA analysis system (Beckman Coulter, USA).

Construction of Mutant Taq Polymerases

The pTV-Taq plasmid was subjected to site-specific mutagenesis to introduce mutations into positions 742 and 743 of Taq polymerase. The sequences of the 14 primers used for mutagenesis are shown in Table 1. The PCR reaction mixture contained 20 ng template DNA, 1× PCR buffer for KOD-Plus-Neo, 1.5 mM Mg_2SO_4, 0.2 mM of each dNTP, 0.3 µM of each primer, and 0.5 unit KOD-Plus-Neo DNA polymerase (TOYOBO, Osaka, Japan) in a final volume of 20 µl. The mixture was heated at 95°C for 30 s and then subjected to thermal cycling (14 cycles of 95°C for 30 s, 55°C for 1 min, and 68°C for 8 min). The PCR product was treated with DpnI at 37°C for 1 h, and introduced into E. coli JM109 cells. For each mutation, the polymerase gene was fully sequenced to ensure that the mutation of interest was present and that no additional mutation was introduced by the PCR.

Table 1: Oligonucletides used to introduce mutations into positions 742 and 743 of Taq polymerase

Name of primer	Sequence (5' to 3')
TaqRR-F	CGGGTGAAGAGCGTGCGCCGCCGCGCCGAGCGCATGGCC
TaqRR-R	GGCCATGCGCTCGGCGCGGCGGCGCACGCTCTTCACCCG
TaqRA-F	CGGGTGAAGAGCGTGCGCCGCGCGGCCGAGCGCATGGCC
TaqRA-R	GGCCATGCGCTCGGCCGCGCGGCGCACGCTCTTCACCCG
TaqAA-F	CGGGTGAAGAGCGTGCGCGCGGCGGCCGAGCGCATGGCC
TaqAA-R	GGCCATGCGCTCGGCCGCCGCGCGCACGCTCTTCACCCG
TaqER-F	CGGGTGAAGAGCGTGCGCGAGCGCGCCGAGCGCATGGCC
TaqER-R	GGCCATGCGCTCGGCGCGCTCGCGCACGCTCTTCACCCG
TaqAR-F	CGGGTGAAGAGCGTGCGCGCGCGCGCCGAGCGCATGGCC
TaqAR-R	GGCCATGCGCTCGGCGCGCGCGCGCACGCTCTTCACCCG
TaqRK-F	CGGGTGAAGAGCGTGCGCCGCAAAGCCGAGCGCATGGCC
TaqRK-R	GGCCATGCGCTCGGCTTTGCGGCGCACGCTCTTCACCCG
TaqKR-F	CGGGTGAAGAGCGTGCGCAAACGCGCCGAGCGCATGGCC
TaqKR-R	GGCCATGCGCTCGGCGCGTTTGCGCACGCTCTTCACCCG
TaqKK-F	CGGGTGAAGAGCGTGCGCAAAAAAGCCGAGCGCATGGCC
TaqKK-R	GGCCATGCGCTCGGCTTTTTTGCGCACGCTCTTCACCCG
TaqQY-F	CGGGTGAAGAGCGTGCGCCAGTATGCCGAGCGCATGGCC
TaqQY-R	GGCCATGCGCTCGGCATACTGGCGCACGCTCTTCACCCG
TaqAH-F	CGGGTGAAGAGCGTGCGCGCGCATGCCGAGCGCATGGCC
TaqAH-R	GGCCATGCGCTCGGCATGCGCGCGCACGCTCTTCACCCG
TaqEH-F	CGGGTGAAGAGCGTGCGCGAGCATGCCGAGCGCATGGCC
TaqEH-R	GGCCATGCGCTCGGCATGCTCGCGCACGCTCTTCACCCG
TaqHA-F	CGGGTGAAGAGCGTGCGCCATGCGGCCGAGCGCATGGCC
TaqHA-R	GGCCATGCGCTCGGCCGCATGGCGCACGCTCTTCACCCG
TaqHH-F	CGGGTGAAGAGCGTGCGCCATCATGCCGAGCGCATGGCC
TaqHH-R	GGCCATGCGCTCGGCATGATGGCGCACGCTCTTCACCCG
TaqHK-F	CGGGTGAAGAGCGTGCGCCATAAAGCCGAGCGCATGGCC
TaqHK-R	GGCCATGCGCTCGGCTTTATGGCGCACGCTCTTCACCCG

Purification of Wild Type and Mutant DNA Polymerases

E. coli JM109 cells carrying the expression plasmid were grown at 37°C, in 1 L of LB medium containing 100 µg/ml ampicillin. The cells were cultured to an A_{600} of 0.2–0.3, and then the expression of the pol gene was induced by further cultivation for 3 h in the presence of 1 mM isopropyl-β-D-thiogalactopyranoside (IPTG). The cells were harvested and disrupted by sonication in the lysis solution (50 mM Tris-HCl, pH 8.0, 1 mM DTT, 1 mM EDTA, 1 mM PMSF). For the preparation of the WT and mutant Taq polymerases, the soluble cell extract, obtained by centrifugation at 12,000 × g for 20 min, was heated at 75°C for 30 min. The heat-stable fraction was obtained by centrifugation, and was treated with 0.15% (w/v) polyethyleneimine in the presence of 1 M NaCl, to remove the nucleic acids. The soluble proteins were precipitated by 80%-saturated ammonium sulfate. The precipitate was resuspended in the buffer [50 mM Tris-HCl, pH 8.0, 0.5 M $(NH_4)_2SO_4$], and was subjected to chromatography on a hydrophobic column (HiTrap Phenyl HP 5 ml, GE Healthcare). The column was washed with 50 mM Tris-HCl, pH 8.0, and the Taq polymerase was eluted with deionized water. An equal volume of 100 mM Tris-HCl, pH 8.0, was added to this fraction, and it was then subjected to affinity chromatography (HiTrap Heparin HP 5 ml, GE Healthcare) with a gradient of 0–2 M NaCl. The proteins that eluted at 0.8 M NaCl were stored at −25°C as the final sample, in 20 mM Tris-HCl, pH 8.0, 100 mM KCl, 0.1 mM EDTA, 1 mM DTT, 0.5% NP-40, 0.5% Tween 20, and 50% (w/v) glycerol.

Measurement of Nucleotide Incorporation Activity

The DNA polymerizing activity was assayed by measuring the incorporation of [methyl-^{3}H]TTP into the acid insoluble materials, basically as described previously (Uemori et al., 1995). To a 50 μl solution, containing 20 mM Tris-HCl, pH 8.8, 5 mM $MgCl_2$, 14 mM 2-mecaptoethanol, 0.2 mM each dATP, dGTP, dCTP, and dTTP, 400 nM [methyl-^{3}H]dTTP, and 20 μg of activated salmon sperm DNA, a constant amount of the enzyme fraction was added, and the reaction was incubated at 74°C for 2.5, 5, and 10 min. After the reaction, a 10 μl portion of each reaction mixture was spotted onto DE81 filters (GE Healthcare Japan, Tokyo, Japan). The filters were washed three times with a 5% Na_2HPO_4 solution, and the radioactivity incorporated into the DNA strands was counted by a scintillation counter. One unit of activity is defined as the amount of enzyme catalyzing the incorporation of 10 nmol of dNTP into DNA per 30 min at 74°C, and the specific activity was calculated as the units per one mg of protein (units/mg) for each DNA polymerase.

Primer Extension Activity

The primer extension ability was investigated by using M13 single-stranded DNA (ssDNA) annealed with a ^{32}P-labeled DNA, as described previously (Cann et al., 1999). M13 ssDNA (0.05 pmol), annealed with a 55 nucleotide long primer (5′-dTCGTAATCATGGTCATAGCTGTTTCCTGTGTGAAATT-GTTATCCGCTCACAATTC-3′), was mixed with each DNA polymerase (0.05 unit) and dNTP (0.2 mM) in a 20 μl solution, containing 20 mM Tris-HCl, pH 8.8, 5 mM $MgCl_2$, and 14 mM 2-mercaptoethanol, and incubated 74°C for 5 min. The reaction mixtures were analyzed by alkaline agarose gel electrophoresis, and the sizes of the products were visualized by autoradiography.

Electrophoretic Mobility-Shift Assay

The electrophoretic mobility-shift assay (EMSA) was performed as described previously (Komori and Ishino, 2000), to measure the DNA binding ability of the DNA polymerases. The ^{32}P-labeled 27mer oligonucleotide (5'-dAGC-TATGACCATGATTACGAATTGCTT-3') was annealed with the 49mer oligo-nucleotide (5'-dAGCTACCATGCCTGCACGAATTAAGCAATTCGTAAT-CATGGTCATAGCT-3') and was used as a DNA substrate. The radiolabeled DNA (3 nM) was mixed with DNA polymerase proteins (0.6–400 nM) in a 20 μl solution, containing 20 mM Tris-HCl, pH 8.8, 10 mM NaCl, 5 mM MgCl$_2$, 14 mM 2-mercaptoethanol, 0.1 mg/ml BSA, and 5%(w/v) glycerol, and incubated at 40°C for 5 min. The DNA-enzyme mixtures were fractionated by 1% agarose gel electrophoresis. The autoradiograms were scanned and the band intensities were quantified using an image analyzer (FLA5000; Fuji Film, Tokyo, Japan). The fraction of bound DNA in each lane was calculated to be: fraction bound DNA = 1 − fraction unbound DNA. The quantitated data for binding (association) was plotted vs. enzyme concentrations. The apparent Kd was determined to be the protein concentration at which the degree of binding equals 0.5.

RESULTS

Amplification of Family a DNA Polymerase Gene

We collected soil samples from various hot spring areas in Japan, and isolated the DNAs within these samples. Using these DNAs as templates, a region of the pol genes encoding family A DNA polymerase-like sequences was amplified. We previously reported the amplification of the pol gene encoding a family A DNA polymerase by a set of mixed primers based on the two conserved sequence motifs (Uemori et al., 1993). The mixed primer set worked for the specific amplification of the pol gene fragments from the hot spring samples in this study. As shown in Figure 1 (a part of the experiments is shown), a 600 bp DNA fragment was amplified as a single band from several samples. We tested 384 different samples, and detected the amplified DNA in 37 samples, obtained at Onikobe, Hachimantai, Nasu, Kirishima, and Beppu, as shown in Table 2. These locations had various environmental conditions, including pH values of 1~7 and temperatures of mostly 70–100°C, and thus were expected to be inhabited by microorganisms with highly diverse genetic resources. The efficiency of DNA isolation was not addressed in this experiment, and thus it is possible that a sufficient amount of DNA was not present in some samples. Therefore, the result that only about 10% of the samples provided target gene amplification is not directly related to the presence or absence of microorganisms in the samples. The amplified gene fragments were excised from the gel, and were cloned into a plasmid vector. Twenty colonies were isolated independently from each cloning of the amplified DNA, and a total of 740 (20 × 37 amplifications) plasmids were isolated.

Table 2: Summary of metagenomic analyses

No.	Sampling place (area)	Temperatue (°C)	pH	Number of clones	Number of different clone
1	Koyasukyo (Tohoku)	98	5	20	5
2	Koyasukyo (Tohoku)	70	7	20	12
3	Koyasukyo (Tohoku)	82	6.5	20	6
4	Koyasukyo (Tohoku)	73	–	20	16
5	Koyasukyo (Tohoku)	74	6.7	20	2
6	Koyasukyo (Tohoku)	35	–	20	10
7	Koyasukyo (Tohoku)	82	7	20	13
8	Koyasukyo (Tohoku)	50	8	20	12
9	Koyasukyo (Tohoku)	57	7	20	11
10	Koyasukyo (Tohoku)	96	4	20	4
11	Onikobe (Tohoku)	94	7	20	7
12	Onikobe (Tohoku)	95	7	20	2
13	Onikobe (Tohoku)	98	8	20	2
14	Onikobe (Tohoku)	89	7	20	9
15	Onikobe (Tohoku)	96	7	20	6
16	Onikobe (Tohoku)	98	8	20	5
17	Onikobe (Tohoku)	56	7	20	9
18	Onikobe (Tohoku)	93	1	20	6
19	Onikobe (Tohoku)	50	2	20	5
20	Onikobe (Tohoku)	85	7	20	4
21	Onikobe (Tohoku)	35	1	20	8
22	Beppu (Kyushu)	96	7	20	2
23	Beppu (Kyushu)	65	6	20	7
24	Beppu (Kyushu)	44	6	20	6
25	Beppu (Kyushu)	77	6.5	20	3
26	Beppu (Kyushu)	97	5	20	6
27	Beppu (Kyushu)	70	6	20	8
28	Beppu (Kyushu)	78	7	20	5
29	Beppu (Kyushu)	–	–	20	10
30	Kirishima (Kyushu)	75	3	20	5
31	Kirishima (Kyushu)	75	5	20	9
32	Kirishima (Kyushu)	–	5	20	6
33	Kirishima (Kyushu)	–	–	20	12
34	Ibusuki (Kyushu)	94	6	20	3
35	Nasu (Kanto)	65	6.5	20	5
36	Nasu (Kanto)	55	7	20	1
37	Nasu (Kanto)	55	6.5	20	8

"–" indicates "not analyzed."

These cloned DNA fragments were subjected to sequencing, and the different sequences were counted (Table 2). In total, we obtained 250 different sequences, which were not present in the public databases. This result suggested that there are still many uncharacterized DNA polymerases in the soil samples, as expected.

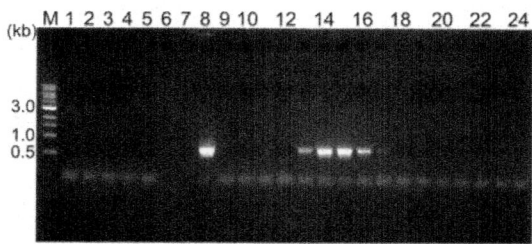

Figure 1: Amplification of a region within the family A DNA polymerase gene from the environmental DNA. PCR reaction mixtures were fractionated by 1% agarose gel electrophoresis, and the DNA bands were visualized by ethidium bromide staining. The lane numbers represent the serial numbers of the samples obtained from the hot spring areas in the Tohoku and Kyushu districts.

Preparation of the Chimeric Taq Polymerases

The amplified gene fragments encode the region in the family A DNA polymerase that is important for the nucleotide connecting reaction. To investigate the structure and function relationships of the family A DNA polymerases, we substituted the corresponding region of Taq polymerase gene with the amplified gene fragments in vitro. To construct the expression plasmids for the chimeric Taq polymerases systematically, restriction sites were created at appropriate sites in the Taq polstructural gene (Figure 2). For this purpose, the BlpI and BglII recognition sequences were suitable, although the substitutions of two amino acids, Leu797Val and Ile798Leu, could not be avoided by the introduction of BglII at the reverse priming site. The mutant Taq polymerase (L797V/I798L) was purified, and we confirmed that its fundamental properties were not affected. Therefore, the mutant Taq, designated as Taq' polymerase, was considered to be equivalent to the WT Taq polymerase. PCR primers containing the recognition sequences for BlpI (forward primer) and BglII (reverse primer) at each 5'-terminus were synthesized and each cloned DNA was re-amplified, and thus 250 chimeric genes were constructed in the Taq polymerase expression plasmid. The total cell extracts of E. coli producing recombinant chimeric Taq polymerases were treated at 75°C for 30 min, and the supernatants were assayed to measure the nucleotide incorporation activity.

About half of the chimeric enzymes were inactivated by the heat treatment. The thermostable chimeric Taq polymerases were further purified to apparent homogeneity by the procedure described in the Materials and Methods, and the specific activity (units/mg protein) was measured by the standard DNA polymerase assay, using activated DNA. Furthermore, the primer extension ability was evaluated for each enzyme, using a constant amount (unit). These results are summarized in Table3, which shows only the chimeric Taq polymerases possessing extension abilities better than 5 kb per 5 min-reaction. As shown in the Table 3, 13 enzymes were superior to WT Taq polymerase. However, the thermostabilities of these high-speed DNA polymerases were not sufficient for PCR applications (data not shown).

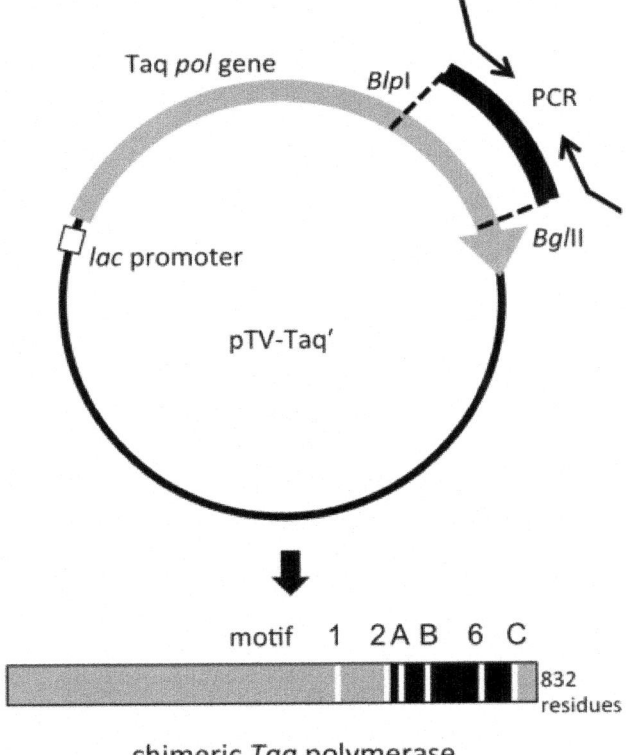

Figure 2: Schematic diagram of the construction of the expression plasmid for chimeric Taq **polymerase**. The recognition sites for BlpI (GCTNAGC) and BglII (AGATCT) were created in the Taq pol structural gene (gray arrow). The introduced gene fragments (black) also have the recognition sequences for these two enzymes from the PCR primers, and the substitution of the DNA fragments can be performed directly on

this plasmid (upper panel). The produced chimeric protein is shown by a bar (lower panel). The replaced region was indicated in black. The motifs conserved in family A DNA polymerases are indicated by white lines (Loh and Loeb, 2005).

Table 3: Properties of the chimeric Taq polymerase

Name	U/mg ($\times 10^5$)	Extention kb/5 min	Kd for DNA (nM)	Sampling pace	
				Temperature (°C)	pH
TaqWT	5.1	0.7	400	–	–
8-1	1.2	5.8	10	50	8
7-1	1.2	5.8	10	82	7
29-1	0.94	6.5	10	–	–
8-2	3.3	6.0	20	50	8
7-2	2.5	5.9	5	82	7
1-1	1.1	5.8	6	98	5
30-1	1.9	6.5	8	75	3
1-2	2.1	6.4	5	98	5
10-1	1.8	6.0	4	96	4
12-1	0.73	6.4	7	95	7
4-1	0.58	6.8	4	73	–
4-2	0.51	7.5	4	73	–
7-3	0.48	7.5	4	82	7

– indicate "not analyzed."

Sequence Comparison of Chimeric Taq Polymerases and Construction of the Mutant Taq Polymerases

The amino acid sequences of the chimeric Taq polymerases with extension rates greater than 1 kb/min (in the condition of 0.0025 unit/µl) are aligned in Figure 3. In this alignment, we focused on the region from amino acids 730 to 745 in the Taq polymerase. One distinct feature is that 3 genes and 4 genes have insertions of 9 amino acids and 3 amino acids, respectively, as compared with WT Taq polymerase. The other characteristic feature is that continuous stretches of basic amino acids were found at residues 741–743 in the chimeric Taq polymerases. The WT Taq polymerase has Glu and Ala at 742 and 743, but many chimeric enzymes showing faster extension have Arg at both positions. The crystal structure of the large fragment of Taq polymerase (Klentaq)-DNA complex revealed that Glu742 directly interacts with the template DNA in the closed conformation, but not in the open conformation (Li et al., 1998). As shown in the right panel of Figure 4, the residues Glu742 and Ala743 (magenta) are located in the finger subdomain and face to the template DNA (blue). The basic amino acid cluster in the chimeric Taq polymerases is supposed to interact with the template DNA. We focused on this finding, and made a series of mutant polymerases by substitutions at positions 742 and 743 in WT Taq polymerase to change the affinity of the enzymes with DNA. The names of the mutant enzymes are as follows: RR (E742R/A743R), RA (E742R), AA (E742A), ER (A743R), AR (E742A/A743R), RK (E742R/A743K), KR (E742K/A743R), KK (E742K/A743K), QY (E742Q/A743Y), AH (E742A/A743H), EH (A743H), HA (E742H), HH (E742H/A743H), and HK (E742H/A743K). Fourteen mutant recombinant enzymes were purified to homogeneity from E. coli cells. The specific activity (units/mg protein) was measured by the standard incorporation assay (Table 4). Thermal stabilities of the mutant Taq polymerases were similar to WT enzyme (data not shown).

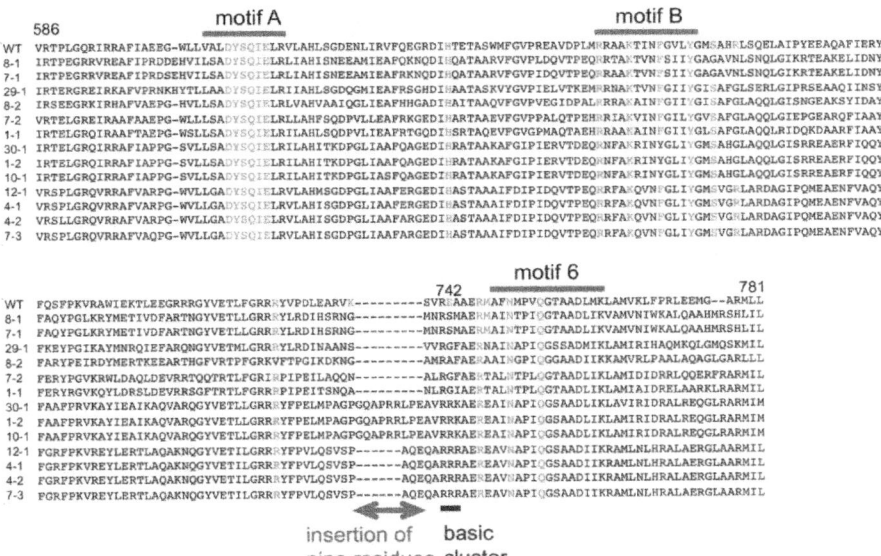

Figure 3: Alignment of the amino acid sequences of chimeric Taq polymerases with extension rates over 1 kb/min. A multiple alignment of the amino acid sequences of the substituted regions in the chimeric Taq polymerases with higher extension rates. The conserved motifs are shown on the top (Loh and Loeb, 2005). The corresponding residues involved in binding the template DNA and ddCTP, identified by the crystal structures of Klentaq (Li et al., 1998), are colored orange. The distinctive region observed in this alignment is indicated by a red line with two arrowheads. The basic cluster is indicated by a blue line.

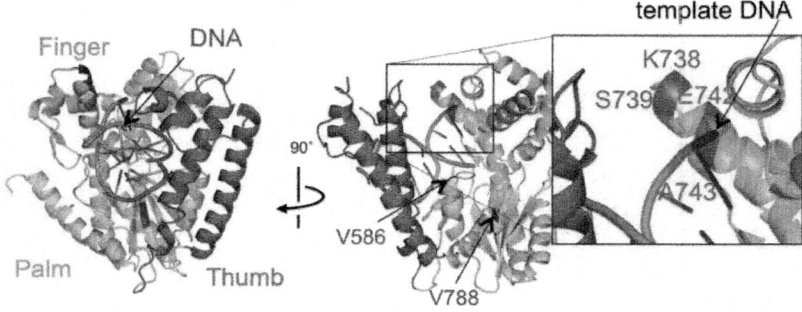

Figure 4: The mutational sites of Taq DNA polymerase. The crystal structure of Taq polymerase with DNA is shown (PDB, 1TAU). The 5' to 3' exonuclease domain is not illustrated for clarity. The polymerase domain is composed of a right hand with finger (green), palm (cyan), and thumb (red) subdomains (Eom et al., 1996). The thumb and finger subdomains hold DNA (backbones are colored orange). The residues E742 and A743 for the mutations are shown in magenta. The site, K738 and S739, where the interesting 9 amino acids were inserted was colored brown. The residues V586 and V788 corresponding to the junctions of the substitutions were shown in gray.

Table 4: Properties of the mutant Taq polymerase

Name	U/mg (× 10^5)	Extention kb/5 min	Kd for DNA (nM)	Charge
TaqEA(WT)	5.1	0.7	400	−1
TaqRR	1.3	>8	9	2
TaqRA	2.3	3.7	8	1
TaqAA	2.8	2	9	0
TaqER	1.4	3.5	8	0
TaqAR	1.1	4.3	8	1
TaqRK	1.5	>8	10	2
TaqKR	1.1	>8	10	2
TaqKK	1.4	>8	10	2
TaqQY	4.7	2.1	10	0
TaqAH	3.7	2.5	6	1
TaqEH	3.2	2.0	6	0
TaqHA	3.8	2.5	6	1
TaqHH	3.8	2.8	6	2
TaqHK	2.2	4.8	6	2

Faster Primer Extension by the Mutant Taq Polymerases

The in vitro primer extension rates were compared for these mutant Taq polymerases, as well as WT Taq polymerase. As shown in Figure 5, all of the mutant Taq polymerases exhibited faster extension reactions compared with that by the WT. The results of these experiments were quantified. The increased extension rate is generally related to the number of positive charges at this site (Table 4). The basic residues gave the varied effects. The positive charge of His appeared to have lower effect than those of Arg and Lys. There is a difference in pKa among the basic residues, and pKa of His, Arg and Lys are 6.8, 12.5, and 11.0, respectively. The relative degree of positive charge of His is estimated to be low. The DNA binding affinity of each enzyme was evaluated by EMSA, using a primed-DNA as a ^{32}P-labeled probe. As shown in Figure 6, the DNA binding ability of all the mutant Taq polymerases was distinctly increased from that of WT Taq polymerase. The increased number of positive charge at the positions 742 and 743 appeared to provide higher binding efficiency of Taq polymerase. Apparent Kd was determined with EMSA (Table 4). All the mutants bound to DNA by up to 2 orders of magnitude more tightly than WT Taq polymerase. Although there is no difference in the apparent Kd among these mutants, the second-shifted bands appeared in the gel images of EMSA in the case of the mutants, which possess Arg or Lys at the positions 742 and 743. The positive charge of Arg or Lys at the positions 742 and 743 might cause a nonspecific binding, in addition to the functional binding, of the enzyme to DNA.

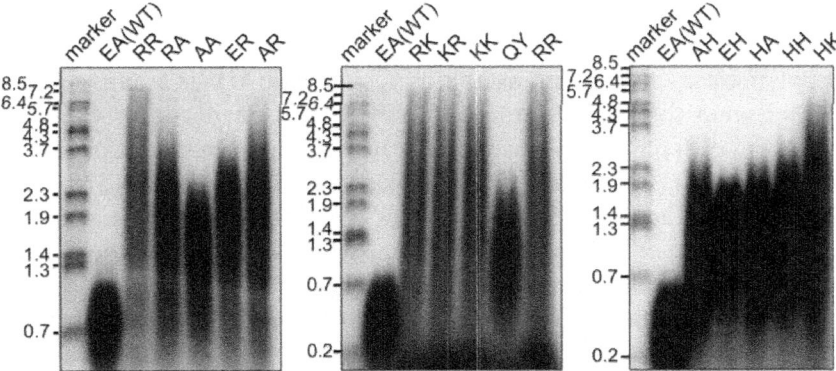

Figure 5: Primer extension activities of WT and mutant Taq polymerases. M13 ssDNA annealed with a ^{32}P-labeled deoxyoligonucleotide (55mer) was used as the substrate. For each DNA polymerase, 0.05 unit was added to 20 μl reaction mixture, containing 2.5 nM primed DNA. The reaction mixtures were incubated at 74°C for 5 min, and

the products were analyzed by 1% alkaline agarose gel electrophoresis, followed by autoradiography. The sizes indicated on the left are from BstPI-digested λ phage DNA labeled with ^{32}P at each 5′ end. The names of the proteins were indicated on the top.

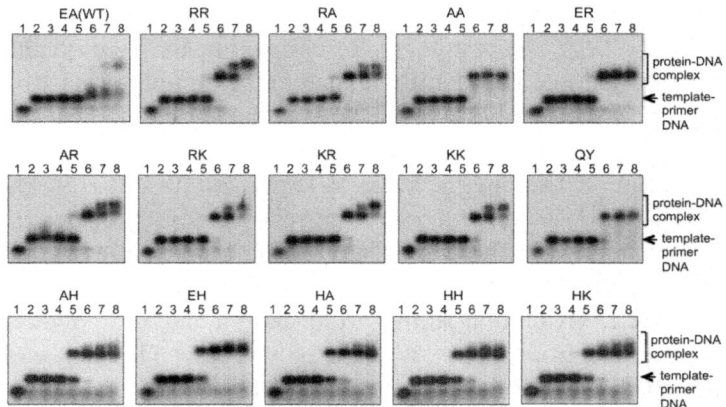

Figure 6: DNA binding ability of mutant Taq polymerases. EMSA was performed using primed DNA (^{32}P-labeled 27mer DNA and 49mer DNA). The names of the mutant proteins were indicated on the top of each panel. Lane 1 was ^{32}P-labeled 27mer ssDNA. Lane 2 had no protein with primed DNA. Lanes 3~8 contained 0.4, 1.6, 6.3, 25, 100, and 400 nM enzyme, respectively.

Better PCR Performance by the Mutant Taq Polymerases

The main goal of this study was to create PCR enzymes with superior performance, as compared to that of WT Taq polymerase. Since the mutant Taq polymerases are as thermostable as WT enzyme, it was promising to apply these enzymes to PCR. Therefore, the PCR performances for several target DNAs with different lengths were compared. A representative example of the PCR experiments is shown in Figure 7. For the amplification of 15 kb of DNA, several mutant Taq polymerases, AA, RA, AH, EH, HA, and HH, successfully amplified the target DNA under conditions where WT Taq polymerase did not function. The other mutant enzymes prepared in this study did not work well in the same conditions. The performances of some of the mutant enzymes, RR, QY, ER, AR, and HK, are shown in Figure 7. These experiments showed inconsistency with the results of primer extension experiment. The target DNA product was not detected by the mutant enzymes that possess primer extension rate with >8 kb/5 min. The enzymes with extension rate of >8 kb/5 min have Arg or Lys at the positions 742 and 743. The observed PCR inhibition by these mutations may be due to the too tight binding to DNA, suggested by the EMSA as shown in the Section Faster Primer Extension by the Mutant

Taq Polymerases. These results indicated that the positions of 742 and 743 in Taq polymerase are important for DNA strand synthesis, and the electrostatic environment of this site severely affects its PCR performance.

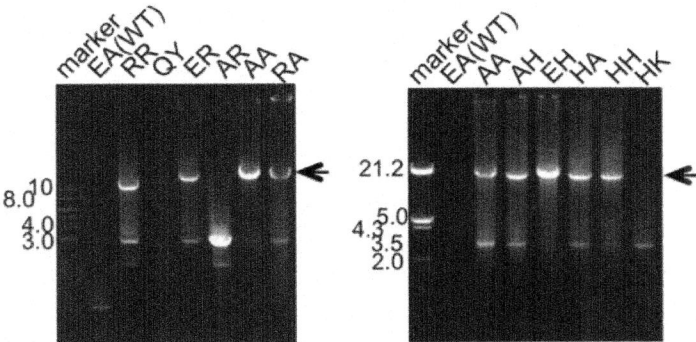

Figure 7: PCR performances of WT and mutant Taq polymerases. Lambda DNA was used as the template. Twenty nanograms of the template DNA and 5 pmol of each primer were added to the standard PCR mixture (total volume 20 μl, with 10 mM Tris-HCl, pH 8.8, 50 mM KCl, 1.5 mM $MgCl_2$, 0.2 mM each dNTP and 1 unit of Taq polymerase) and 30 cycles were run (99°C for 5 s and 66°C for 5 min) in a DNA thermal cycler. A 10 μl portion of each reaction mixture was analyzed by 1% agarose gel electrophoresis. The primers designed to amplify 15 kb fragment were 15-F (5'-dGAGTTCGTGTCCGTACAACTGGCGTAATCATGGCC-3'), and 15-R (5'-dGAATATCTGGCGGTGCAATATCGGTACTGTTTGC-3').

Discussion

DNA polymerase is an important enzyme for both fundamental living phenomena (DNA replication/repair) in cells and applications to genetic engineering in vitro. Therefore, numerous structural and functional investigations of DNA polymerase have been reported to date. In this study, we developed PCR enzymes that provide a superior extension reaction as compared to Taq polymerase, the standard enzyme for PCR. As compared to the PCR performance of Taq polymerase, these enzymes achieved the amplification of either the same length of DNA in a shorter time or a longer DNA in the same reaction time.

Metagenomic analysis is a revolutionary technique for microbiological ecology. The amplification of target genes from metagenomic DNA is a very powerful method to investigate many different DNA polymerases from uncultivated microbes. In this study, we focused on thermophilic bacteria as useful genetic resources for new thermostable family A DNA polymerases.

We obtained many new sequences encoding a region of a family A DNA polymerase from the hot spring soil samples. These results suggested that our strategy to amplify a specific region of the family A DNA polymerase genes is actually applicable to the analysis of microbial populations in any habitat. We employed the same strategy to search for new family B DNA polymerases, and some of this work was published previously (Matsukawa et al., 2009).

We constructed chimeric enzymes between Taq polymerase and the products of the various polgenes amplified from the metagenomic DNAs, and their primer extension abilities were compared. Many chimeric polymerases possessing excellent extension ability were obtained by this experiment. However, none of the chimeric enzymes were sufficiently thermostable for PCR use. The microbial sources of the gene fragments used for the construction of the chimeric genes are not necessarily extreme thermophiles, and some moderate thermophiles and mesophiles may be included among the amplified genes. The chimeric Taq polymerases showing faster extension ability than WT Taq polymerase would have gene fragments from the organisms, which are not extreme thermophiles. However, the amino acid sequence comparison of the chimeric Taq polymerases provided an important clue to design a mutant Taq polymerase with superior speed for the primer extension reaction, by site-specific mutagenesis. We focused on positions 742 and 743 in this study. The positions 742 and 743 are located in the finger subdomain and affected the interaction with DNA. The PCR performances of some of the mutant Taq polymerases showed reliable improvement, and they are useful for faster PCR and also for longer target DNAs. The conversion of the electrostatic environment at this position, from a negative charge to a positive charge, will affect the stabilization of the DNA binding near the active site of Taq polymerase. It is important to check whether the mutations in this site affect the fidelity of Taq polymerase. Our preliminary data revealed that the fidelities of these enzymes are not different from that of WT Taq polymerase (data not shown). We will confirm this with more experiments and provide statistical data in the future.

In addition to positions 742 and 743, we found one more remarkable feature in the sequences of the chimeric enzymes. These enzymes have an insertion of either 9 or 3 amino acids between positions 738 and 739 of Taq polymerase. It will be interesting to investigate the effects of these insertions in the finger subdomain on the PCR performance of Taq polymerase. Characterizations of the mutant Taq polymerases with the different inserted sequences are now underway.

In conclusion, we designed a method for engineering Taq polymerase to improve its primer extension rate, by using information obtained from the metagenomic analysis of soil samples from various hot-spring areas. The created enzymes showed robust PCR performances that were better than that of Taq polymerase. Since Taq polymerase is the standard enzyme used for PCR, an abundance of PCR data using this enzyme has been accumulated to date. The enzymes created in this study basically retain the properties of Taq polymerase, and therefore, they are applicable to many uses that have already been optimized with Taq polymerases.

CONFLICT OF INTEREST STATEMENT

The authors declare that the research was conducted in the absence of any commercial or financial relationships that could be construed as a potential conflict of interest.

ACKNOWLEDGMENTS

We thank Drs. Masaaki Takahashi and Yukiko Miyashita for valuable discussions and encouragement. This work was supported by grants from the Ministry of Education, Culture, Sports, Science and Technology of Japan [grant numbers 21113005, 23310152, and 26242075 to Yoshizumi Ishino]. This work was partly supported by Institute for fermentation, Osaka (IFO).

REFERENCES

1. Baar, C., d'Abbadie, M., Vaisman, A., Arana, M. E., Hofreiter, M., Woodgate, R., et al. (2011). Molecular breeding of polymerases for resistance to environmental inhibitors. Nucleic Acids Res. 39, e51. doi: 10.1093/nar/gkq1360

2. Barns, W. M. (1994). PCR amplification of up to 35-kb DNA with high fidelity and high yield from lambda bacteriophage templates. Proc. Natl. Acad. Sci. U.S.A. 91, 2216–2220. doi: 10.1073/pnas.91.6.2216

3. Braithwaite, D. K., and Ito, J. (1993). Compilation, alignment, and phylogenetic relationships of DNA polymerases. Nucleic Acids Res. 21, 787–802. doi: 10.1093/nar/21.4.787

4. Brakmann, S. (2005). Directed evolution as a tool for understanding and optimizing nucleic acid polymerase function. Cell. Mol. Life Sci. 62, 2634–2646. doi: 10.1007/s00018-005-5165-5

5. Cann, I., Ishino, S., Hayashi, I., Komori, K., Toh, H., Morikawa, K., et al. (1999). Functional interactions of a homolog of proliferating cell nuclear antigen with DNA polymerases in Archaea. J. Bacteriol. 181, 6591–6599.

6. Cann, I., and Ishino, Y. (1999). Archaeal DNA replication: identifying the

pieces to solve a puzzle. Genetics 152, 1249–1267.

7. d›Abbadie, M., Hofreiter, M., Caisman, A., Loakes, D., Gasparutto, D., Cadet, J., et al. (2007). Molecular breeding of polymerases for amplification of ancient DNA. Nat. Biotechnol. 25, 939–943. doi: 10.1038/nbt1321

8. Eckert, K. A., and Kunkel, T. A. (1991). DNA polymerase fidelity and the polymerase chain reaction. PCR Methods Appl. 1, 17–24. doi: 10.1101/gr.1.1.17

9. Eom, S. H., Wang, J., and Steitz, T. A. (1996). Structure of Taq ploymerase with DNA at the polymerase active site. Nature 382, 278–281. doi: 10.1038/382278a0

10. Ghadessy, F. J., Ong, J. L., and Holliger, P. (2001). Directed evolution of polymerase function by compartmentalized self-replication. Proc. Natl. Acad. Sci. U.S.A. 98, 4552–4557. doi: 10.1073/pnas.071052198

11. Ghadessy, F. J., Ramsay, N., Boudsocq, F., Loakes, D., Brown, A., Iwai, S., et al. (2004). Generic expansion of the substrate spectrum of a DNA polymerase by directed evolution. Nat. Biotechnol. 22, 755–759. doi: 10.1038/nbt974

12. Henry, A. A., and Romesberg, F. E. (2005). The evolution of DNA polymerases with novel activities. Curr. Opin. Biotechnol. 16, 370–377. doi: 10.1016/j.copbio.2005.06.008

13. Holmberg, R. C., Henry, A. A., and Romesberg, F. E. (2005). Directed evolution of novel polymerases. Biomol. Eng. 22, 39–49. doi: 10.1016/j.bioeng.2004.12.001

14. Ishino, S., and Ishino, Y. (2013). "DNA polymerases and DNA ligases," in Thermophilic Microbes in Environmental and Industrial Biotechnology, eds J. Litterchild, T. Satyanarayana, and Y. Kawarabayasi (Dordrecht: Springer Science; Business Media), 429–457.

15. Ishino, S., and Ishino, Y. (2014). DNA polymerase as a useful reagent for genetic engineering ~History of developmental research on DNA polymerase~. Front. Microbiol. 5:465. doi: 10.3389/fmicb.2014.00465

16. Ishino, S., Kawamura, A., and Ishino, Y. (2012). Application of PCNA to processive PCR by reducing the stability of its ring structure. J. Jap. Soc. Extremophiles 11, 19–25.

17. Ishino, Y., and Cann, I. (1998). The euryarchaeotes, a subdomain of archaea, survive on a single DNA polymerase: fact or farce?Genes Genet. Syst. 73, 323–336.

18. Ishino, Y., Ueno, T., Miyagi, M., Uemori, T., Imamura, M., Tsunasawa,

S., et al. (1994). Overproduction of Thermus aquaticusDNA polymerase and its structural analysis by ion-spray mass spectrometry. J. Biochem. 116, 1019–1024.

19.   Joyce, C. M., and Steitz, T. A. (1994). Function and structure relationships in DNA polymerases. Annu. Rev. Biochem. 63, 777–822. doi: 10.1146/annurev.bi.63.070194.004021

20.   Kermekchiev, M. B., Kirilova, L. I., Vail, E. E., and Barnes, W. M. (2009). Mutants of Taq DNA polymerase resistant to PCR inhibitors allow DNA amplification from whole blood and crude soil samples. Nucleic Acids Res. 37, e40. doi: 10.1093/nar/gkn1055

21.   Kermekchiev, M. B., Tzekov, A., and Barnes, W. M. (2003). Cold-sensitive mutants of Taq DNA polymerase provide a hot start for PCR. Nucleic Acids Res. 31, 6139–6147. doi: 10.1093/nar/gkg813

22.   Komori, K., and Ishino, Y. (2000). Functional interdependence of DNA polymerizing and $3'\rightarrow 5'$ exonucleolytic activities inPyrococcus furiosus DNA polymerase I. Protein Eng. 13, 41–47. doi: 10.1093/protein/13.1.41

23.   Li, Y., Kong, Y., Korolev, S., and Waksman, G. (1998). Crystal structures of open and closed forms of binary and ternary complexes of the large fragment of Thermus aquaticus DNA polymerase I: structural basis for nucleotide incorporation. EMBO J. 17, 7514–7525. doi: 10.1093/emboj/17.24.7514

24.   Lipps, G., Röther, S., Hart, C., and Krauss, G. (2003). A novel type of replicative enzyme harbouring ATPase, primase and DNA polymerase activity. EMBO J. 22, 2516–2525. doi: 10.1093/emboj/cdg246

25.   Loh, E., and Loeb, L. A. (2005). Mutability of DNA polymerase I: implications for the creation of mutant DNA polymerases. DNA Repair 4, 1390–1398. doi: 10.1016/j.dnarep.2005.09.006

26.   Matsukawa, H., Yamagami, T., Kawarabayasi, Y., Miyashita, Y., Takahashi, M., and Ishino, Y. (2009). A useful strategy to construct DNA polymerases with different properties by using genetic resources from environmental DNA. Genes Genet. Syst. 84, 3–13. doi: 10.1266/ggs.84.3

27.   Ohmori, H., Friedberg, E. C., Fuchs, R. P., Goodman, M. F., Hanaoka, F., Hinkle, D., et al. (2001). The Y-family of DNA polymerases. Mol. Cell 8, 7–8. doi: 10.1016/S1097-2765(01)00278-7

28.   Ong, J. L., Loakes, D., Jaroslawski, S., Too, K., and Holliger, P. (2006). Directed evolution of DNA polymerase, RNA polymerase and reverse transcriptase activity in a single polypeptide. J. Mol. Biol. 361, 537–550. doi: 10.1016/j.jmb.2006.06.050

29. Perler, F. B., Kumar, S., and Kong, H. (1996). Thermostable DNA polymerases. Adv. Protein Chem. 48, 377–435. doi: 10.1016/S0065-3233(08)60367-8

30. Suzuki, M., Avicola, A. K., Hood, L., and Loeb, L. A. (1997). Low fidelity mutants in the O-helix of Thermus aquaticus DNA polymerase I. J. Biol. Chem. 272, 11228–11235. doi: 10.1074/jbc.272.17.11228

31. Suzuki, M., Yoshida, S., Adman, E. T., Blank, A., and Loeb, L. A. (2000). Thermus aquaticus DNA polymerase I mutants with altered fidelity. Interacting mutations in the O-helix. J. Biol. Chem. 275, 32728–32735. doi: 10.1074/jbc.M000097200

32. Takagi, M., Nishioka, M., Kakihara, H., Kitabayashi, M., Inoue, H., Kawakami, B., et al. (1997). Characterization of DNA polymerase from Pyrococcus sp. strain KOD1 and its application to PCR. Appl. Environ. Microbiol. 63, 4504–4510.

33. Terpe, K. (2013). Overview of thermostable DNA polymerases for classical PCR applications: from molecular and biochemical fundamentals to commercial systems. Appl. Microbiol. Biotechnol. 97, 10243–10254. doi: 10.1007/s00253-013-5290-2

34. Tosaka, A., Ogawa, M., Yoshida, S., and Suzuki, M. (2001). O-helix mutant T664P of Thermus aquaticus DNA polymerase I: altered catalytic properties for incorporation of incorrect nucleotides but not correct nucleotides. J. Biol. Chem. 276, 27562–27567. doi: 10.1074/jbc. M010635200

35. Uemori, T., Ishino, Y., Doi, H., and Kato, I. (1995). The hyperthermophilic archaeon Pyrodictium occultum has two α-like DNA polymerases. J. Bacteriol. 177, 2164–2177.

36. Uemori, T., Ishino, Y., Fujita, K., Asada, K., and Kato, I. (1993). Cloning of the Bacillus caldotenax DNA polymerase gene and characterization of the gene product. J. Biochem. 113, 401–410.

37. Villbrandt, B., Sobek, H., Frey, B., and Schaumburg, D. (2000). Domain exchange: chimeras of Thermus aquaticus DNA polymerase, Escherichia coli DNA polymerase I and Thermotoga neapolitana DNA polymerase. Protein Eng. 13, 645–654. doi: 10.1093/protein/13.9.645

38. Wang, Y., Prosen, D. E., Mei, L., Sullivan, J. C., Finney, M., and Vander Horn, P. B. (2004). A novel strategy to engineer DNA polymerases for enhanced processivity and improved performance in vitro. Nucleic Acids Res. 32, 1197–1207. doi: 10.1093/nar/gkh271

Chapter 2

A GENETIC REPLACEMENT SYSTEM FOR SELECTION-BASED ENGINEERING OF ESSENTIAL PROTEINS

Sonja Billerbeck and Sven Panke

Department for Biosystems Science and Engineering (D-BSSE), ETH Zürich

ABSTRACT

Background

Essential genes represent the core of biological functions required for viability. Molecular understanding of essentiality as well as design of synthetic cellular systems includes the engineering of essential proteins. An impediment to this effort is the lack of growth-based selection systems suitable for directed evolution approaches.

Results

We established a simple strategy for genetic replacement of an essential gene by a (library of) variant(s) during a transformation.

The system was validated using three different essential genes and plasmid combinations and it reproducibly shows transformation efficiencies on the order of 10^7 transformants per microgram of DNA without any identifiable false positives. This allowed for reliable recovery of functional variants out of at least a 10^5-fold excess of non-functional variants. This outperformed selection in conventional bleach-out strains by at least two orders of magnitude, where recombination between functional and non-functional variants interfered with reliable recovery even in recA negative strains.

Conclusions

We propose that this selection system is extremely suitable for evaluating large libraries of engineered essential proteins resulting in the reliable isolation of functional variants in a clean strain background which can readily be used for in vivoapplications as well as expression and purification for use in in vitro studies.

BACKGROUND

About eight percent of E. coli genes are essential for the cell [1]. Essential genes are of particular scientific interest as they encode proteins required for important biological functions, thereby building the minimal core of cellular viability which tends to be conserved across species. Knowledge about essential genes and their protein products is important for drug design [2, 3], biotechnological applications [4], minimal genome approaches [5–8] and, in general, crucial for understanding and engineering the basic cellular functions required for life [9]. While the construction of the Keio-collection, a collection of single gene knock-outs in E. coli[1], enormously facilitated the systematic investigation of the physiology of E. coli as well as protein and strain-engineering approaches, it is still restricted to non-essential genes and their protein products. Engineering approaches involving essential genes and proteins are complicated because knock-outs cause lethality. This means that phenotypes of engineered proteins cannot be easily evaluated in vivo as suitable clean strain backgrounds are not available.

In the last decade several approaches have been investigated to identify essential genes and to study their function in vivoby conditional elimination of the protein from the cell. This was achieved by triggering interference of the synthesis of the target protein on either the transcriptional or translational level [10–15]. However, these "bleach-out" methods rely on conditional protein elimination rather than elimination of the target gene itself and thus retain a wild-type copy of the essential gene in the cell. This sets limitations for the utility of these systems as ready-to-use selection systems for directed evolution experiments since recombination of library members with the chromosomal wild-type gene or mutations in the system regulating the expression of the wild-type protein can lead to the selection of false positive variants. This is particularly true when using a library for which only a small fraction of variants is expected to be functional. In this case recombination events are preferentially selected over functional library members, which results in every selection effort turning into a laborious screen for bona fide functional library members.

Besides evaluation of large libraries, another desire during protein engineering of essential genes is to replace the wild-type gene by a single engineered or heterologous variant for in vivo functional studies or for the construction of specialized strains which can be used to purify the mutant protein free of wild-type protein. Phage P1-mediated transduction of a chromosomal knock-out into a strain expressing a variant of the essential gene of interest from a plasmid is the current method of choice to achieve genetic replacement of an essential target gene by a variant (e.g [16]). The knock-out was thereby created while complementing the chromosomal gene loss by a plasmid-encoded version of the essential gene. Although P1-transduction is widely used, the protocol is time-consuming and restricted to a few variants at a time as efficiencies of successful transductions are low, often requiring empirical testing for the proper phage concentration followed by re-plating and PCR-screening for correct genotypes. Therefore, it was our aim to develop a general genetic set-up which turns working with essential genes and the engineering of their gene products into a straight-forward approach as facile as working with non-essential genes. Here, we present a simple transformation-based system. Establishment of the system begins with the chromosomal replacement of the essential gene of interest by a PCR-derived selection marker [17], in conjunction with a complementary vector-encoded version of the target. A central element of the method is that the complementation vector carries an I-Sce I nuclease recognition site and can thus be rapidly and conditionally eliminated in the presence of an I-Sce I nuclease-expressing helper plasmid. During elimination of the complementation vector, cells are made electrocompetent and transformed with a vector-encoded library or a specific variant of the essential gene. Thus, the actual step of gene exchange is reduced to a single transformation plus preparation of a suitable knock-out strain.

RESULTS

Overview of the Replacement System

The selection system can be flexibly assembled in various user-defined ways, but will be discussed in its simplest version first. It relies on two vectors: the knock-out and complementation vector pKOCOMP and the helper plasmid pI-Sce I (Figure1A and Table1). Plasmid pKOCOMP is a derivative of pKD46 [17] and encodes the arabinose-inducible λ red recombination system (genes β γ and exo) as well as the essential gene of interest. It also contains an 18 bp I-Sce I cleavage site which allows for the conditional elimination of pKOCOMP in the presence of I-Sce I. I-Sce I is expressed from the helper plasmid pI-Sce I under the control of the rhamnose-inducible RhaSR/$P_{rha\ BAD}$-system [18]. A variant of this helper plasmid, pP$_{ara}$I-Sce I, carries an arabinose-inducible regulatory system AraC/$P_{ara\ BAD}$[19] instead of the rhamnose-based system. After testing various set-ups, these promoters were chosen as they exhibited high expression level in the presence of rhamnose or arabinose but could be efficiently switched off in the presence of glucose due to catabolite repression [19, 20].

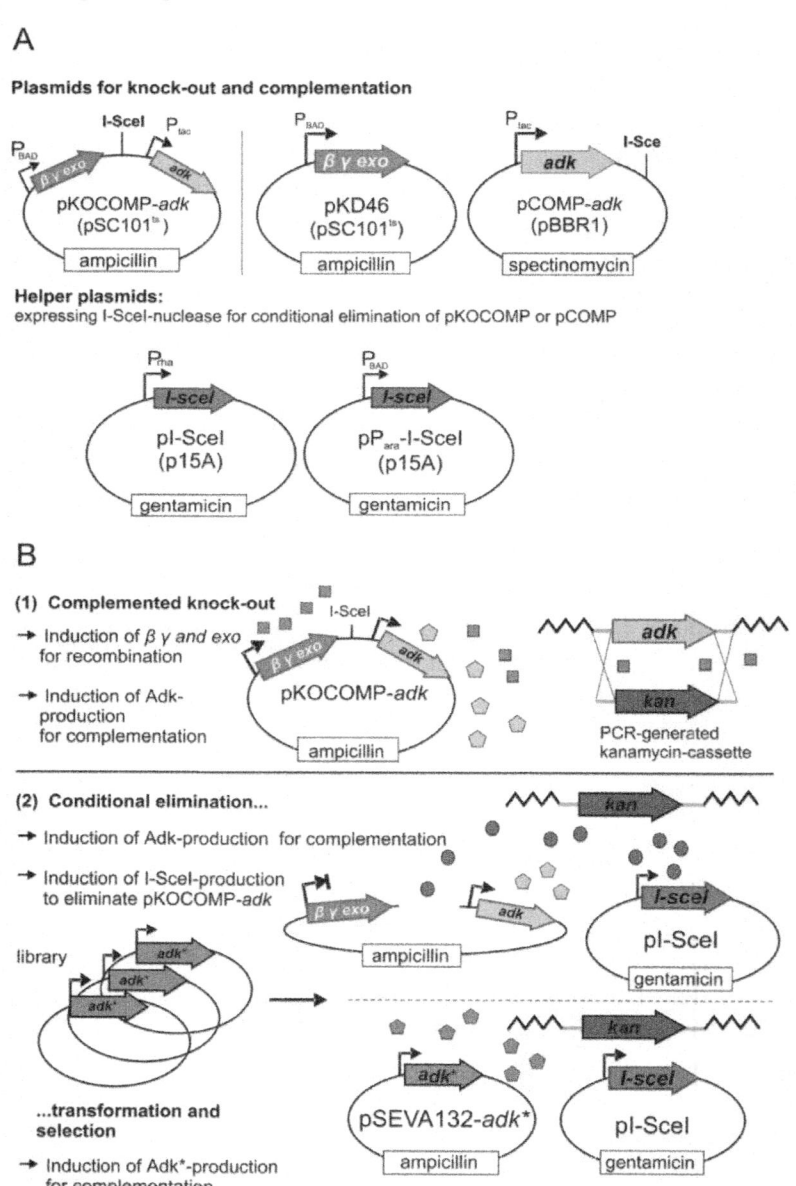

Figure 1: Overview on the transformation-based replacement of essential genes. A:

Plasmids required for establishment of the selection system B: General procedure: The essential target gene adk is replaced by an antibiotic resistance cassette while pKOCOMP-adkcomplements for the chromosomal loss. Plasmid pKOCOMP can be conditionally eliminated by co-expression of I-Sce I nuclease from helper plasmid pI-Sce I. During elimination, cells are made electrocompetent and are transformed with a variant or library under investigation to evaluate functionality or to select for functional library members which can complement for the loss of pKOCOMP.

It should be noted that the specific set-up of the complementation vector is flexible as long as it carries the I-Sce I site, which can easily be introduced by PCR, is compatible with pKD46, which is required for λ red-mediated knock-out of the target gene and can be cured by growth at 42°C. The system can then be completed by subsequent transformation with pI-Sce I or pP_{ara} I-Sce I.

Table 1: Bacterial strains and plasmids used in this study

Bacterial strain or plasmid	Characteristics	Source
Strains		
W3110	F⁻ λ⁻rph-1 INV(rrnD, rrnE)	[33], internal strain collection
DH10B	F⁻endA1 recA1 galE15 galK16 nupG rpsL ΔlacX74 Φ80lacZΔM15 araD139 Δ(ara,leu)7697 mcrA Δ(mrr-hsdRMS-mcrBC) λ⁻	[34], internal strain collection
W3110 adk::kan	Chromosomal adk replaced by a Km^R cassette. The strain is only viable if the adk deletion is complemented.	This work
W3110 groE::kan	Chromosomal groS and groL replaced by a Km^R cassette. The strain is only viable if the gro E deletion is complemented.	This work
W3110 secB-gpsA::kan	Chromosomal secB and gpsA replaced by a Km^R cassette. The strain is only viable if the gps A deletion is complemented.	This work
SBΔrecA	lacI q rrnB3 ΔlacZ4787 hsdR514 Δ (araBAD) 567 Δ (rhaBAD) 568 rph-1 groE::P_araBAD-groE-Km ^RrecA::FRT	This work
Plasmids		
pKD46	ori pSC101^ts, Ap^R. Encodes λ red recombination genes γ, β and exo under control of the arabinose-responsive promoter P_ara BAD.	[1]

pKOCOMP-adk	pKOCOMP-derived vector with adk under control of an IPTG-responsive tac promoter	This work
pSEVA432	ori pBBR1, Spec^R, multiple cloning site	provided by Victor de Lorenzo
pCOMP-adk	pSEVA432 encoding for adk under control of an IPTG-responsive tac promoter, contains an I-Sce I cleavage site	This work
pCOMP-ESL	pSEVA432 encoding for groS and groL control of an IPTG-responsive tac promoter, contains an I-Sce I cleavage site	This work
pCOMP-sec B-gps A	pSEVA432 encoding for the natural secB-gpsA transcriptional unit, contains a I-Sce I cleavage site	This work
pSEVA132	ori pBBR1, Ap^R, multiple cloning site	provided by Victor de Lorenzo
pSEVA132-adk	pSEVA132 encoding for adk under control of its natural promoter	This work
pSEVA132-adk^{Stop}	pSEVA132-adk with an internal stop codon	This work
pSEVA132-adk^{watermark}	pSEVA132-adk with a peptide insertion behind position P140	This work
pSEVA132-gro E	pSEVA132 coding for groS and groL under control of their natural promoter	This work
pSEVA132-gro E^{Stop}	pSEVA132-groE with an internal stop codon	This work
pSEVA132-groE^{watermark}	pSEVA132-groE with a peptide insertion behind site I301	This work
pSEVA132-sec Bgps A	pSEVA132 encoding for secB and gpsA under control of their natural promoter	This work
pSEVA132- sec BgpsA^{Stop}	pSEVA132- sec Bgps A with an internal stop codon	This work
pSEVA671	ori p15A, Gm^R, multiple cloning site	provided by Victor de Lorenzo

pI-Sce I	pSEVA671, with I-Sce I nuclease under control of the rhamnose inducible promoter P_{Rha} and the response regulators RhaS and RhaR, derived from the rhammnose metabolizing transcriptional unit of E. coli.	This work
pP_{ara}I-Sce I	pI-Sce I with P_{rhaBAD} and regulators RhaS and RhaR exchanged by the arabinose promoter P_{araBAD} and the regulator araC	This work

Validation and characterization of the system using the essential gene product adenylate kinase (Adk)

The first target gene chosen for validation and characterization of the system was adk, encoding E. coli' s Adk, an essential gene product required for the biosynthesis of purine ribonucleotides and for the regulation of intracellular nucleotide availability [1, 21, 22]. For complementation, adk under control of P_{tac} was inserted into pKOCOMP, giving rise to pKOCOMP-adk. Upon induction of pKOCOMP-adk with IPTG, the chromosomal copy of adk was replaced by a PCR-generated kanamycin resistance cassette [17]. The genotype of the resulting strain E. coli adk::kan [pKOCOMP-adk was confirmed by PCR using primers binding to chromosomal regions up- and downstream of the adk-locus (Additional file 1: Table S1 and Additional file 2 Figure S1). To complete the selection system, strain adk::kan [pKOCOMP-adk was transformed with helper plasmid pI-Sce I. The resulting strain adk::kan [pKOCOMP-adk; pI-Sce I] was grown in the presence of glucose during maintenance to repress I-Sce I nuclease production. However, when cells were transferred to glucose-free LB medium, I-Sce I production was apparently efficiently induced by the addition of rhamnose, as no colony-forming units could be recovered after an induction-8period of 180 minutes (Figure 2A). Importantly, even without a copy of the adkgene, cells remained viable for about two generations, presumably until all mRNA and protein was depleted. This was expected as linear DNA is rapidly degraded by cellular exonucleases and has a half-life in the range of minutes [23] whereas most proteins are relatively stable with half-lives in the range of hours [24].

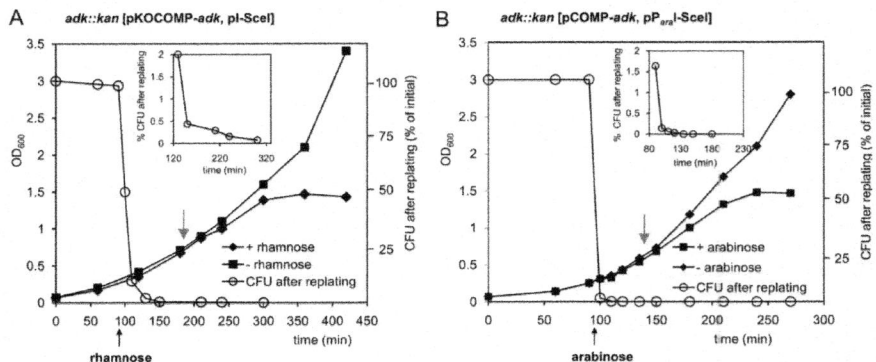

Figure 2: Characterization of the pKOCOMP- adk /pI- Sce I and the pCOMP- adk / pP $_{ara}$ I- Sce I systems. A: Elimination of pKOCOMP-adk from adk::kan due to rhamnose-induced expression of I-Sce I nuclease from helper plasmid pI-Sce I. Red arrow: Time point for harvest and competent cell preparation. The inlet gives a better resolution of the elimination dynamics. The chromosomal adk-replacement was established using pKOCOMP-adk. B: Elimination of pCOMP-adk from adk::kan due to arabinose-induced expression of I-Sce I nuclease from helper plasmid pP $_{ara}$ I-Sce I. The chromosomal adk-replacement was established using pKD46 and pCOMP-adk.

During this intermediate period cells were made electrocompetent by washing with H$_2$O and glycerol and transformed with various test plasmids (Table 1) in order to determine transformation efficiencies and the frequency of false positive variants. Vector pSEVA132-adk encoding the wild-type adenylate kinase under control of its natural promoter was used for determination of transformation efficiencies. Transformation efficiencies of 10^7 µg^{-1} DNA were routinely achieved. The system was intensively characterized regarding the frequency of false positive variants which could arise due to recombination of the library plasmid with residual linearized pKOCOMP-adk or due to an uninduced subpopulation of cells that maintained pKOCOMP (Table 2). Recombination-based false positives were tested by transformation with pSEVA132-adk ^{stop}, harboring a stop codon in the adk gene that could be repaired by recombination allowing growth of the corresponding carrier cell. False positives due to incomplete induction were examined by transformation with the empty vector pSEVA132. In both cases no false positive colonies were detected when I-Sce I expression was induced and pKOCOMP was eliminated before competent cell preparation (Table 2), corresponding to a frequency of less than 2×10^{-4} false positives per transformed cell. However, there was a detectable frequency of recombination events when pKOCOMP was eliminated only after transformation with the test plasmids or when we tried to eliminate pKOCOMP solely by growth at the non-permissive temperature. In the former case, I-Sce I nuclease expression was induced only after transformation by plating on arabinose-supplemented agar plates, such that both plasmids would be simultaneously present in the cells for a short period of time. Here we found illegitimate events with a frequency of about 3×10^{-3} per transformed cell (Table 2). This demonstrates the importance of careful management of the plasmid elimination step.

Table 2: Transformation efficiencies and frequency of false positive variants for the Adk-specific selection systems

Strain	I-Sce I induction during competent cell preparation	Number of transformants (frequency CFU per transformant)		
		pSEVA132-adk	pSEVA132-adk^{stop}	pSEVA132
W3110 adk::kan [pKO-COMP- adk , pI- Sce I]	No	20 200±2900	66±21 (~3×10^{-3})	28±13 (~1×10^{-3})
	Yes	21 000±6000	0 (<2×10^{-4})	0 (<2×10^{-4})
W3110 adk ::kan [pCOMP- adk, pP$_{ara}$-I- Sce I]	No	32 100±3700	61±17 (2×10^{-3})	35±11 (1×10^{-3})
	Yes	28 000±2200	0 (<2×10^{-4})	0 (<2×10^{-4})

Flexibility of the Replacement System

To verify that the selection system can be set up with alternative combinations, we constructed pCOMP-adk. This vector is based on the pBBR1 ori with an expected copy number of 10-20 per cell [25], carrying a P$_{tac}$ promoter-controlled adk gene, the lac-repressor LacI, and an I-Sce I cleavage site. While the previous complementation vector pKOCOMP-adk used a temperature-sensitive pSC101 ori with about 2-3 copies per cell when grown at 37°C [26], the increased copy number of pCOMP-adk allowed for examination of whether plasmid elimination was sufficiently efficient at higher intracellular plasmid concentration. We used pCOMP-adk in combination with pKD46 to replace adk by a kanamycin cassette. After curing cells of pKD46 by growth at 43°C, the resulting strain adk::kan [pCOMP-adk was transformed with pP$_{ara}$I-Sce I and used for gene replacement as described before using arabinose to induce I-SceI production. Despite the higher copy number of pCOMP, the plasmid was again rapidly eliminated from the cells in the presence of arabinose (Figure 2B). Probably due to the faster on-set and the possibly higher expression levels of the arabinose responsive promoter P_{araBAD}[19, 20], elimination-dynamics of the pCOMP/pP$_{ara}$ I-Sce I system were faster than those of the pKOCOMP/pI-Sce I system. Although the elimination data of the two systems are difficult to compare due to differences in copy number and promoters, these results indicate that the described approach can be set up in multiple ways, making it easy to adapt to plasmid strategies for specific purposes. Transformation efficiencies of the pCOMP/pP$_{ara}$ I-Sce I system (~10^7 colonies µg^{-1}DNA) were comparable to the pKOCOMP/pI-Sce I system (Table 1) and we could not identify false positive transformants.

Generality of the System

To confirm that the utility of the system was not limited to adk but could be easily extended to other essential genes, we constructed in vivo selection systems for other essential gene products: the chaperonin GroEL and its co-chaperonin GroES (encoded by the groE operon containing the genes groL and groS), and glycerol-3-phosphate dehydrogenase (GpsA encoded by gpsA). For establishment of the GroEL-specific system we introduced an I-Sce I cleavage site into the vector pSEVA431-groE by PCR giving rise to vector pCOMP-groE. Plasmid pSEVA431-groE encodes the groE-operon under control of the IPTG-inducible P$_{tac}$ promoter. It also harbors the lacI gene, a spectinomycin resistance cassette and replicates with a pBBR1 ori. Plasmid pCOMP-groE was used in combination with pKD46 to replace the chromosomal groE-operon by a kanamycin resistance cassette (Additional file 2 Figure S1). The resulting strain groE::kan [pCOMP-groE] was cured from pKD46 at 43°C and then transformed with the helper plasmid pP$_{ara}$ I-Sce I.

The GpsA-specific system was constructed by cloning the natural gpsA transcriptional unit (consisting of genes secB andgpsA under control of their natural promoter) into vector pSEVA431 using primers encoding for an I-Sce I restriction site, resulting in vector pCOMP-secBgpsA. Chromosomally encoded secB and gpsA were then replaced by a kanamycin resistance cassette using pKD46 (Additional file 2 Figure S1). To complete the set-up, strain secBgpsA::kan [pCOMP-secBgpsA] was cured of pKD46 and transformed with helper plasmid pP$_{ara}$ I-Sce I. Both systems were characterized regarding elimination dynamics of the complementing plasmids pCOMP-groE and pCOMP-secBgpsA after I-Sce I induction, as well as regarding transformation efficiencies of electrocompetent cells prepared during pCOMP-elimination and frequency of false positive variants. Both pCOMP-type plasmids were lost at a comparable rate to the pCOMP-plasmid carrying adk (Figure 3). After that, both systems routinely showed transformation efficiencies of 10^6-10^7 colonies µg^{-1} DNA when transformed with the positive control vectors pSEVA132-groE or pSEVA132-secBgpsA. No false positive variants could be detected after transformation with the test plasmids pSEVA132-groE[stop] and pSEVA132-secBgpsA[stop], constructed in analogy to pSEVA132-adk[stop] before. Importantly, it seems to be a general feature that cells stay viable - as judged by the doubling time in comparison to a non-induced control culture that did not produce I-Sce I - for one or more generations, depending on the target gene, after loss of the complementing plasmid. This is an important characteristic of the system as cells can be made competent for transformation with a library or variant of an essential gene during a period where the complementing plasmid has already been lost and can no longer contribute to recombination.

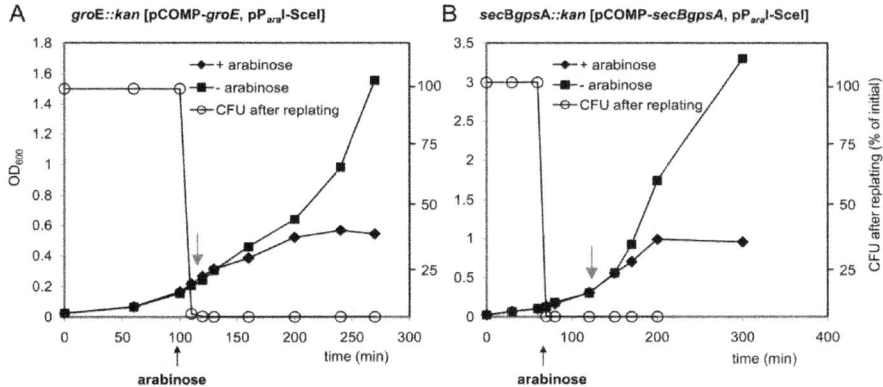

Figure 3: Characterization of the pCOMP- groE /pP$_{ara}$ I- Sce I and the pCOMP-sec BgpsA /pP$_{ara}$ I- Sce I system. Elimination of (A) pCOMP-groE from groE::kan and (B) pCOMP-secBgpsA from secBgpsA::kan in the presence of arabinose and the helper plasmid pP$_{ara}$I-SceI. Elimination is induced by arabinose-induced expression of I-Sce I nuclease from helper plasmid pP$_{ara}$I-Sce I. Red arrow: Time point for harvest and competent cell preparation.

Performance of the System During Selection

To test the system regarding its performance in reliably identifying functional variants from a large library without contamination by false positive variants we challenged the GroEL- and the Adk-specific selection system with mock libraries with various excesses of non-functional variants. As there is a conventional bleach-out system available for GroEL (the E. coli MGM100 strain [10]), we also compared the performance of the replacement system developed here to the bleach-out system. In strain MGM100, the groE promoter has been replaced by the araBAD promoter and the regulatory gene araC. The strain can therefore be maintained in the presence of arabinose but GroEL production from the chromosomal locus can be fully repressed in the presence of glucose. This way, functional variants can be selected from a library in the presence of glucose. However, as a copy of the wild-type gene is still present during selection, recombination can lead to selection of false positive variants, which can dominate even in stringent selection conditions. To improve the strain regarding the expected recombination frequency we deleted recA leading to the strain SBΔrecA (Additional file 2: Figure S1).

The created mock libraries contained decreasing ratios of functional to non-functional gene variants from $1:10^2$ to $1:10^5$(functional:non-functional) mimicking libraries where a decreasing number of variants is expected to be functional within a large pool of non-functional variants, as it is typical e.g. for libraries created by error-prone PCR with different error rates. As non-functional variants we used pSEVA132-groE ^{stop} or pSEVA132-adk ^{stop} . As functional variants we used pSEVA132-groE ^{watermark} or pSEVA132-adk ^{watermark} . These plasmids encode for a GroEL or Adk- variant with an in-frame insertion of a short peptide at previously identified permissive sites. Both variants are fully functional and can be identified by PCR using watermark-specific primers. To compare the GroEL-specific selection system with the GroEL bleach-out system SBΔrecA, electrocompetent W3110 groE::kan [pCOMP-groE, pP$_{ara}$I-Sce I] cells or electrocompetent SBΔrecA cells were transformed with the different mock libraries in separate experiments. As positive and negative controls, cells were transformed with only functional or only non-functional variants. After plating and incubation overnight on LB-agar plates supplemented with arabinose, the corresponding antibiotics, and, in the case of W3110 groE::kan [pCOMP-groE, pP$_{ara}$I-Sce I], glucose, colonies were scored and a subset of the colonies was genotyped with watermark-specific primers (Table 3).

Table 3: Recovery of functional GroEL-variants and Adk-variants from mock libraries with increasing excess of non-functional variants using the established selection system in comparison with a conventional bleach-out system

Strain	Library (functional: non-functional)	Number of colonies after transformation[a]	Number of genotyped variants	Number of detected false positive variants

W3110 groE::kan [pCOMP-groE, pP$_{ara}$ I –Sce I]	Only functional	~ 27 000	20	0
	1:10²	231	20	0
	1:10³	33	20	0
	1:10⁴	4	4	0
	1:10⁵	1[a]	1	0
	Only non-functional	0	0	0
SBΔrecA	Only functional	~ 25 000	20	0
	1:10²	253	20	0
	1:10³	25	20	1
	1:10⁴	7	7	5
	1:10⁵	6	6	5
	Only non-functional	7	7	7
W3110 adk::kan [pCOMP-adk, pP$_{ara}$ I-Sce I]	Only functional	~ 46 000	20	0
	1:10²	459	20	0
	1:10³	36	20	0
	1:10⁴	3	3	0
	1:10⁵	1	1	0
	Only non-functional	0	0	0

[a]Typically, 10^9 cells were used in the transformation, which would in some cases not have allowed to isolate a positive colony in the high-stringency case. In these cases, the number of transformed cells was increased to $3*10^9$ cells.

In the GroEL-specific system, the numbers of colonies correlated well with the numbers expected from transformation efficiencies and the decreasing number of functional variants in each library. Even from the most stringent library (functional variants at a frequency of 10^{-5}) only functional variants carrying the watermark were recovered after transformation of ~ 81,000 cells. No colony was detected after transformation of approximately the same number of cells with only non-functional variants. In contrast when using SBΔrecA for the same experiment, false positive variants were detected after transformation with the negative control (transformation with only non-functional variants) with a frequency of ~ 2 x 10^{-4}. In addition, we identified false positive variants after genotyping a subset of the colonies which had been selected from the different mock libraries (Table 3). Selection stringency positively correlated with the false positive rate, thus requiring laborious orthogonal assays to differentiate true from false positives. In a directed evolution experiment this would result in the requirement for intensive post-screening of selected variants for true functional library members. Plasmids of three of the false positive variants, which had been isolated after transformation with only non-functional variants, were further analyzed by sequencing. For two of them the stop codon had been reverted to the wild-type codon, probably due to recombination with the chromosomal groL copy. The third analyzed variant still had the stop codon withingroL indicating that the ability to grow must have arisen from a mutation in the araBAD promoter preventing full repression. This phenomenon had been described before for strain MGM100 [16].

The same experiment was performed with the Adk-specific selection system using mock libraries with decreasing ratios of functional Adk-variants to non-functional Adk-variants. Also for this set-up we did not identify any false positive variant and after transformation with the most stringent library we could reliably identify a functional variant containing the watermark after transformation of~ 138,000 cells (Table 3).

Discussion

In this study we present a facile and efficient set-up for a (within the tested boundaries) background-free selection system for functional, engineered, essential proteins. It is based on the conditional elimination of a complementary plasmid-based copy of an essential gene in a knock-out strain in order to replace the essential gene by genes from a library (for example).

We show that I-Sce I nuclease-based cleavage proves to be a suitable strategy for the fast and efficient elimination of a complementing plasmid from a knock-out strain which can occur while the strain is being made transformation-competent. Efficient plasmid elimination is an essential prerequisite for the high performance of the system during evaluation of large libraries under stringent conditions because elimination of the wild-type gene from the cells prior to introduction of variants prevents recombination-based background growth. The major shortcoming of conventionally used bleach-out systems is indeed due to recombination with a silenced wild-type gene during selection. Even in a recA strain, RecA-independent recombination between homologous regions occurs with frequencies between 10^{-3}-10^{-4}[27]. As soon as the number of expected functional variants within a library drops below 10^{-3} selection is primarily for recombination events instead of functional library members. This can clearly be seen in our selection experiment using the GroEL bleach-out strain SBΔrecA and mock libraries with decreasing ratio of functional variants. Even though our strain was deficient forrecA, we frequently isolated false positive variants. Besides being recombination-free, another advantage of the presented set-up over bleach-out systems is that selected variants are directly expressed in a clean strain-background circumventing laborious post-transformation work such as P1-transductions. This allows for subsequent purification of engineered proteins for in vitro characterization and user-definedin vivo applications without running the risk of wild-type contamination.

We also show that even after elimination of the complementing vector from the knock-out strain, cells remain viable for one or more generations - depending on the gene - and electrocompetent cells prepared after elimination yield up to 10^7transformants μg^{-1} DNA.

As in our set-up the essential target genes are expressed from an inducible promoter during competent cell preparation, appropriate bleach-out times can be adjusted for individual gene products by tuning the inducer concentration.

The system was validated with three different essential E. coli proteins: Adk, glyceraldehyde-3-phosphate dehydrogenase and the chaperonin GroEL. Transformation efficiencies and the absence of detectable recombination events proved to be independent of the essential target gene.

CONCLUSIONS

In the current work we present a straightforward, transformation-based system which enables the genetic replacement of a wild-type essential gene of interest by a library or variant. It thereby directly results in the isolation of functional variants in a clean strain background with considerably reduced effort.

Furthermore, it substantially facilitates working with and engineering of essential genes and their protein products making it an experimentally easy, fast and scalable task.

Finally, it should be possible to adapt the here introduced replacement strategy to other hosts – like e.g. yeast, Bacillus subtilis or Clustridium spec. - for which homologous recombination-based knock-out strategies are available [28–30].

MATERIALS AND METHODS

Chemical and Enzymes

Restriction enzymes and ligase were obtained from New England Biolabs (Ipswich, MA, USA) and used according to manufacturers' instructions. Chemicals were purchased in the highest purity available from Sigma-Aldrich, Fluka (Buchs, Switzerland) or Roth (Lauterbourg, France). Trypton and yeast extract were from BD Bioscience (Basel, Switzerland). Oligonucleotides and Sanger-sequencing service were purchased from Microsynth (Balgach, Switzerland).

Strains and Plasmids

E. coli DH10B was used for general cloning procedures. E. coli W3110 was used as the chassis for all chromosomal deletions (see Table 1 for an overview on strains used in this study). SBΔrecA is a derivative of strain MGM100 [10]. It was constructed from BW25113 recA:: FRT [1] by P1-phage transduction using a lysate from MGM100 and selecting for Km ^{R}. The final clone was confirmed by PCR analysis of the groE and recA::FRT locus and by its inability to grow on glucose after GroEL-bleach-out. Plasmid pKOCOMP-adk is a derivative of pKD46 [17] and was constructed by first cloning the adk gene into the multiple cloning site of expression vector pACT3 [31] via restriction sites Bam HI and Hin dIII using primers pKOCOMP-adk-fw and pKOCOMP-adk-rv (see Table S1 for primer sequences). The resulting vector pACT-adk was used as template to amplify the Ptac promoter-controlled adk gene with primers pACT-forward and pACT-reverse, encoding for a I-Sce I recognition site, and cloned into pKD46 via its unique Nco I-site. For construction of pCOMP-adk the P$_{tac}$ promoter-controlled adk gene was amplified with primers pACT-SceI-Spe and pACT-Pac and the PCR product was cloned into the unique Spe I and Pac I sites of pSEVA432 (ori pBBR1, Spec^R resistance). Plasmid pSEVA132-adk is a derivative of pSEVA132 (ori pBBR1, Ap^R) and encodes adk controlled by its natural promoter and fused to a C-terminal 6xHis-tag. It was amplified from genomic E. coli DNA using primers adk-forward and adk-reverse and cloned into pSEVA132 via restriction sites Xma I and Sac I. Plasmid pCOMP-gro E was constructed by amplification of the P$_{tac}$-controlled groE-operon from pACT-ESL using primers pACT-SceI-Spe and pACT-Pac and cloned into the unique SpeI and PacI sites of pSEVA431. pACT-ESL is pACT3 derived and encodes the P$_{tac}$ controlled groE operon. groS was amplified from W3110 by PCR with primers groS-fw andgroS-rv and groL was PCR-amplified with primers groL-fw and groL-rv and then sequentially cloned into the Kpn I and HindIII sites of pACT3. Plasmid pSEVA132-groE was constructed by amplification of the natural groE operon from E. coligenomic DNA using primers groE-forward and groE-reverse and cloning them into the unique sites Xma I and Xba I. Plasmid pCOMP-secBgpsA was constructed by cloning the natural secB-gpsA transcriptional unit, amplified with primerssecBgpsA-forward and gpsA_I-Sce I-reverse, into pSEVA431 via restriction sites Xma I and Xba I. Plasmids pSEVA132-adk ^{stop}, pSEVA132-adk ^{watermark}, pSEVA132-groE ^{stop}, pSEVA132-groE ^{watermark} and pSEVA132-secBgps A^{stop} were constructed by amplification and re-ligation of pSEVA132-adk, pSEVA132-groE or pSEVA132-secBgps A using primers adk-stop-fw/adk-stop-rv, adk-watermark-fw/adk-watermark-rv,   groE-watermark-fw/groE-watermark-rv, groE-stop-fw/groE-stop-rv and secBgpsA-stop-fw/secBgpsA-stop-rv.

Helper plasmids pI-Sce I and pP$_{ara}$ I-Sce I are derivatives of pSEVA671 (ori p15A, Gm^R). The gene for I-SceI nuclease was amplified from plasmid pSTKST [32] using primers I-SceI-fw and I-SceI-rv and cloned into pSEVA671 via Pac I and Eco RI restriction sites. The RhaR-RhaS/P$_{rhaBAD}$ regulatory system was amplified from E. coli genomic DNA using primers Rha-forward and Rha-reverse and cloned in front of I-SceI via restriction sites Nsi I and Spe I. The AraC/P$_{ara BAD}$ regulatory system was amplified from E. coli genomic DNA using primers ParaBAD-fw and ParaBAD-rv and used to exchange RhaR-RhaS/P$_{rha}$ using sites Nsi I and Spe I.

Preparation of Competent Cells and Transformation

E. coli adk::kan [pKOCOMP-adk, pI-Sce I] or [pCOMP-adk, pP$_{ara}$ I-Sce I] were grown overnight in LB liquid broth supplemented with 50 μg mL^{-1} kanamycin, 10 μg mL^{-1} gentamicin, 100 μM IPTG and 0.5% (wt/vol) glucose (to efficiently repress I-Sce I production) at 30°C or 37°C. Cells were pelleted, washed once with LB and diluted 1:100 in fresh LB broth, supplemented with the same antibiotics as before but without glucose. Cells were grown at 37°C. At an OD$_{600}$ of 0.2, 10 mM rhamnose or 0.2% arabinose (wt/vol) was added to induce I-Sce I nuclease production. When reaching an OD$_{600}$ of 0.4-0.5, cells were chilled on ice for 30 min, harvested and washed twice with chilled water and once with 10% glycerol as described [26]. For transformation with the test plasmids and the mock libraries, 50 μL cells (OD$_{600}$ around 100) were mixed with 1.5 ng DNA, exposed to an electrical pulse of 1.3 kV and recovered in 1 mL LB broth supplemented with 10 mM rhamnose or 0.2% arabinose for 1 h at 37°C. Selection was done overnight at 37°C on LB agar plates containing 50 μg mL^{-1}kanamycin, 10 μg mL^{-1} gentamicin, 100 μg mL^{-1} ampicillin and 10 mM rhamnose or 0.2% arabinose. The GroEL-specific selection system based on W3110 groE::kan [pCOMP-groE, pP$_{ara}$ I-Sce I] and the GpsA-specific selection system based on W3110 secBgpsA::kan [pCOMP-secBgpsA, pP$_{ara}$ I-Sce I] were treated the same way. For preparation of competent SBΔrec A, cells were grown overnight in LB supplemented with 50 μg mL^{-1} kanamycin and 0.2% arabinose. Cells were washed twice with water and diluted 1:100 in fresh LB medium. When cells reached an OD$_{600}$ of 0.1 0.5% glucose was added to repress chromosomal GroEL production. Cellular GroEL was bleached for two generations before cells were harvested for competent cell preparation (at OD$_{600}$ of 0.4).

Determination of Plasmid Loss

Loss of pKOCOMP-adk or pCOMP-adk, pCOMP-groE and pCOMP-secBgpsA was determined as follows: The corresponding knock-out strains adk::kan, groE::kan and secBgpsA::kan containing helper plasmid pI-Sce I or pP_{ara} I-Sce I were grown in LB liquid broth supplemented with 50 μg mL^{-1} kanamycin and 10 μg mL^{-1} gentamicin until exponential growth was reached. Next, I-Sce I nuclease production was induced with 10 mM rhamnose or 0.2% arabinose. After induction aliquots were taken after 0, 10, 20, 40, 60, 120 and 180 min and normalized to OD_{600}. Serial dilutions were subsequently plated on LB agar supplemented with 0.5% glucose. The next day the number of colony forming units (CFU) was counted. The number of colonies resulting from aliquots which had been taken from a control culture grown in the absence of rhamnose or arabinose were set to 100% CFU.

Chromosomal Knock-out of Essential Genes

Knock-outs were done by λ red-based recombination with a PCR-encoded kanamycin resistance cassette as described earlier [17]. The kanamycin-cassettes were generated with pKD13 as a template and primers adk- H1 and adk-H2, groE-H1 and groE-H2 or secBgpsA- H1 and secBgpsA-H2. Knock-outs were PCR-verified with primers P1-P6 as indicated in Additional file2 Figure S1 and Additional file 1 Table S1.

ACKNOWLEDGMENT

This work was funded by the EU FP6 projects Eurobiosyn and NANOMOT and the ESF project Nanocell. The authors are indebted to Tania Roberts for proofreading of the manuscript.

ELECTRONIC SUPPLEMENTARY MATERIAL

12934_2012_723_MOESM1_ESM.docx Additional files file 1 : Table S1. Primers used in this study. (DOCX 17 KB)

12934_2012_723_MOESM2_ESM.tiff Additional files file 2 : Figure S1. PCR verification of knock out strains. (TIFF 1 MB)

AUTHORS' ORIGINAL SUBMITTED FILES FOR IMAGES

Below are the links to the authors' original submitted files for images.

12934_2012_723_MOESM3_ESM.tiff Authors' original file for figure 1

12934_2012_723_MOESM4_ESM.tiff Authors' original file for figure 2

12934_2012_723_MOESM5_ESM.tiff Authors' original file for figure 3

AUTHORS' CONTRIBUTIONS

SB performed the experiments, SB and SP designed the experimental approach and wrote the manuscript, SP supervised the research. All authors read and approved the final manuscript.

REFERENCES

1.   Baba T, Ara T, Hasegawa M, Takai Y, Okumura Y, Baba M, Datsenko KA, Tomita M, Wanner BL, Mori H: Construction of Escherichia coli K-12 in-frame, single-gene knockout mutants: the Keio collection. Mol Syst Biol. 2006, 2: 2006-0008.

2.   Rosamond J, Allsop A: Harnessing the power of the genome in the search for new antibiotics. Science. 2000, 287: 1973-1976. 10.1126/science.287.5460.1973

3.   Haselbeck R, Wall D, Jiang B, Ketela T, Zyskind J, Bussey H, Foulkes JG, Roemer T: Comprehensive essential gene identification as a platform for novel anti-infective drug discovery. Curr Pharm Des. 2002, 8: 1155-1172. 10.2174/1381612023394818

4.   Johnson DBF, Xu JF, Shen ZX, Takimoto JK, Schultz MD, Schmitz RJ, Xiang Z, Ecker JR, Briggs SP, Wang L: RF1 knockout allows ribosomal incorporation of unnatural amino acids at multiple sites. Nat Chem Biol. 2011, 7: 779-786. 10.1038/nchembio.657

5.   Gil R, Silva FJ, Pereto J, Moya A: Determination of the core of a minimal bacterial gene set. Microbiol Mol Biol Rev. 2004, 68: 518-537. 10.1128/MMBR.68.3.518-537.2004

6.   Posfai G, Plunkett G, Feher T, Frisch D, Keil GM, Umenhoffer K, Kolisnychenko V, Stahl B, Sharma SS, de Arruda M, et al: Emergent properties of reduced-genome Escherichia coli. Science. 2006, 312. 10.4-1046.

7.   Trinh CT, Unrean P, Srienc F: Minimal Escherichia coli cell for the most efficient production of ethanol from hexoses and pentoses. Appl Environ Microbiol. 2008, 74: 3634-3643. 10.1128/AEM.02708-07

8.   Lee JH, Sung BH, Kim MS, Blattner FR, Yoon BH, Kim JH, Kim SC: Metabolic engineering of a reduced-genome strain of Escherichia coli for L-threonine production. Microb Cell Fact. 2009, 8: 2- 10.1186/1475-2859-8-2

9.   Danchin A: Natural selection and immortality. Biogerontology. 2009. 10.

503-516. 10.1007/s10522-008-9171-5

10. McLennan N, Masters M: GroE is vital for cell-wall synthesis. Nature. 1998, 392: 139-139. 10.1038/32317

11. Herring CD: Introduction of conditional lethal amber mutations in Escherichia coli. Methods Mol Biol. 2008, 416: 323-334. 10.1007/978-1-59745-321-9_21

12. Jin Y, Watt RM, Danchin A, Huang JD: Use of a riboswitch-controlled conditional hypomorphic mutation to uncover a role for the essential csrA gene in bacterial autoaggregation. J Biol Chem. 2009, 284: 28738-28745. 10.1074/jbc.M109.028076

13. Ji Y, Zhang B, Van SF, Horn , Warren P, Woodnutt G, Burnham MK, Rosenberg M: Identification of critical staphylococcal genes using conditional phenotypes generated by antisense RNA. Science. 2001, 293: 2266-2269. 10.1126/science.1063566

14. Tang YC, Chang HC, Chakraborty K, Hartl FU, Hayer-Hartl M: Essential role of the chaperonin folding compartment in vivo. EMBO J. 2008, 27: 1458-1468.

15. Herring CD, Blattner FR: Conditional lethal amber mutations in essential Escherichia coli genes. J Bacteriol. 2004, 186: 2673-2681. 10.1128/JB.186.9.2673-2681.2004

16. van der Vies SM, Lund PA: Determination of chaperonin activity in vivo. Methods Mol Biol. 2000, 140: 75-96.

17. Datsenko KA, Wanner BL: One-step inactivation of chromosomal genes in Escherichia coli K-12 using PCR products. Proc Natl Acad Sci U S A. 2000, 97: 6640-6645. 10.1073/pnas.120163297

18. Tobin JF, Schleif RF: Transcription from the rha operon psr promoter. J Mol Biol. 1990, 211: 1-4. 10.1016/0022-2836(90)90003-5

19. Guzman LM, Belin D, Carson MJ, Beckwith J: Tight regulation, modulation, and high-level expression by vectors containing the arabinose PBAD promoter. J Bacteriol. 1995, 177: 4121-4130.

20. Egan SM, Schleif RF: A regulatory cascade in the induction of rhaBAD. J Mol Biol. 1993, 234: 87-98. 10.1006/jmbi.1993.1565

21. Cousin D, Buttin G: Thermosensitive mutants of K12 Escherichia coli. 3. A lethal mutation of E. coli affecting the activity of the adenylate kinase. Ann Inst Pasteur (Paris). 1969, 117: 612-630.

22. Glaser M, Nulty W, Vagelos PR: Role of adenylate kinase in the regulation of macromolecular biosynthesis in a putative mutant of Escherichia coli defective in membrane phospholipid biosynthesis. J Bacteriol. 1975,

123: 128-136.

23. Kuzminov A, Schabtach E, Stahl FW: Chi sites in combination with RecA protein increase the survival of linear DNA in Escherichia coli by inactivating exoV activity of RecBCD nuclease. EMBO J. 1994, 13: 2764-2776.

24. Tobias JW, Shrader TE, Rocap G, Varshavsky A: The N-end rule in bacteria. Science. 1991, 254: 1374-1377. 10.1126/science.1962196

25. Antoine R, Locht C: Isolation and molecular characterization of a novel broad-host-range plasmid from Bordetella bronchiseptica with sequence similarities to plasmids from gram-positive organisms. Mol Microbiol. 1992, 6: 1785-1799. 10.1111/j.1365-2958.1992.tb01351.x

26. Sambrook J, Russell DW: Molecular cloning : a laboratory manual. 2001, Cold Spring Harbor Laboratory Press, Cold Spring Harbor, N.Y., 3

27. Bi X, Liu LF: RecA-independent and RecA-dependent intramolecular plasmid recombination - Differential homology requirement and distance effect. J Mol Biol. 1994, 235: 414-423. 10.1006/jmbi.1994.1002

28. Wang Y, Weng J, Waseem R, Yin X, Zhang R, Shen Q: Bacillus subtilis genome editing using ssDNA with short homology regions. Nucleic Acids Res. 2012, 40: e91- 10.1093/nar/gks248

29. Kuehne SA, Minton NP: ClosTron-mediated engineering of Clostridium. Bioengineered. 2012, 3: 245-252.

30. Wach A, Brachat A, Pohlmann R, Philippsen P: New heterologous modules for classical or PCR-based gene disruptions in Saccharomyces cerevisiae. Yeast. 1994. 10. 1793-1808.

31. Dykxhoorn DM, St Pierre R, Linn T: A set of compatible tac promoter expression vectors. Gene. 1996, 177: 133-136. 10.1016/0378-1119(96)00289-2

32. Kolisnychenko V, Plunkett G, Herring CD, Feher T, Posfai J, Blattner FR, Posfai G: Engineering a reduced Escherichia coli genome. Genome Res. 2002, 12: 640-647. 10.1101/gr.217202

33. Hayashi K, Morooka N, Yamamoto Y, Fujita K, Isono K, Choi S, Ohtsubo E, Baba T, Wanner BL, Mori H, Horiuchi T: Highly accurate genome sequences of Escherichia coli K-12 strains MG1655 and W3110. Mol Syst Biol. 2006, 2: 0007-

34. Durfee T, Nelson R, Baldwin S, Plunkett G, Burland V, Mau B, Petrosino JF, Qin X, Muzny DM, Ayele M, et al: The complete genome sequence of Escherichia coli DH10B: insights into the biology of a laboratory workhorse. J Bacteriol. 2008, 190: 2597-2606. 10.1128/JB.01695-07

Chapter 3

SQPRIMER: THE UTILITY OF DESIGNING HOMOLOGOUS PRIMERS FOR THE GENETIC ANALYSIS BASED ON THE PCR

Julio Perez-Marquez

Department of Biomedicine and Biotechnology, University of Alcalá de Henares, Spain

INTRODUCTION

Primers are a short chain of nucleotides chemically synthesized in orientation 5' to 3' that are used for amplification of DNA by the polymerase chain reaction (PCR). The PCR technique includes three steps in one cycle: the DNA denaturation, the alignment of two primers to the extremes of the sequence of DNA to be amplified, and the synthesis of DNA using polymerases; the repetition of this cycle (c) produces 2c molecules of DNA named amplicons. Many genetic applications use the PCR; among them, those that detect variations in the nucleotide sequence of genes. For instance, PCR is used in the analysis of mutations involved in genetic diseases, in forensics and in the studies of differences amongst species. The design of specific and selective primers is a critical factor for a successful DNA amplification by PCR [1,2].

There are several software programs in the web that design specific primers for the PCR [3-8]. Some applications are devoted to specialized tasks such as primer-BLAST [9] that designs primers that do not match to any other DNA, apart from the one of interest. By contrast, there are software applications that obtain primers from a group of nucleotide sequences; among others the Primaclade and BatchPrimer3 software which are based on Primer3 [10,11]. That particular software has an array of different applications; for instance, it designs degenerated primers, finds primers that recognize microsatellites of nucleotide repeats (or SSR, simple sequence repeat) or detects primers that include single nucleotide polymorphism (SNP), a kind of nucleotide variation in the DNA. The existing software has in common that they serve to design primers that are different and specific for one or various DNA templates.

Some areas of the genetic analysis that use the PCR require the design of homologue primers in divergent nucleotide sequences. There are two possible strategies to distinguish by PCR different DNAs with a degree of homology: one is to use primers that are unique to each sequence, the other is to use a combination of primers that are unique to each sequence and homologous to the sequences in the analysis. In the case of sequences that vary in one single nucleotide the second strategy is the only alternative: one of the pair of primers that will be used in the PCR is homologous to all sequences in the analysis. Here I show the design of one application named SQPrimer that is particularly useful in produce the primers that are identical in multiple DNAs together with primers that are unique to each sequence.

In the year 2002 we cloned a cDNA from rat tissue that codifies a putative protein (CLRP) with complex leucine repeats [12]. With concrete examples that use this and other sequences of nucleotides [13], I show that SQPrimer designs primers at conserved regions among different species and also serves to find and to design mutation specific primers in sequences with SNPs by using different strategies of the PCR analysis. SQPrimer is also a tool to design primers to detect length polymorphisms: insertions, deletions or expansion of triplet nucleotide repeats.

METHODS

Algorithms and Conditions for the Design of Primers in SQPrimer

The most important function of SQPrimer contains one algorithm that finds short strings of nucleotides (primers) in the inputted sequence templates. The process to design primers for the PCR requires setting several variables; thus, this function depends on the values of the primer conditions that are indicated by the user in one panel of the application. The variables in the function are: the primer length and the number of guanines+cytosines/primer length (%GC), and the number of repeats of single nucleotide (i.e. AAA...) or dinucleotide (i.e. ATATAT...) that are allowed in its sequence. The algorithm also evaluates the melting temperature (Tm) of the alignment of the primer to the DNA template; regarding to this variable, three different options can be selected by the user: the basic, the simple and the salt 50mM, with equations that have previously described [2,15]:

Basic: $nAT \times 2° + nGC \times 4°$.

Simple: $64.9° + 41° \times (nGC-16.4)/\text{primer length}$

Salt: $81.5° + 16.6° \times (\log_{10} [0.05]) + 0.41° \times (\%GC) - 675/\text{primer length}$.

(nAT: number of adenine + thymine; nGC: number of guanine + cytosine;

%GC: nGC/primer length; °: Celsius degrees). Basically, the algorithm starts with the declarations of two arrays of the size of the length of the primer; one array is for the sense and the other for the antisense. Another two arrays serve to memorize the position of the primers in the template. Then, the algorithm progressively takes the primers from the sequence template up to its full length. For each primer, the algorithm determines the presence of single or double nucleotide repeats; if the selected primer fits the conditions indicated by the user, then the algorithm determines whether the primer also fits with the Tm. Primers that fulfill all conditions are stored, together with its position, in the arrays. Finally, the algorithm evaluates whether each pair of sense and antisense primers in the arrays are separated in the template the distance that has been specified by the user; if they do, both primers are displayed in the application

Software Design

I applied object-oriented programming using the C++ Builder 2009 application from Embarcadero technologies to produce the SQPrimer software that runs in the Windows environment. This programming tool has been previously tested for the design of bioinformatics applications [14]. The SQPrimer software is open access at http:// www2.uah.es/biologia_celular/JPM/SQPP/SQPrimer. html.

Applications within SQPrimer and input of nucleotide sequences: SQPrimer contains two main interconnected windows: the multi-sequence application and the tool that design pairs of sense and antisense primers for single DNA templates. Additionally, there is one primer analysis tool that analyzes the nucleotide composition and the self-complementarity of the primers and there are also two graphical displays (Figure 1). The multi-sequence tool requires the input of various nucleotides sequences in the text box; the nucleotide sequences can be also pasted by the user and, alternatively, the application can open *.txt files or *.SQP files that can previously be saved with the single sequence application. Once the template sequences are included in the interfaces, the production of the primers by the software requires two steps: first, to indicate the values primer conditions and then to click the buttons of function. The functions in the two main applications will produce lists of primers that meet the conditions established by the user; those lists can be exported to excel (Microsoft) and thus, SQPrimer features connectivity to other functions and applications of the Windows environment. The design of primers is accompanied with the display of a graphical representation of the position of the primers in the nucleotide templates. Additional features of SQPrimer that help the user to familiarize with the application are instruction menus in each window as well as examples of the nucleotide sequences that are explained in the results. Compared to other software, SQPrimer offers simplicity, as it is required for a didactic tool; for instance, if the software does not produce results using a particular set of primer conditions, a new search can be done in few steps by changing the variables of primer length and/or GC content followed by a click on the button of the function

The multi-sequence application: Two different functions can be run in this application clicking in the respective buttons: one that designs the primers that are different or identical in various sequences and another that produces primers that detect length polymorphisms. For the first function, the software produces two groups of results: primers that can be found in all the templates (identical) and primers that are unique to each inputted sequence; all of them meet the conditions specified by the user. If the application finds primers that are unique, SQPrimer will automatically display the primers that have one distinctive nucleotide at the extreme 3'. As shown in the results, the multi-sequence application serves to clone orthologous DNAs and is also useful for the detection of SNPs. The function that designs primers to analyze length polymorphisms of the DNA produces pairs of primers that flank nucleotide sequences that differ in either deletions or insertions, including the expansion of trinucleotide repeats.

Cloning of the cDNA of CLRP from different species: To design primers to clone the CLRP cDNA of CHO cells I followed one strategy based in the design with SQPrimer of primers that are homologues in the rat and human nucleotide sequences. We had previously cloned the CLRP gene of Rattus norvergicus [GenBank: AF406814.1]) and blasted the sequence to obtain information of a similar nucleotide sequence from the Homo sapiens chromosome 5, BAC clone from the database [GenBank: AC005214.1]. One segment of the human cDNA containing CLRP was isolated from human prostate Marathon-Ready cDNA (Clontech) using a PCR strategy based on the sequence homology between these two species. I tested a batch of homologous primers for the rat CLRP cDNA that were designed with SQPrimer to amplify the human cDNA from human prostate and obtained a positive PCR reaction. The human cDNA of CLRP isolated from human prostate was 2642 nt long (nt: nucleotide).

One positive PCR reaction on human DNA was obtained with the homologous sense 5'-AGGGCATCAGCAGTATTG-3' and antisense 5'-GAGGAAGAGGTTCTGAAG-3' primers from the published rat cDNA (nt 1-18 and 1174-1191, respectively) for 30 cycles of denaturation at 94°C, annealing at 55°C for 30 sec, each and extension at 72°C for 2 min. After sequencing, the extremes of the human cDNA were amplified by rapid amplification of cDNA extremes (RACE) using the nested and polyT adaptor primers included in the Marathon cDNA amplification kit (Clontech laboratories) using the following rat primers: the 5' was amplified with the human antisense primer 5'-CCATCCTCTACACTCATAC-3' (nt 720-438) and the 3' extreme with the sense primer 5'-CATTCTGTACTGCCTCATC-3' (nt 1297- 1315) at 94°C for 30 sec, 55°C for 30 sec and 68°C for 3 min. Finally The PCR products obtained in the reactions were gel purified, subcloned in the pGEM-T vector (Promega) and sequenced. The human CLRP cDNA obtained was 2642 nt long.

Results

Cloning of orthologous cDNAs using homologous primers designed by SQPrimer

The applicability and accuracy of SQPrimer was tested in the laboratory in concrete examples of genetic analysis. The steps to clone the CLRP cDNA of CHO cells was as follows: 1-To obtain the human cDNA (described in the methods), 2-To use of SQPrimer to design the primers that are identical in the rat and human sequences, 3-To use different combinations of those homologous primers in PCRs using the cDNA of CHO cells as template. 4-To isolate and purify the PCR products in the agarose gels and clone that cDNA.

Having obtained the CLRP cDNA from rat and human I used these two sequences as template and run one of the functions included in the multi-sequence application in order to find both the primers that are identical in the two templates as well as those primers that are different (unique to each template). The application was used with the following fixed values of the primer conditions: GC=50% ± 2 and basic Tm of 54 ± 2. As expected, the SQPrimer application designed more primers if the restrictive conditions of the variables of the primer are more permissive: no homologous primers were found in the templates at any primer length selected if nucleotide repeats was 0. By contrast, if 2 repeats of one single nucleotide and 2 repeats of dinucleotide were allowed the software designed 4 primers of 18 nt, 2 of 20 nt but none over 21 nt long. Thus, the number of homologous primers designed by the application also increased with the decrease of the primer length

I purified the total cDNA of CHO cells and made different PCR amplifications using different combinations of these four homologous sense and antisense primers (Figure 2). It should be warned that this strategy may not always produce results since only three, out of the four primers that are identical in rat and human, produced positive PCR reactions. One 315 nt long PCR product was obtained with two sense and antisense primers and was subcloned and sequenced. After confirming nucleotide homology to the rat CLRP cDNA I proceeded to amplify the 5' and 3' extremes to obtain the full cDNA sequence by RACE. The final cDNA product was 2306 nt long and displays 94.7 % homology with the rat ortholog CLRP (data not shown). I conclude that the multi-sequence function of SQPrimer is useful to design primers at conserved regions of nucleotides among different species or DNAs sequences that display a degree of homology.

Detection of Alleles by PCR Using the Unique/Identical Function of SQPrimer

In one analysis of human DNA samples I found one abundant allele of CLRP (allele A) and one individual that had one allelic sequence with two single nucleotide differences in the open reading frame (allele B). In allele A, the codon at position 950 of the cloned DNA was GCC which encodes Ala and one second codon starting at position 962 was GCA, which also encodes Ala. By contrast, the allele B had ACC at the first codon and GTA in the second codon, which is translated to Thr and Val, respectively (Figure 2). The two allelic sequences of CLRP are included in the examples of the SQPrimer application.

All primers designed by SQPrimer that are unique to one sequence should hybridize differentially the SNPs in the alleles A and B. Because it is generally accepted that among all possible primers the best ones to recognize the SNPs are those that have the differential nucleotide at the extreme 3', I selected two 18 nt long reverse (or antisense) primers, each unique to one of the alleles, that have that feature and that were provided by the function included in SQPrimer. The two reverse primers that were selected are located at the same position in the two DNA templates (Figure 2). To recognize the alleles in agarose gels by their lengths, a combination of 18 nt long primers was used in two PCR reactions per individual. One reaction was carried out with one of the reverse primers and one forward primer that is identical in both sequences; the second PCR reaction was performed with the second reverse primer and a different forward primer at a different position than the one used in the previous reaction, which is also present in both sequences (Figure 2). As shown in the gel of the figure, one individual is homozygous for the allele A and the other is heterozygote and has the alleles A and B of the CLRP gene. In conclusion, the multi-sequence application of SQPrimer was useful in the design of specific primers that detect nucleotide sequences with SNPs.

The functionality of SQPrimer was tested by changing the variables of the design of primers with these two sequences as templates. For instance, the software was run with a fixed value of the primer length of 20 nt. The application found no differential primers if the repetition of one single nucleotide was not allowed; conversely, if that repetition is allowed in the primer sequence, the number of unique primers designed increases with increasing variation in the proportion of GCs and if the presence of two nucleotide repeats is admitted (Table 1). I conclude that SQPrimer produces results that are consistent with the effect of restricting the conditions of the primers.

Using the SQPrimer application to design identical primers that detect length polymorphisms in various sequences

Insertions and deletions of nucleotides are forms of genetic mutations in the DNA; they range from one to a large amount of nucleotides. There is one function in SQPrimer that detects homologous primers in several sequences that may serve to distinguish length polymorphisms. Two different examples were included in the application to test this functionality; starting from the sequence of the CLRP cDNA, I artificially created two sequences using a word processor: one sequence with a single nucleotide insertion and another with one nucleotide deletion. A second example is an insertion that consists on the extension of triplet nucleotide repeats; a kind of mutation that is shared by a group of genes that cause genetic diseases. In this example, the sequences included in the application were: the real coding sequence of the Huntingtin mRNA and one sequence with a repeat of nucleotide triplets that expands 63 nt, which was also constructed artificially. As shown in the Figure 3, the application finds pairs of sense and antisense primers that are identical in the sequences of the two Huntingtin alleles and also displays the different lengths of expected PCR products for each sequence using those homologous primers. In conclusion, this application of SQPrimer designs primers that flank the sites at where the length polymorphism occurs and serve to detect insertions, deletions or expansions of trinucleotide repeats

Discussion

The PCR method has a large applicability in very different fields such as the detection of pathogens, drug discovery, genetic engineering, genetic diagnoses, mutagenesis, molecular anthropology, genetic phylogeny, etc. In all these fields of research the products of the PCR are used to clone and to sequence DNAs. For any application, the technique of the PCR largely depends on the design of primers that are used in the reaction.

SQPrimer might be useful in fields such as anthropology or phylogeny. I believe that the finding of identical primers in a large number of sequences from different species is one strategy to find the same sequence in new species. With this approach, SQPrimer showed to be useful to clone the CLRP cDNA of CHO cells; the results indirectly show that CLRP is a conserved gene in vertebrates because the homology of sequence of their primers SQPrimer proved useful in the design of identical primers for the analysis of different types of polymorphisms. The PCR can determine directly the presence of

SNPs between individuals by using a combination of primers that are unique to each sequence and primers that identical in the templates; this application of the PCR does not require additional steps of DNA purification, cloning and sequencing. There are two methods to try to avoid mispriming in the research of sequences with high homology by PCR; one is at the step of the primer design using applications as SQPrimer: it would be advisable to design primers with the highest possible values of the Tm and to select primers with the distinctive nucleotide at the extreme 3'. The second method can be carried out at the laboratory: test those primers in different PCR reactions with increasing melting temperatures

Several genetic human diseases are caused by deletions and insertions in genes; there is also a group of diseases that are determined by the extension of the number of triplet repeats in particular genes [16]. SQPrimer is focused in the design of identical primers in different DNA sequences that display these kinds of polymorphisms; therefore, the application may be useful in the studies of human genetic mutations that use molecular techniques based on the PCR.

The functionality of SQPrimer can be compared with some existing primer design tools. Primer3 (http://biotools.umassmed.edu/bioapps/ primer3_www. cgi) [3], Primer3Plus (http://www.bioinformatics.nl/ cgi-bin/primer3plus/ primer3plus.cgi/) [6] or Primer-Blast (http:// www.ncbi.nlm.nih.gov/tools/ primer-blast/) are tools for finding unique primers in single PCR templates; in contrast, the multisequence application of SQPrimer designs the differential and also the homologous primers in several templates. Primaclade (http:// primaclade.org/index.html) admits a group of nucleotide sequences; the software is developed to design minimally degenerated primers to work reliably and specifically on a number of species [10]; in contrast, SQPrimer does not produce degenerated primers but primers that can directly be used in PCR analysis. Some of the indicated applications require the input of a previously formatted sequence, others include a considerable amount of variables that need to be set in the design of primers; in the case of SQPrimer, any sequence can be pasted in the application and the simplicity in the number of the most important variables of the primer design may facilitate an educational use of the application. Finally, the function of SQPrimer that localizes the primers that are identical in the inputted sequences is a tool that may help researchers that do not know where to expect the length polymorphism in the sequences or to uncover unexpected insertions or deletions of nucleotides in batches of sequences. In conclusion, a common feature of existing software is that it is developed to design primers that are unique for each template; SQPrimer adds the capability of designing primers that are identical in a batch of sequences and this an utility that can be used in some strategies of the PCR analysis.

Present research in my laboratory is focused on developing bioinformatics software that covers different aspects of genetic engineering and has educational utility. SQPrimer is related to bioinformatics software named SQRestriction that serves for various types of restriction analysis of nucleotide sequences [17]. The main limitation of SQPrimer is that it is an executable (*.exe) application and it is not a multi-platform; therefore it only runs in the Windows OS environment. Having that limitation in mind, future work includes the translation of the C++ code in SQR to Java and implementing its interfaces in Html5

CONCLUSION

SQPrimer is a bioinformatics tool that designs the primers that are identical in different DNAs together with primers that are unique to each sequence, as it is required for several types of genetic analysis that use the PCR technology. Because its usability, the application may also be an educational tool for teaching the requirements of the design of primers for the PCR. The application was tested in the cloning of orthologous genes and in finding of SNP by PCR. SQPrimer is particularly interesting to design homologous primers that detect length polymorphisms, including the expansion of triplet repeats

ACKNOWLEDGEMENTS

Thanks to Daniel Pérez Grande for the scientific review of the manuscript. This work was supported by Ministerio de Economía y Competitividad, Spain (grant number: BFU2011-30217-C03-01).

REFERENCES

1. Saiki RK, Scharf S, Faloona F, Mullis KB, Horn GT, et al. (1985) Enzymatic amplification of beta-globin genomic sequences and restriction site analysis for diagnosis of sickle cell anemia. Science 230: 1350-1354.

2. Murphy WJ, O'Brien SJ (2007) Designing and optimizing comparative anchor primers for comparative gene mapping and phylogenetic inference. Nat Protoc 2: 3022-3030.

3. Rozen S, Skaletsky H (2000) Primer3 on the WWW for general users and for biologist programmers. Methods Mol Biol 132: 365-386.

4. van Baren MJ, Heutink P (2004) The PCR suite. Bioinformatics 20: 591-593.

5. Arányi T, Váradi A, Simon I, Tusnády GE (2006) The BiSearch web

server. BMC Bioinformatics 7: 431.

6.   Untergasser A, Nijveen H, Rao X, Bisseling T, Geurts R, et al. (2007) Primer3Plus, an enhanced web interface to Primer3. Nucleic Acids Res 35: W71-74.

7.   Qu W, Zhou Y, Zhang Y, Lu Y, Wang X, et al. (2012) MFEprimer-2.0: a fast thermodynamics-based program for checking PCR primer specificity. Nucleic Acids Res 40: W205-208.

8.   San Millán RM, Martínez-Ballesteros I, Rementeria A, Garaizar J, Bikandi J1 (2013) Online exercise for the design and simulation of PCR and PCR-RFLP experiments. BMC Res Notes 6: 513.

9.   Ye J, Coulouris G, Zaretskaya I, Cutcutache I, Rozen S, et al. (2012) Primer-BLAST: a tool to design target-specific primers for polymerase chain reaction. BMC Bioinformatics 13: 134.

10.   Gadberry MD, Malcomber ST, Doust AN, Kellogg EA (2005) Primaclade--a flexible tool to find conserved PCR primers across multiple species. Bioinformatics 21: 1263-1264.

11.   You FM, Huo N, Gu YQ, Luo MC, Ma Y, et al. (2008) BatchPrimer3: a high throughput web application for PCR and sequencing primer design. BMC Bioinformatics 9: 253.

12.   Pérez-Márquez J, Reguillo B, Paniagua R (2002) Cloning of the cDNA and mRNA expression of CLRP, a complex leucine repeat protein of the Golgi apparatus expressed by specific neurons of the rat brain. J Neurobiol 52: 166-173.

13.   (1983) A novel gene containing a trinucleotide repeat that is expanded and unstable on HD chromosomes. The Huntington Disease Collaborative Research Group. Cell 72: 971-983.

14.   Garcia-Segura LM, Perez-Marquez J2 (2014) A new mathematical function to evaluate neuronal morphology using the Sholl analysis. J Neurosci Methods 226: 103-109.

15.   Dieffenbach CW, Lowe TM, Dveksler GS (1993) General concepts for PCR primer design. PCR Methods Appl 3: S30-37.

16.   Goldberg YP, Andrew SE, Clarke LA, Hayden MR (1993) A PCR method for accurate assessment of trinucleotide repeat expansion in Huntington disease. Hum Mol Genet 2: 635-636.

17.   Perez-Marquez J (2014) SQRestriction: Bio-Informatics Software for Restriction Fragment Length Polymorphism of Batches of Sequences. J Comput Sci Syst Biol 7: 186-192.

Chapter 4

DEVELOPMENT OF ENDOTHELIALSPECIFIC SINGLE INDUCIBLE LENTIVIRAL VECTORS FOR GENETIC ENGINEERING OF ENDOTHELIAL PROGENITOR CELLS

GuanghuaYang[1,2], M. Gabriela Kramer[1], Veronica Fernandez-Ruiz[1], Milosz P. Kawa[1], Xin Huang[3], Zhongmin Liu[2], Jesus Prieto[1] & Cheng Qian[4]

[1]Division of Gene Therapy and Hepatology, Center for Applied Medical Research (CIMA), University of Navarra, Pamplona, Spain

[2]Research Center for Translational Medicine, East Hospital, School of 1Medicine, Tongji University, Shanghai 200120, China

[3]Present address: Department of Biotechnology, Instituto de Higiene, Facultad de Medicina, Universidad de la República, Av. A. Navarro 3051, 11600 Montevideo, Uruguay

[4]Magee-Womens Research Institute, University of Pittsburgh School of Medicine, Pittsburgh, PA 15213, USA

ABSTRACT

Endothelial progenitor cells (EPC) are able to migrate to tumor vasculature. These cells, if genetically modified, can be used as vehicles to deliver toxic material to, or express anticancer proteins in tumor. To test this hypothesis, we developed several single, endothelial-specific, and doxycycline-inducible self-inactivating (SIN) lentiviral vectors. Two distinct expression cassettes were inserted into a SIN-vector: one controlled by an endothelial lineage-specific, murine vascular endothelial cadherin (mVEcad) promoter for the expression of a transactivator, rtTA2S-M2; and the other driven by an inducible promoter, TREalb, for a firefly luciferase reporter gene. We compared the expression levels of luciferase in different vector constructs, containing either the same or opposite orientation with respect to the vector sequence. The results showed that the vector with these two expression cassettes placed in opposite directions was optimal, characterized by a robust induction of the transgene expression (17.7- to 73-fold) in the presence of doxycycline in several endothelial cell lines, but without leakiness when uninduced. In conclusion, an endothelial lineage-specific single inducible SIN lentiviral vector has been developed. Such a lentiviral vector can be used to endow endothelial progenitor cells with anti-tumor properties.

INTRODUCTION

For gene therapy-based anti-tumor treatment, therapeutic genes need to be specifically introduced and highly expressed in neoplastic cells, which remains a challenge in the field. Although some lentiviral vectors and replication deficient recombinant adenovirus vectors carrying specific transgenes demonstrate clear therapeutic benefits in a variety of animal tumor models, clinical trials show that these gene therapy systems possess very low anti-tumor capability because of their low specificity in the transduction of neoplastic cells[1]. An alternative strategy has been developed for gene therapy of solid tumors, based on the observation that tumor growth depends on the number of recruited endothelial cells, which contribute to the generation of functional neo-vasculature. Endothelial progenitor cells (EPCs) are considered functional platforms for gene therapy because of their ability to home to the tumor vasculature and to develop new vessels. Bone marrow–derived EPCs have also been frequently detected both in the circulation of cancer patients and in lymphoma-bearing mice. In addition, tumor-targeted migration of EPC from the bone marrow is correlated with tumor volume and the production of VEGF by tumor cells[2,3]. The homing of EPCs to the tumor vasculature may lead to their incorporation throughout the tumor mass — up to 95% of the tumor vasculature in the peripheral region[4,5]. Transduction of these endothelial cells with therapeutic genes holds the potential to retard the tumor growth—even to eradicate it.

Lentiviral vectors are unique tools for gene delivery into the hematopoietic system because of their biological properties and the relatively easy manipulations required for ex vivo gene transfer[1]. In addition to differentiated cells, lentiviral vectors can efficiently transduce committed progenitors and primitive hematopoietic stem cells[6,7]. One study has shown that lentiviral vectors can be used for the in vitro transduction of human umbilical vein endothelial cells (Huvec) and human bone marrow–derived mesenchymal stem cells with high efficiency[8]. The angiogenic potential of EPCs genetically modified by lentiviruses may be particularly useful in anti-angiogenic therapies of cancer; e.g., in attenuated tumor growth, induced tumor apoptosis and increased survival in vivo and in vitro. It has been proposed that delivering these genes by EPC has the advantage of generating high concentration of proteins that is specific to tumors, avoiding systemic toxicity. However, safety also requires that the expression of the therapeutic genes to be well controlled.

For selective gene expression in endothelial cells, endothelial cell–specific promoters are required9. Of note, the vascular endothelial cadherin (VEcad) gene is exclusively and constitutively expressed in endothelial cells9 and may be a good candidate for genetic manipulation to achieve specific destruction of solid tumor vascular system.

For controlled gene expression in vivo, the Tet-on/off system has been widely used10. In Tet-on/off system, a reverse repressor of tetracycline operon, rtetR, is fused to a herpes simplex virus transcriptional factor VP16, to generate a reverse tetracycline-controlled transactivator (rtTA), which interacts with the inducible promoter in the presence of doxycycline (Dox) and activates transcription1,10,11. The inducible The tetracycline-responsive element (TRE) promoter is composed of seven copies of the Tet operator (tetO) fused to a cytomegalovirus minimal promoter region (CMVm). For better performance, one mutant derived from rtTA, named rtTA2S-M2, was introduced into the Tet-on/off system. This mutant transactivator binds with much lower efficiency to the tetO regions than rtTA in the uninduced state, and its VP16 domain was shortened to avoid cell toxicity11,12. This mutant transactivator is highly sensitive to Dox; it can induce the same gene expression levels at 10% of Dox dose required with the original rtTA12. The fused CMVm promoter, due to its self-activation, may cause a basal activation of the Tet-on system13; i.e., expression leakage. Thus, a tTS(Kid) repressor was used in the all-in-one inducible lentiviral vector containing CMVm inducible promoter14. Although the repressor protein had no effect on the already minimal uninduced expression level, it had a negative influence on the induction process in the presence of Dox14,15. To further minimize expression leakage without using repressor molecules in the Tet-on system, a Tet-responsive promoter for albumin (TREalb) was generated10.

Tet-regulated transgene expression in a host cell requires the delivery of two expression cassettes, one cassette contains the transactivator gene, and the other contains the transgene of interest1,10,11. However, combining the two expression cassettes (Tet-on system) in a single vector can reduce the potential integration sites in the host cells. The quantity of the vector administered to the patient should also be as small as possible in order to minimize the risk of oncogene activation by the integration of the vector to the host genome16. In the case of lentiviral vectors, which belong to the RNA vectors family, the possibilities of inserting more than one expression cassette appear to be limited because of the generic use of the termination of transcription by poly (A) sequences. For the expression of small regulable transgenes, such as green fluorescent protein (GFP), an overlapping expression cassette, placed in the same orientation, is used to minimize the size of the inserted gene17.

Alternatively, opposite cassette orientations with respect to the vector RNA can also be used in lentiviral vectors in which a monodirectional poly (A) signal has been added to the antisense cassette14. The cassette placed in the opposite orientation has shown a higher transgene expression level14. In this study, we developed single endothelial-specific and Dox-inducible self-inactivating (SIN) lentiviral vectors, which was constructed by deleting 133 bp promoter and enhancer elements from the U3 region of the 3' long terminal repeat (LTR)18. The deletion is transferred to the 5' LTR after reverse transcription and integration in infected cells, resulting in the transcriptional inactivation of the LTR in the proviruses and thus, the risk of oncogene activation due to viral insertional mutagenesis can be eliminated using this safety feature of the SIN lentiviral vectors. We compared the expression level of luciferase as a report gene in different vector constructs in endothelial cell lines. Our goal was to generate a single inducible lentiviral vector for specific transduction of endothelial cells. This vector should have three optimal features: (1) no leakiness of the transgene expression in the uninduced state, (2) a high level of gene expression in the presence of doxycycline as an inducer in the Tet-on system, and (3) a high titer of lentiviral vector production. Our results demonstrated a single endothelial lineage-specific inducible SIN lentiviral vector with opposite orientation of two expression cassettes that meets the above requirements. This lentiviral vector was then used to genetically modify endothelial progenitor cells, endowing them with anti-tumor properties.

Results

Construction of inducible lentiviral vectors for endothelial-specific regulable gene expression. We constructed several lentiviral vectors based on a pRRLsin-derived self-inactivating backbone19, in which the different elements of the Tet-on system were introduced sequentially (Fig. 1). Two expression cassettes were inserted into the lentiviral vector. Firefly luciferase was used as a reporter gene under the control of TRE/alb, consisting of a TRE and mouse albumin promoter in the first cassette (vector: TRELuc, Fig. 1A). The second cassette consisted of the endothelial specific promoter (VEcad) for driving the expression of the Tet-responsive transactivator (rtTA2s-M2) in the endothelial cells (vector: VEcadrtTA, Fig. 1B). The second cassette was oriented in the antisense direction with respect to the vector RNA sequence. A monodirectional poly (A) (60PA) sequence was inserted into the second cassette for polyadenylation of the rtTA2s-M2 transcript without disturbing the transcription of the genome during the process of the lentiviral vector production (vector: SindLuc-A1, Fig. 1C). The promoters of the two cassettes were separated by a 2.9-Kb stuffer sequence to suppress the transactivation effects between them (vector: SindLuc-A2, Fig. 1D). In the other construction, where the two cassettes were oriented in the sense direction with respect to the vector RNA, the two cassettes shared the same poly (A) signal of the vector in the 3'long terminal repeat (LTR) (vector: SindLuc-S, Fig. 1F).

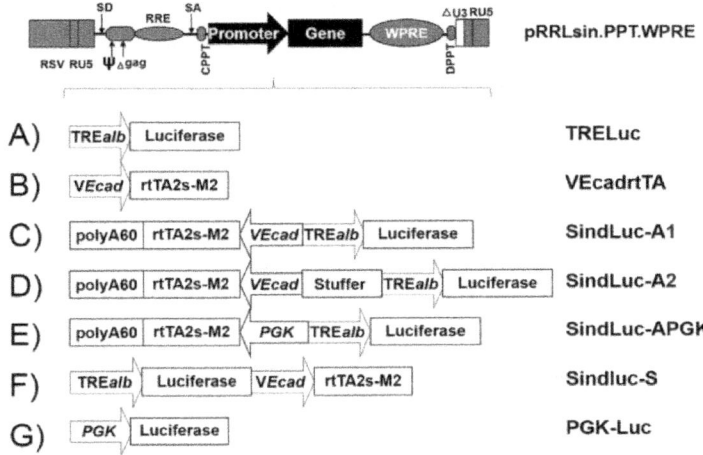

The constructed Lentiviral vectors

Figure 1: The schematic representations of the lentiviral vectors designed for this study.

Vectors are based on the third generation self-inactivated lentiviral vector backbone (pRRLsin.cPPT.WPRE). (A) TRE-luc: vector containing luciferase as a reporter gene under the control of the Tetracycline-regulated albumin promoter (TRE/alb). (B) VEcadrtTA: a vector containing an endothelial-specific promoter (VEcad) for driving the expression of the Tet-responsive transactivator (rtTA2s-M2). (C) SindLuc-A1: a vector containing two cassettes oriented in the antisense direction with respect to the vector RNA. (D) SindLuc-A2: a vector containing the stuffer DNA sequence between the two promoters (based on SindLuc-A1). (E) SindLuc-APGK: a vector containing human phosphoglycerate kinase (hPGK) promoter replacing VEcad promoter in the SindLuc-A1 vector. (F) SindLuc-S: a vector containing two cassettes oriented in the sense direction with respect to the vector RNA sequence. (G) PGK-Luc: vector containing human phosphoglycerate kinase (hPGK) promoter driving luciferase gene expression and used as a positive control in the experiments. Abbreviations: RSV, the Rous Sarcoma Virus promoter driving viral mRNA; SD, major splice donor site; ψ, encapsidation signal including the 5′ portion of the gag gene (Δ gag); RRE, Rev-response element; SA, splice acceptor site; cPPT, central polypurine tract; WPRE, the post-transcriptional regulatory element of woodchuck hepatitis virus.

Comparison of Production titer of Different Inducible Vectors

In order to compare the production titer of different inducible vectors in Fig. 1, we examined the lentiviral titers in three independent viral production experiments. The lentiviral production was performed on five 15-cm dishes each time, concentrated by ultracentrifugation, and titrated by qPCR. The results indicated that the viral titer was not affected by inserting the antisense expression cassette and 60PA in the lentiviral vector (compared to the SindLuc-S vector) (Table 1). In addition, there is no significant difference between those single inducible vectors containing opposite expression cassette.

Table 1: The lentiviral titers from different vectors

Constructed Lentiviral vectors	Final Titer (TU/mL)	Size of Inserted Fragment (bps)
TRELuc	7.3×10^8	1900
VEcadrtTA	8.1×10^8	1075
SindLuc-A1	4.0×10^8	3428
SindLuc-A2	5.3×10^8	6129
SindLuc-APGK	3.8×10^8	3642
SindLuc-S	2.6×10^8	3387
PGK-Luc	7.3×10^8	2312

Efficacy of transgene expression mediated by inducible vectors in endothelial and non-endothelial cell lines

The induction levels of different inducible vectors were analyzed in six endothelial and non-endothelial cell lines, including Humevc, Huvec, mEPC, Svr, Hep3B, and HeLa cells. Each cell line was transduced with different inducible lentiviral vectors at a multiplicity of infection (MOI) of 5, induced with 1 μg/mL Dox for another 68 hours, and harvested for luciferase activity measurement (Fig. 2A–D). We detected basal luciferase expression in the case of the regulable cassette vector, TRELuc, infected together with transactivator expression cassette vector, VEcadrtTA, indicating a high leakage level of luciferase (50–2600 RLU/μg protein) in the absence of Dox. In the case of a single inducible vector, SindLuc-A1, the leakage level of luciferase expression was relatively low (50–1000 RLU/μg protein). Moreover, we observed a high leakage level of luciferase expression (60–4000 RLU/μg protein) using the SindLuc-S vector. A strong luciferase expression was observed when the rtTA2S-M2 transactivator was included in the expressing cassettes (vectors:

SindLuc-A1, SindLuc-A2 and SindLuc-S). The activity of the reporter gene in the examined vectors varied, depending on the relative orientation of the transgene- and rtTA2s-M2-containing cassettes. Although the expression level of the same vector in the presence of Dox varied among the cell lines, opposite orientation of the cassettes (vector: SindLuc-A1) resulted in higher expression than did identical orientation (vector: SindLuc-S) in the same cell lines. At the same time, gene induction was observed to be more potent in both Huvec and SVR cells transduced with SindLuc-A1 than in the cells cotransfected with TRELuc and VEcadrtTA (Fig. 2C,D). Insertion of a 2.9-Kb stuffer DNA sequence between the two promoters VEcad and TREalb had a negative effect on the induction in the presence of Dox (Fig. 2A–D). In the non-endothelial cell lines, only a basal level of luciferase expression (8–78 RLU/μg protein) was detected.

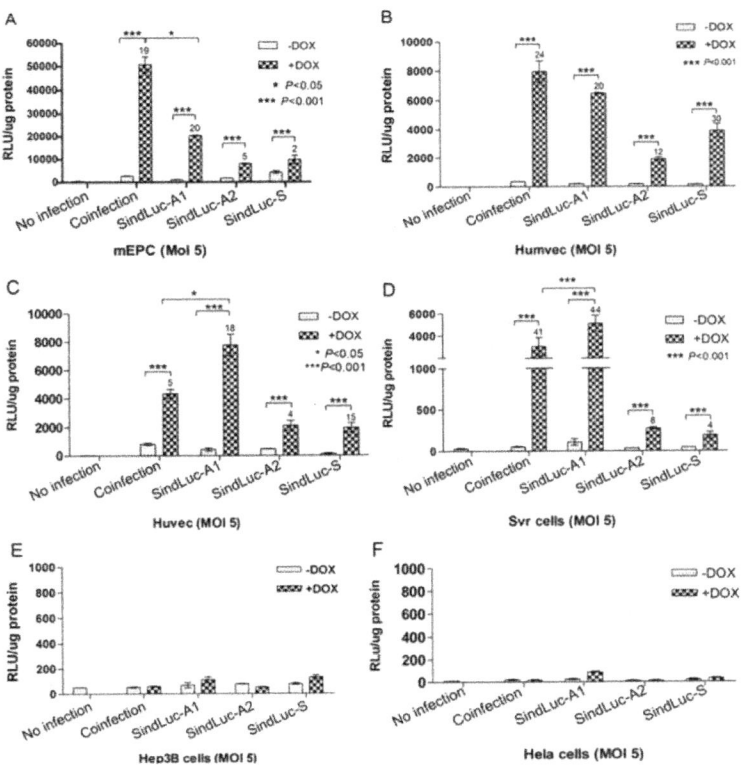

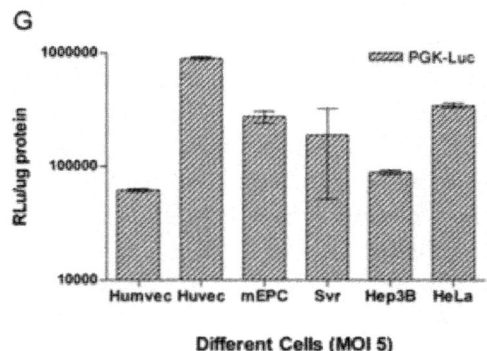

Figure 2: Quantification of doxycycline-dependent gene expression in endothelial cells and non-endothelial cells infected by the different lentiviral vectors encoding luciferase transgene.

The expression level was similar to that detected in the cells without infection which served as negative control. Moreover, a PGK-Luc vector was used as a positive control to test the infectivity of different cell lines in all the experiments. Huvec and HeLa cells presented higher accessibility to lentiviral vector infection compared with the other cell lines (Fig. 2G). Taken together, the SindLuc-A1 vector was considered as an optimal construction because of its minimal leakage in the uninduced state and its higher level of transgene induction in the presence of Dox among all the vectors compared in this study.

Cells were transduced (MOI 5) and grown in the presence or absence of Dox (1 µg/mL). Coinfection: TRELuc and VEcadrtTA vectors were used. Luciferase activity was measured 72 hours post-transduction. Fold induction is indicated at the top of each bar. (A) Luciferase detection in murine endothelial progenitor cells (mEPC). (B) Luciferase detection in human microvascular endothelial cells (Humvec). (C) Luciferase detection in human umbilical vein endothelial cells (Huvec). (D) Luciferase detection in SVR cells. (E) Luciferase detection in Hep3B cells. (F) Luciferase detection in HeLa cells. (G) Luciferase detection in different cell lines infected with PGK-Luc vector as a positive control. (*P < 0.05; ***P < 0.001). Results are presented as mean ± SD of data from each group, n = 3.

Dose-Dependent Dox-Induced Transgene Expression

Tet-on is a doxycycline-dependent system[11]. SVR cells, which are human neoplastic endothelial cells, were transduced with SindLuc-A1 vector (MOI 5) and cultured for 68 hours with various concentrations of Dox (ranging from 0 to 10 µg/mL) before being harvested for luciferase activity measurement.

The maximal transgene activation can be reached with 4 µg/mL or higher concentrations of Dox (Fig. 3A), agreeing with published results14,20. Therefore, an optimal concentration of Dox (4 µg/mL) was selected for all the subsequent experiments.

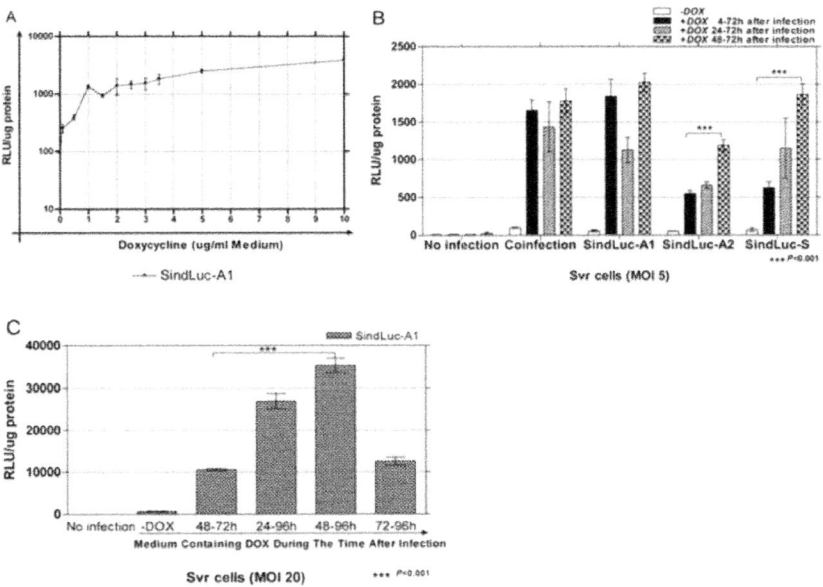

Figure 3: Quantification of doxycycline dose-dependent gene expression in SVR cells infected with SindLuc-A1 vector encoding luciferase transgene controlled by an inducible system.

(A) SVR cells transduced with SindLuc-A1 vector (MOI 5) and cultured for 68 hours with different concentrations of Dox (0.01; 0.5; 1; 1.5; 2; 2.5; 3; 3.5; 5; 10 µg/mL) were subjected to luciferase activity measurement. (B) SVR cells transduced with all the designed inducible vectors (MOI 5) and cultured in the presence of Dox (4 µg/mL) at different time points (4, 24 or 48 hours after infection) were subjected to luciferase activity measurement at 72 hours post transduction. Coinfection: TRELuc and VEcadrtTA vectors were used. (C) SVR cells transduced with SindLuc-A1 vector (MOI 5 or 20) and cultured with Dox (4 µg/mL) added to the culture at different time points (4, 24 or 48 hours after infection) were subjected to luciferase activity measurement at 72 or 96 hours post transduction. (***P<0.001). Results are presented as mean ± SD of data from each group, n = 3.

To optimize the duration of Dox induction, we infected SVR cells with the inducible vectors (MOI 5) and added Dox (4 µg/mL) to the culture medium at different time points (4, 24 or 48 hours after infection). The cells were harvested for measurement of luciferase activity 72 hours after transduction. The highest luciferase expression was observed when the Dox was added at 48 hours after lentivirus infection (Fig. 3B). Coinfection of SVR cells with the vectors encoding the luciferase-inducible expression cassette (TRELuc) and transactivator expression cassette (VEcadrtTA) showed no significant differences in luciferase expression among the various Dox-induction time points tested. Infection with the SindLuc-A1 vector reached the highest induction level at the presence of Dox for 4 or 48 hours after infection. However, both SindLuc-A2 and SindLuc-S vectors achieved the highest luciferase expression with 48 hours of Dox induction following lentiviral infection.

To determine whether longer Dox induction could further enhance transgene expression, we compared the 24-, 48-, and 72-hour Dox induction of infected endothelial cells. When the SVR cells were infected with the single inducible vector SindLuc-A1 at MOI 20 and harvested for luciferase expression measurement at 72 hours or 96 hours post transduction, we observed the highest luciferase expression when cells were harvested 96 hours post transduction with 48 hours of Dox induction (Fig. 3C).

Influence of Orientation of Expression Cassettes on Transgene Expression

Different endothelial cell lines were transduced with the Tet-inducible lentiviral vectors (MOI 5) in the presence or absence of Dox (4 μg/mL). Luciferase expression was measured at 96 hours post-transduction. We detected a low luciferase expression in all inducible vectors in the absence of Dox, indicative of leaky expression in the uninduced state (Fig. 4A–D). A robust Dox-dependent luciferase expression was detected when the rtTA2s-M2 transactivator was present (vector: SindLuc-A1). Placing the two cassettes in the opposite orientations resulted in lower uninduced levels of transgene expression and higher induction efficiency in the presence of the inducer. The PGK-Luc vector was used as a positive control in all the experiments. Lentivirus had the highest infection efficiency in Huvec cells among the tested endothelial cell lines (Fig. 4E). Based on the above data, we considered the SindLuc-A1 as an optimal viral vector because it had the lowest level of leaky expression in the uninduced state and the highest luciferase expression after induction in the selected endothelial cell lines (Fig. 4A–D).

Different cell lines transduced with various vectors (dose: MOI 5) were cultured in the presence or absence of Dox (4 μg/mL) and subjected to luciferase activity measurement at 96 hours post transduction. Coinfection: TRELuc and VEcadrtTA vectors were used. (A) Luciferase detection in murine endothelial progenitor cells (mEPC). (B) Luciferase detection in human microvascular endothelial cells (Humvec). (C) Luciferase detection in human umbilical vein endothelial cells (Huvec). (D) Luciferase detection in SVR cells. (E) Luciferase detection in different cell lines infected with PGK-Luc vector as a positive control. (***P<0.001). Results are presented as mean±SD of data from each group, n=3.

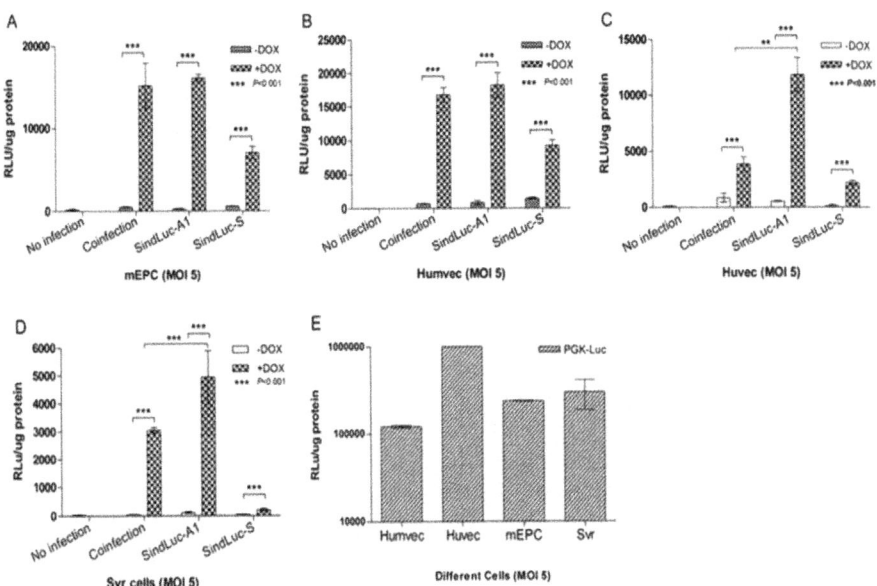

Figure 4: Quantification of doxycycline-dependent gene expression in endothelial cells infected with different lentiviral vectors carrying luciferase transgene.

Expression of the Tet-On System Driven By Endothelial-Specific or Non-Endothelial Promoters

To compare the function of promoters with endothelial-specific or constitutive activity, we generated a new single inducible lentiviral vector with a human phosphoglycerate kinase (hPGK) constitutive promoter that replaced the endothelial specific promoter VEcad used in the SindLuc-A1 construct (vector: SindLuc-APKG, Fig. 1E). In this case, the transactivator (rtTA2S-M2) was driven by the constitutive promoter hPGK. Different endothelial cell lines were transduced with SindLuc-A1 or SindLuc-APGK lentiviral vectors (MOI 5). The cells transduced with single inducible vector, SindLuc-APGK, produced high luciferase expression (~100.000 RLU/μg protein), which was 88- to 345-fold higher in the presence of Dox (4 μg/mL) when compared to the basal expression (<800 RLU/µg protein) in the absence of Dox. Additionally, we investigated the relation between MOI and luciferase expression level in SVR cells. As expected, increasing the vector dose led to augmented luciferase expression in the presence of Dox (Fig. 5E).

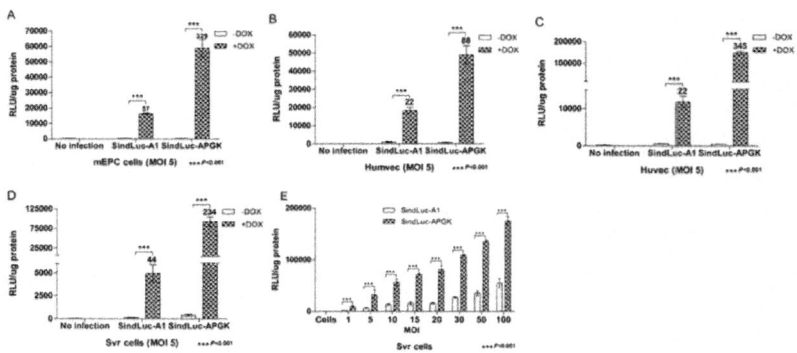

Figure 5: Quantification of doxycycline-dependent gene expression in endothelial cells infected with different lentiviral vectors carrying luciferase transgene.

Different cell lines transduced with SindLuc-A1 or SindLuc-APGK vector (MOI 5) were cultured in the presence or absence of Dox (4 µg/mL) and subjected to luciferase activity measurement at 96 hours post transduction. Fold induction is indicated at the top of each bar. (A) Luciferase detection in murine endothelial progenitor cells (mEPC). (B) Luciferase detection in human microvascular endothelial cells (Humvec). (C) Luciferase detection in human umbilical vein endothelial cells (Huvec). (D) Luciferase detection in SVR cells. (E) Luciferase detection in SVR cells infected with SindLuc-A1 or SindLuc-APGK vector in a dose-growing scale (MOI 0; 1; 5; 10; 15; 20; 30; 50; and 100). Results are presented as mean ± SD of data from each group, n = 3.

In Vivo Biological Function of Single Inducible Lentiviral Vectors

Analysis of the biological function of the single inducible lentiviral vectors (SindLuc-A1 and SindLuc-APGK) was carried out in a gastrointestinal cancer model (MC38, murine colon carcinoma). The MC38 cells were injected subcutaneously into the dorsal area of mice. Intra-tumor injection of different lentiviruses was performed 10 days later. Dox was administrated in drinking water for 10 days and the animals were sacrificed at day 21 of the experiment.

The transgene expression was evaluated by luciferase activity measurement and in situ detection of luciferase protein in tumor tissue. Luciferase activity was highest in those tumors collected from mice treated with unregulable PGK-Luc vector. This group served as a positive control in this experiment (Fig. 6A). The detected luciferase signal in SindLuc-APGK (+Dox) mice was 10- to 22-fold higher than in the uninduced group (Fig. 6A). SindLuc-A1 (+DOX) mice reached two-fold higher expression of luciferase compared to the SindLuc-A1 (-DOX) mice. SindLuc-A1 (-DOX) and SindLuc-APGK (-DOX) mice had nearly undetectable levels of luciferase activity; bioluminescence signal level detected in the tumors of these animals was comparable to that in mice without any injection (negative control). Indeed, such levels of luciferase expression were considered as a background signal (Fig. 6A).

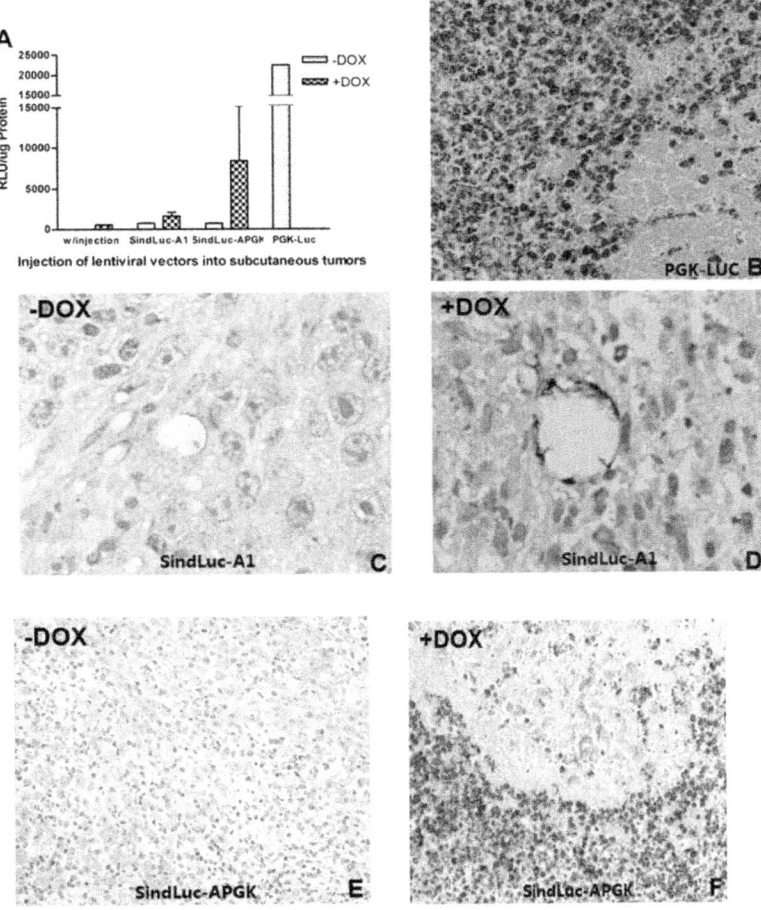

Figure 6: Quantification of luciferase expression in vivo and localization of luciferase protein in tumor tissues by immunohistochemistry.

(A) Luciferase activity measurement in tumors collected from each experimental group. (B) A 40× magnification of the tumor treated with unregulable PGK-Luc vector. (C) A 200× magnification of the tumor treated with regulable SindLuc-A1 vector without Dox induction. (D) A 200× magnification of the tumor treated with regulable SindLuc-A1 vector with Dox induction. (E) A 40× magnification of the tumor treated with regulable SindLuc-APGK vector without Dox induction. (F) A 40× magnification of the tumor treated with regulable SindLuc-APGK vector with Dox induction. Immunohistochemistry was performed on paraffin-embedded tumor sections. The presence of intracellular luciferase was visualized in brown due to peroxidase reaction. The tissues were counterstained with hematoxylin, which stains the nuclei blue. Results are presented as mean ± SD of data from each group, n = 3.

The intra-tumor expression of luciferase detected by immunohistochemistry was consistent with the luciferase activity measured above. We observed that the majority of tumor cells were positive for luciferase expression in the tissue samples collected from animals in PGK-LUC mice (Fig. 6B). Furthermore, we were able to detect the luciferase-positive cells only in the endothelium lining the vessels of the tumor nodules excised from SindLuc-A1 (+DOX) mice (Fig. 6D). However, no luciferase-positive cells were detected in the tumor tissue collected from SindLuc-A1 (-DOX) mice (Fig. 6C). Similarly, the tumors that have been grown in the SindLuc-APGK (-DOX) mice were negative for luciferase expression as well (Fig. 6E). In contrast, ubiquitous luciferase-positive cells were easily detected in tumors isolated from the SindLuc-APGK (+Dox) mice (Fig. 6F). In conclusion, our results demonstrated that the endothelial lineage-specific single regulable lentiviral vector can limit the reporter gene expression in tumor endothelial cells in vivo following doxycycline induction.

In summary, we believe that the single inducible lentiviral vector with an endothelial lineage-specific promoter as demonstrated in our study could be used for efficient targeting of endothelium at the sites of tumor angiogenesis. The induced gene expression within the tumor would reach therapeutic levels through this novel cancer treatment strategy.

Discussion

In a previously published study21, an increased dose of inducible vector produced high background expression of a transgene. This could result from the dose-dependent accumulation of the vector genomes integrating into chromosomal locations in the transduced cell populations, where the TRE/min was constitutively activated. It will be necessary, for future clinical trials, to reduce the amount of vectors used to avoid the risk as a result of viral integration into the host genome16.

The aim of this study was to develop an endothelial lineage-specific single inducible lentiviral vector that permits doxycycline to control the expression of therapeutic transgenes in endothelial cells. The importance of this project attaches to the possibility it raises of targeting tumor vascular tissue for therapeutic purposes using lentiviral vectors directly or stable endothelial progenitor cell lines, which have been recombined with a single endothelial-specific, Dox-inducible expression of therapeutic genes. Thus, we generated a single lentiviral vector containing both the transgene-inducible expression cassette and the transactivator expression cassette. Compared with other research groups14,19,22,23, we generated an optimal single inducible vector construct by using TREalb inducible promoter. Next, we cloned a minimal mono-directional poly (A) site into this structure. Due to limitations of the lentivirus packaging capacity, this minimal poly (A) site could be used widely for other vector designs in the future. Finally, we orientated the transactivator expression cassette in the antisense direction with respect to the vector RNA sequence without affecting the viral production capability.

A two-vector and several single-vector approaches were compared in our study for delivery of the transactivator rtTA2S-M2 and regulable transgene expression cassettes. We obtained reproducible, regulable and efficient transgene expression in endothelial cells. Moreover, in non-endothelial cells, there were nearly undetectable levels of transgene expression. In a previously described study14, the tetracycline-regulated CMV promoter (TRE/CMVmini), composed of a tetracycline-responsive element (TRE) and a CMV minimal promoter, was used. Placing the cassettes in the opposite orientation resulted in a slightly higher induction level not only in presence of Dox, but also higher in the uninduced state, compared with placing the cassettes in the same orientation. The major factor responsible for the distinct expression level was the minimal promoter used in those studies, which was also supported by another finding24; a tetracycline-regulated albumin promoter produced a lower gene expression in the uninduced state than TRE/CMVmini and could achieve a high induction level10. Rather than using an all-in-one vector14,19,24, we chose this tetracycline-regulated albumin-inducible promoter for our study to reduce basal expression levels without using repressor molecules.

During viral production, all the SindLuc-A vectors presented higher titer than the SindLuc-S vector. This result indicated that consistently oriented transcriptional products, as observed in the case of the SindLuc-S vector, can reduce the yield of viral mRNA. Both transcription products were terminated by the same Poly (A) signal which was located in the R region of 3' LTR.

In addition, in the set of SindLuc-A vectors, there are no any transcriptional products interfering with viral mRNA composition during the viral production process.

There was a very low transgene expression level in the uninduced state in all of our experiments (MOI 5–20). The action of the vectors was different depending on the relative orientation of the transgene and rtTA2S-M2 transactivator cassettes. Additionally, the same inducible vector possesses the different induction level in various endothelial cell lines. This might be attributed to a difference in the activity of the VEcad endothelial promoter in different endothelial cells (Fig. 2A–D). From all of the vectors tested, we found SindLuc-A1 to be optimal because transgene expression was absent in the state of Dox deficiency and was induced around 80-fold when Dox (1 μg/mL) was added to the medium. These data indicate that there was no influence between the two promoters (VEcadh/TREalb). Based on the same structure, a non-lineage-specific single inducible vector was constructed by replacing an endothelial specific VEcad promoter with an hPGK promoter (vector: SindLuc-APGK). In the cells transduced with SindLuc-APGK at MOI 5, luciferase expression was undetectable in the absence of Dox, and there was a very high induction efficiency (~345-fold) in the presence of Dox (Fig. 5A–C). Likewise, the maximal luciferase expression level obtained with this vector was only 10 times lower than that achieved with PGK-Luc unregulable vector (Fig. 4E). Furthermore, we also generated a vector, SindLuc-A2, which contained a 2.9-Kb stuffer DNA between the two promoters; a longer fragment was not allowed due to the maximal packaging capacity (8–9 K bps) of lentiviral vector. It seemed that the 2.9-Kb stuffer DNA had a negative influence on Dox-dependent induction (Fig. 2A–D). This stuffer DNA sequence plays an unknown role in the function of this vector. However, in the viral production process, the titer of this particular virus was similar to the titer of SindLuc-A1 vector (Table 1).

In our study, when the two expression cassettes were placed in the same orientation (vector: SindLuc-S) with respect to viral RNA sequence, it seemed that the transgene expression was affected in the doxycycline induced state. Previous studies also indicated that the same orientation of cassettes in a vector affected transcription when using the Tet-responsive promoter[25,26]. Although this effect resulted in the suppression of its basal activity, it strongly reduced the expression of transgenes in the "on" state[17]. In another study[14], a similar result was also reported; the same orientation of cassettes in a vector involved overlapping expression cassettes.

The cassette was required for driving the production of the genome of the vector from the precursor plasmid, and the cassettes used to express the transactivator and the transgene share the same poly (A) sequence, which is in the 3'LTR of the vector. In our study, we found that the action of the SindLuc-S vector was not satisfactory in the Dox-induction state. It produced only around 4-to 30-fold induction with Dox, and the maximal transgene expression level was much lower than that from the SindLuc-A1 vector (Fig. 2A–D).

In our work, we tested the kinetics of Dox induction and harvesting time in relation to induced gene expression by adding the Dox (4 µg/mL) to the culture medium at 4, 24 or 48 hours after infection and harvesting cells for transgene activity measurement at 72 or 96 hours post transduction. The highest luciferase expression was observed, when the Dox was added at 48 hours (Fig. 3B), which is higher than previously published results14,20. Compared to a previously described lentiviral infection protocol27, in which luciferase activity was measured at 72 hours post transduction, our induction protocol, in which we kept the cells for 96 hours in the presence of Dox, achieved higher transgene expression. Thus, we have demonstrated that we could increase the luciferase expression by 2- to 3- fold with our newly established protocol, compared to previously described method27.

For optimal doxycycline induction, we found that cells should be infected with inducible vectors during the first 48 hours and cultured for the subsequent 48 hours in the presence of Dox at 4 µg/mL. The cells are then harvested for transgene detection after 96 hours of culturing.

In distinction from previous studies14,17, we developed a minimal monodirectional poly (A) sequence for a single inducible vector. Lentiviral vectors have a less-than-8-Kb site for foreign cassette insertion. Our intent was to minimize the vector size without affecting its function in order to reduce the genome integration sequence. It is also a new design of a minimal lentiviral vector. In a previous study28, researchers generated a minimal Rev-Response Element and packaging signals for HIV virus. In future clinical trials, a minimal lentiviral vector will be required to avoid the production of replicatively competent recombinants—one that can be modified by reducing the congenetic sequence from wide-type HIV virus. This minimal synthetic monodirectional poly (A) site should be very useful for the novel vector designs.

We have chosen the murine vascular endothelial cadherin promoter (VEcad) for our study because of its highly specific activity in endothelial cells with no reported activity in non-endothelial cells. In 2003, Dancer et al. used VEcad promoter to drive the expression of thymidine kinase and successfully inhibit tumor growth of Lewis lung carcinoma cells in transgenic mice29. In 2005, the human VEcad promoter transcriptional activity was also characterized, and the hVEcad promoter was subjected to organ-specific regulation and was activated in tumor angiogenesis30,31. The inducible endothelial-specific lentiviral system we have developed is expected to have a broad application in future tumor therapy. For tumors that are easily accessible, such as in melanoma or head and neck tumors, the lentiviruses could be injected directly into the tumor mass. However, for internal tumors the lentiviruses could be used to transduce isolated EPCs from patients ex vivo and then the transduced EPCs can be infused to patients. We believe our novel lineage-specific single inducible lentiviral vector will provide a new and effective resource in the future treatment of human tumors.

METHODS

General Methods

All methods described in this manuscript were carried out in accordance with the approved guidelines in the University of Navarra, Pamplona, Spain.

Animals and Cell Lines

Five- to six-week-old female C57BL76J mice were purchased from Harlan Laboratories (Barcelona, Spain). Animals were maintained under standard conditions and all procedures were approved by the institutional ethical committee. The viral vector was administered by intratumoral injection of 70 μL of saline to evaluate its biological function in the Tet-on system. Doxycycline (Dox) induction started 48 hours after the vector injection. Dox (Sigma-Aldrich, St Louis, USA) was given in drinking water (1 mg/mL with 10% sucrose). The following cell lines were obtained from the American Type Culture Collection: human embryonic kidney 293 (HEK293) cells, human embryonic kidney 293 T (HEK293T) cells, human cervical carcinoma (HeLa), murine colon carcinoma (MC38), human angiosarcoma (SVR), human umbilical vein endothelial cells (Huvec) and human microvascular endothelial cells (Humvec). Murine endothelial progenitor cells (mEPC) were obtained from animals as primary culture cells.

The cell lines were cultured in an incubator at 37 °C with 5% CO_2 and various media. Specifically, HEK293, HEK293T, HeLa, and MC38 cells were cultured in standard DMEM supplemented with 10% fetal bovine serum (FBS), 100 IU/mL penicillin, 100 μg/mL streptomycin and 2 mM glutamine (all from Life Technologies, Carlsbad, CA). Huvec cells were cultured in M131 medium (Life Technologies) supplemented with 25 mM HEPES (Life Technologies), 10% FBS, 100 IU/mL penicillin, 100 μg/mL streptomycin, 2 mM glutamine, 50 μg/mL heparin (Mayne Pharma, Madrid, Spain) and 25 μg/mL of Endothelial Cell Growth Supplement (Sigma). Cells were cultured on human fibronectin (Sigma) coated 75- or 125-cm2 flasks and plates (Sarstedt, Wexford, Ireland). Humvec were cultured in M131 medium supplemented with 5% (v/v) Microvascular Endothelial Cells Growth Supplement (MVGS, Life Technologies), 10% FBS, 100 IU/mL penicillin, 100 μg/mL streptomycin, 2 mM glutamine and 50 μg/mL heparin. Cells were cultured on fibronectin-coated flasks and plates.

EPCs were cultured according to the protocol as previously described32. Briefly, bone marrow-derived mononuclear cells (MNC) were isolated by density gradient centrifugation with Ficoll Histopaque (Sigma). After purification, 8×106 MNCs were plated on fibronectin-coated six-well plates (Cellstar, Greiner Bio-One, Frickenhausen, Germany). Cells were cultured in EBM-2 medium (Clonetics, Walkersville, MD) with supplements: 10 ng/mL recombinant rat VEGF, 1 ng/mL bovine basic FGF, 10 ng/mL recombinant mouse IGF-I, 10 ng/mL recombinant human EGF (all from R&D Systems, Minneapolis, MN), 1 μg/mL Hydrocortisone (Sigma) and 5% FBS. The cultured cells were replated on day 4 and used for particular experiments.

Construction of INDUCIBLE LENTIVIRAL VECTORS

The third generation self-inactivating and inducible lentiviral vectors were generated on the basis of the backbone pRRL.cPPT.PGK.eGFP.WPRE as described23. A BamH1-SaL1 digest of the synthetic DNA linker containing Poly-cloning sites (BamH1-Xho1-Xma1-Sma1-Xba1-Mlu1-Nhe1-BglII-SaL1) was performed for the ligation into a BamH1-SaL1 digest of pRRL.cPPT. PGK.eGFP.WPRE to obtain the particular construct of pRRL.cPPT.PGK.linker. WPRE. The firefly luciferase was released from pGl3-basic by Xma1-Xba1 digestion and was ligated into Xma1-Xba1 digested pRRL.cPPT.PGK.linker. WPRE to obtain a lentiviral vector encoding PGK-Luc. The PGK promoter from pRRL.cPPT.PGK.linker.WPRE was removed by Xho1 digestion to get pRRL. cPPT.linker.WPRE. The Nhe1-BamH1 digestion of pmVEcad-rtTA (No PA) released the VEcad-rtTA fragment for subsequent ligation into an Nhe1-BglII digest of pRRL.cPPT.Linker.WPRE to achieve the lentiviral vector expressing the mVEcadrtTA fragment. TREalb-Luc sequence was isolated from pTREalb-Luc with Xho1-Xba1 and placed into the Xho1-Xba1 digested pRRL.cPPT. linker.WPRE to create the TREalb-Luc lentiviral vector. The S60PA element is derived from published result33 with slight modifications. The sequence of our modified S60PA is 5'-GATCCAATAAAAGATCTTAAGTTTCATT AGATCTGTGTGTTGGTTTTTTGTGTG-3', where the two underlined nucleotides were mutated to prevent the transcriptional stop function of the polyA. To construct SindLuc-A1 lentiviral vector, the mVEcad-rtTA-S60PA cassette was excised with Xho1-Acc1, and digestion of pmVEcad-rtTA (S60

PA) was done for the ligation into the pTRELuc vector with Xho1-ClaI. A 2.9 Kbp stuffer DNA fragment was excised from STK120 with Xho1 digestion and was ligated into Xho1 digest site of SindLuc-A1 to gain the SindLuc-A2 lentiviral vector. To construct the SindLuc-S lentiviral vector, the mVEcad-rtTA fragment was released with Nhe1-BamH1 from pmVEcad-rtTA (No PA) and finally ligated into TRELuc with Nhe1-BglII. Digestion of pUHrT62-1 with BamH1-EcoR1 (blunted) excised the rtTA2S-M2 fragment that was linked into pRRL.cPPT.PGK.eGFP.WPRE with BamH1-Sal1 (blunted) to obtain the pRRL.cPPT.PGK.rtTA.WPRE construct. The PGK-rtTA (half gene) fragment was generated from pRRL.cPPT.PGK.eGFP.WPRE with Xho1-Pml1 digestion and was ligated into a pmVEcad-rtTA (S60 PA) plasmid, digested with Xho1-Pml1 to acquire phPGK-rtTA (S60 PA). SindLuc-APGK lentiviral vector was obtained by excision of the hPGK-rtTA-60PA expression cassette from the phPGK-rtTA (S60 PA) plasmid with Xho1-Acc1 digestion and ligated into a TRELuc lentiviral vector with Xho1-Cla1.

Lentiviral Vector Production

Dr Antonia Follenzi (University of Torino Medical School, Italy) kindly provided all the plasmids for lentiviral vector production. Vectors were produced by transfection of HEK293T cells as described34, with the following modifications. HEK293T cells (7×106/plate) were growing in 15-cm-diameter dishes for 24 hours prior to transfection in DMEM supplemented with 10% FBS, penicillin (100 IU/mL) and streptomycin (100 μg/mL) at 37 °C in a 5% CO2 incubator. The culture medium was exchanged 2 hours before transfection (DMEM with 2% FBS). Transient transfection of HEK293T cells with the plasmids was obtained by the calcium phosphate precipitation method. Briefly, a total of 40 μg of plasmid DNA was used for the transfection of one culture dish, including 12 μg of the envelope plasmid (pMD.G) encoding VSV-G, 5 μg of the packaging plasmid (pMDLg/pRRE), 3 μg of the plasmid producing Rev regulatory protein (pRSVrev) and 20 μg of the transfer vector plasmid. The calcium phosphate precipitate was formed by adding the plasmids to a final volume of 1000 μl of filtered dH2O and 200 μl of 2.5 M CaCl2 solution. Next, 1000 μl of 2×-concentrated HEPES-buffered saline (281 mM NaCl, 100 mM HEPES, 1.5 mM Na2HPO4, pH 7,05) was added drop-wise with brief vortexing for 10 seconds. After 15 minutes, the precipitate was added to cell cultures. Post-transfection culture medium (16 hours) was replaced with DMEM (10% FBS). The conditioned medium was collected after 24 hours, cleared by low-speed centrifugation and filtered through a 0.45-μm-pore filter (500 mL) (Dominique Dutscher S.A., Brumath, France). All the vectors were concentrated by ultracentrifugation at 19,500 rpm for 2 hours at 12 °C, aliquoted, and stored at −80 °C until use.

Lentiviral Vector Titration

Infectious viral particles were determined by transduction of 5×104 HeLa cells with serial dilutions of the vector preparation in a 24-well plate in the presence of $8\,\mu g/mL$ Polybrene (Sigma). 72 hours later, genomic DNA from transduced HeLa cells was extracted using a DNA Blood Mini Kit 50 (Qiagen, Santa Clara, CA). The transducing unit titer (TU/mL) was determined by quantitative PCR (qPCR) as described35. Real-time PCR was used for the quantitative analysis of proviral DNA copies. Probes were labeled at the 5′end with the reporter dye molecule FAM (emission wavelength: 518 nm) and at the 3′end with the quencher dye TAMRA (emission wavelength: 582 nm). The 3′end of the probe was additionally phosphorylated to prevent extension during PCR. For detection of the lentivirus WPRE sequence in HeLa cells, the following primers and probe were used: forward primer (1277 F): 5′-CCGTTTCAGGCAACGTG-3′; reverse primer (1361 R): 5′-AGCTGACAGGTGGTGGCAAT-3′; probe (1314 P): 5′-FAM-TGCTGACGCAACCCCCACTGGT-TAMRA-3′ [52]. For human β-actin gene copy numbers in HeLa cells, the following primers and probe were used: forward primer, 5′-GCGAGAAGATGACCCAGCTC-3′; reverse primer: 5′-CCAGTGGTACGGCCAGAGG-3′; probe: 5′-FAM-CCAGCCATGTACGTTGCTATCCAGGC-TAMRA-3′ [53]. For the PCR reaction, the universal PCR Master Mix ($4\,\mu L$; Promega, Madison, WI), with a 1-μM concentration of each primer, and the probe were combined. Finally, genomic DNA was added to each reaction, and the total reaction volume was adjusted to $10\,\mu L$. Standard conditions were used for the PCR reaction (2 minutes at 50 °C, 10 minutes at 95 °C, and then 40 cycles of 15 seconds at 95 °C and 1 minute at 60 °C).

Cell Transduction

Cells (HEK293, HEK293T, HeLa and SVR cell lines) growing in 24-well plates to 50–60% confluence were transduced with lentiviral vector preparations in a total volume of $300\,\mu L$ of DMEM (10% FBS) supplemented with Polybrene ($8\,\mu g/mL$). After 16 hours, the cells were washed extensively with PBS to remove lentiviral genomic RNA and fresh medium was added. The cells were maintained in culture for another 72 hours. Other cell lines were incubated with lentiviral vector preparations in a total volume of $300\,\mu L$ of specific medium (according to the each endothelial cell type) supplemented with Polybrene ($8\,\mu g/mL$). For inducible lentiviral vectors, various Dox dosages and induction times were tested to optimize the induction conditions.

Cell Transfection

The cells were cotransfected with $1\,\mu g$ of each transfer plasmid and 20 ng of pRL-SV40 plasmid (Promega). The Huvec and Humvec cells were transfected with Lipofectamine Plus reagent (Life Technologies). HEK293 and HEK293T were transfected with polyethylenimine (Polysciences, Warrington, PA). SVR cells were transfected with FuGENE HD transfection reagent (Promega). Finally, HeLa cells were transfected by the Calcium phosphate method. Cells were cultured in 24-well plates for transfection. The post-transfection medium was exchanged according to the protocol for each transfection reagent. The cells were incubated for 48 hours in the presence or absence of Dox and then harvested for the luciferase activity measurement.

Luciferase Detection in Vitro

The cells transduced with the lentiviral vector were measured with the Dual-Luciferase Reporter Assay System (Promega) for luciferase expression. Following the manufacturer's instructions, the cells were recovered at different time points, washed twice with PBS, and lysed in 250 µL Passive Lysis Buffer (Promega) by 10 minutes of shaking. The cell extracts were serially diluted, and 20 µL of each dilution were mixed with 100 µL of the luciferase assay reagent in luminometer tubes. The firefly luciferase activity was measured using a single tube luminometer (Berthold Detection Systems, Pforzheim, Germany). Protein concentration was calculated using the Bio-Rad protein assay (Bio-Rad, Hercules, CA). Experiments were repeated three times.

Animal Model of Gastrointestinal Cancer

For the gastrointestinal cancer model, 1×106 of MC38 cells (murine colon carcinoma) in 100 µL of saline were injected subcutaneously into the dorsal area of mice. Ten days after the tumor cell inoculation, the tumor-bearing mice were assigned to five groups of three animals each. Intratumoral injections of different lentiviral vectors (5×107 TU/mL/tumor) were performed. Dox was administrated in drinking water for a period of 10 days. Group A remained unmedicated and served as a negative control group. Group B was injected with SindLuc-A1 vector and drank water without Dox. Group C was injected with SindLuc-A1 vector and drank water treated with Dox. Group D was injected with SindLuc-APGK vector and drank water without Dox. Group E was injected with SindLuc-APGK vector and drank water treated with Dox. Group F was injected with PGK-Luc vector and drank water without Dox. The survival of the animals was checked daily. Finally, the animals were sacrificed by cervical dislocation at day 21 of the experiment. The subcutaneous tumors were collected, and each one was divided into two parts, one part fixed in 4% paraformaldehyde for histological examination and the other part frozen and stored at $-80\,°C$ for luciferase activity measurement.

Immunohistochemistry

Three-μm-thick sections of 4% paraformaldehyde-fixed and paraffin-embedded tumor tissues were processed for immunohistochemistry. Staining with the following antibodies was performed: goat anti-firefly luciferase (1:50; Cortex Biochemical, San Leandro, CA, CR2029GAP), polyclonal rabbit anti-goat immunoglobulins biotinylated antibody (1:600; Dako, Glostrup, Denmark, E0466) and streptavidin- biotinylated horseradish peroxidase (1:100; Amersham, RPN1051V). For detection, a diaminobenzidine reagent (Dako) was used and counterstaining performed with Mayer's hematoxylin solution (Merck-KGaA, Darmstadt, Germany). Slices were mounted in DePex Mounting Solution. Histological samples were visualized with a Nikon microscope, and images were acquired using a Leica camera (Leica Microsystems, Wetzlar, Germany) and analyzed with Aquacosmos acquisition software (Hamamatsu Photonics, Hamamatsu City, Japan).

Luciferase Detection in Tumor Tissue

Tumor tissue was cut and lysed with 250 μL of Passive Lysis Buffer (Promega) and homogenized in Eppendorf tubes using the manual method with plastic sticks. The tissue extracts were collected and processed according to the protocol described above in the section on luciferase detection in vitro.

Statistical Analysis

All analyses were done using SPSS version 9.0 software (Chicago, IL) with $P < 0.05$ considered to be statistically significant. Data were analyzed by Mann-Whitney nonparametric tests due to the sample size being less than 10.

Additional Information

How to cite this article: Yang, G. et al. Development of Endothelial-Specific Single Inducible Lentiviral Vectors for Genetic Engineering of Endothelial Progenitor Cells. Sci. Rep. 5, 17166; doi: 10.1038/srep17166 (2015).

REFERENCES

1.   Liechtenstein, T., Perez-Janices, N. & Escors, D. Lentiviral vectors for cancer immunotherapy and clinical applications. Cancers 5, 815 (2013).

2.   Pirro, M. et al. Baseline and post-surgery endothelial progenitor cell levels in patients with early-stage non-small-cell lung carcinoma: impact on cancer recurrence and survival. Eur. J. Cardiothorac. Surg. 44,

3.   De la Puente, P., Muz, B., Azab, F. & Azab, A. K. Cell trafficking of

endothelial progenitor cells in tumor progression. Clin. Cancer Res. 19, 3360–3368 (2013).

4.   Rodrigues, C. G. et al. VEGF 165 gene therapy for patients with refractory angina: mobilization of endothelial progenitor cells. Arq. Bras. Cardiol. 101, 141–148 (2013).

5.   Ha, X. et al. Identification and clinical significance of circulating endothelial progenitor cells in gastric cancer. Biomarkers 18, 487–492 (2013).

6.   Case, S. S. et al. Stable transduction of quiescent CD34+CD38− human hematopoietic cells by HIV-1-based lentiviral vectors. Proc. Natl Acad. Sci. USA 96, 2988–2993 (1999).

7.   Piacibello, W. et al. Lentiviral gene transfer and ex vivo expansion of human primitive stem cells capable of primary, secondary, and tertiary multilineage repopulation in NOD/SCID mice. Blood 100, 4391–4400 (2002).

8.   Vidoni, S., Zanna, C., Rugolo, M., Sarzi, E. & Lenaers, G. Why mitochondria must fuse to maintain their genome integrity. Antioxid. Redox Signaling 19, 379–388 (2013).

9.   Gory, S. et al. Requirement of a GT Box (Sp1 Site) and two ets binding sites for vascular endothelial cadherin gene transcription. J. Biol. Chem. 273, 6750–6755 (1998).

10.   Zabala, M. et al. Optimization of the tet-on system to regulate interleukin 12 expression in the liver for the treatment of hepatic tumors. Cancer Res. 64, 2799–2804 (2004).

11.   Lamartina, S. et al. Stringent Control of Gene Expression In Vivo by Using Novel Doxycycline-Dependent Trans-Activators. Hum. Gene Ther. 13, 199–210 (2002).

12.   Urlinger, S. et al. Exploring the sequence space for tetracycline-dependent transcriptional activators: novel mutations yield expanded range and sensitivity. Proc. Natl Acad. Sci. USA 97, 7963–7968 (2000).

13.   Kramer, M. G. et al. In vitro and in vivo comparative study of chimeric liver-specific promoters. Mol.

14.   Barde, I. et al. Efficient control of gene expression in the hematopoietic system using a single tet-on inducible lentiviral vector. Mol. Ther. 13, 382–390 (2006).

15.   Shi, Q. et al. Anti-arthritic effects of FasL gene transferred intra-articularly by an inducible lentiviral vector containing improved tet-on system. Rheumatol. Int. 34, 51–57 (2014).

16. Hacein-Bey-Abina, S. et al. A serious adverse event after successful gene therapy for X-linked severe combined immunodeficiency. N. Engl. J. Med. 348, 255–256 (2003).

17. Vigna, E. et al. Robust and efficient regulation of transgene expression in vivo by improved tetracycline-dependent lentiviral vectors. Mol. Ther. 5, 252–261 (2002).

18. Miyoshi, H., Blömer, U., Takahashi, M., Gage, F. H. & Verma, I. M. Development of a self-inactivating lentivirus vector. J. Virol. 72, 8150–8157 (1998).

19. Dull, T. et al. A third-generation lentivirus vector with a conditional packaging system. J. Virol. 72, 8463–8471 (1998).

20. Markusic, D., Oude-Elferink, R., Das, A. T., Berkhout, B. & Seppen, J. Comparison of single regulated lentiviral vectors with rtTA expression driven by an autoregulatory loop or a constitutive promoter. Nucleic Acids Res. 33, e63, 10.1093/nar/gni062 (2005).

21. Recchia, A. et al. Site-specific integration mediated by a hybrid adenovirus/adeno-associated virus vector. Proc. Natl Acad. Sci. USA 96, 2615–2620 (1999).

22. Bai, J., Li, J. & Mao, Q. Construction of a single lentiviral vector containing tetracycline-inducible Alb-uPA for transduction of uPA expression in murine hepatocytes. PLoS ONE 8, e61412, 10.1371/journal.pone.0061412 (2013).

23. Follenzi, A., Ailles, L. E., Bakovic, S., Geuna, M. & Naldini, L. Gene transfer by lentiviral vectors is limited by nuclear translocation and rescued by HIV-1 pol sequences. Nat. Genet. 25, 217–222 (2000).

24. Hoffmann, A., Villalba, M., Journot, L. & Spengler, D. A novel tetracycline-dependent expression vector with low basal expression and potent regulatory properties in various mammalian cell lines. Nucleic Acids Res. 25, 1078–1079 (1997).

25. Debowski, A. W., Verbrugghe, P., Sehnal, M., Marshall, B. J. & Benghezal, M. Development of a tetracycline-inducible gene expression system for the study of helicobacter pylori pathogenesis. Appl. Environ. Microbiol. 79, 7351–7359 (2013).

26. Fiorini, E. et al. Inducible gene expression in fetal thymic epithelium: a new BAC transgenic model. Genesis 51, 717–724 (2013).

27. Sastry, L., Johnson, T., Hobson, M., Smucker, B. & Cornetta, K. Titering lentiviral vectors: comparison of DNA, RNA and marker expression methods. Gene Ther. 9, 1155–1162 (2002).

28. Clever, J. L., Miranda, D. & Parslow, T. G. RNA structure and packaging signals in the 5′ leader region of the human immunodeficiency virus type 1 genome. J. Virol. 76, 12381–12387 (2002).

29. Dancer, A. et al. Expression of thymidine kinase driven by an endothelial-specific promoter inhibits tumor growth of Lewis lung carcinoma cells in transgenic mice. Gene Ther. 10, 1170–1178 (2003).

30. Prandini, M.-H. et al. The human VE-cadherin promoter is subjected to organ-specific regulation and is activated in tumour angiogenesis. Oncogene 24, 2992–3001 (2005).

31. Lopez, D., Niu, G., Huber, P. & Carter, W. B. Tumor-induced upregulation of twist, snail, and Slug represses the activity of the human VE-cadherin promoter. Arch. Biochem. Biophys. 482, 77–82 (2009).

32. Griese, D. P. et al. Isolation and transplantation of autologous circulating endothelial cells into denuded vessels and prosthetic grafts: implications for cell-based vascular therapy. Circulation 108, 2710–2715 (2003).

33. Levitt, N., Briggs, D., Gil, A. & Proudfoot, N. J. Definition of an efficient synthetic poly(A) site. Genes Dev. 3, 1019–1025 (1989).

34. Kutner, R. H., Zhang, X.-Y. & Reiser, J. Production, concentration and titration of pseudotyped HIV-1-based lentiviral vectors. Nat. Protocols 4, 495–505 (2009).

35. Lizée, G. et al. Real-time quantitative reverse transcriptase-polymerase chain reaction as a method for determining lentiviral vector titers and measuring transgene expression. Hum. Gene Ther. 14, 497–507 (2003).

Chapter 5

REDUCTION IN C-TERMINAL AMIDATED SPECIES OF RECOMBINANT MONOCLONAL ANTIBODIES BY GENETIC MODIFICATION OF CHO CELLS

Mihaela Škulj[1], Dejan Pezdirec[1], Dominik Gaser[1], Marko Kreft[2,3,4] and Robert Zorec[2,3]

[1]Sandoz Biopharmaceuticals, Mengeš, Lek Pharmacetucals d.d
[2]Celica Biomedical Centre
[3]Laboratory of Neuroendocrinology – Molecular Cell Physiology, Institute of Pathophysiology, Faculty of Medicine, University of Ljubljana
[4]Biotechnical Faculty, University of Ljubljana

ABSTRACT

Background

During development of recombinant monoclonal antibodies in Chinese hamster ovary (CHO) cells, C-terminal amidated species are observed. C-terminal amidation is catalysed by peptidylglycine α-amidating monooxygenase (PAM), an enzyme known to be expressed in CHO cells. The significant variations between clones during clone selection, and the relatively high content of amidated species (up to 15%) in comparison to reference material (4%), led us to develop a cell line with reduced production of C-terminal amidated monoclonal antibodies using genetic manipulation.

Results

Initial target validation was performed using the RNA interference approach against *PAM*, which resulted in a CHO cell line with C-terminal amidation decreased to 3%. Due to the transient effects of small-interfering RNAs, and possible stability problems using short-hairpin RNAs, we knocked-down the *PAM* gene using zinc finger nucleases. Plasmid DNA and mRNA for zinc finger nucleases were used to generate a *PAM* knock-out, which resulted in two CHO cell lines with C-terminal amidation decreased to 6%, in CHO *Der2* and CHO *Der3* cells.

Conclusion

Two genetically modified cell lines were generated using a zinc finger nuclease approach to decrease C-terminal amidation on recombinant monoclonal antibodies. These two cell lines now represent a pool from which the candidate clone with the highest comparability to the reference molecule can be selected, for production of high-quality and safe therapeutics.

BACKGROUND

The production of biopharmaceuticals for human use began in 1982 with recombinant insulin, and the development of new biopharmaceuticals has grew almost exponentially ever since. Over the past two decades, Chinese hamster ovary (CHO) cells have become the standard mammalian host cell line, with the expression and production of nearly 70% of all biopharmaceuticals [1, 2]. CHO cells provide efficient post-translational modifications, which allow the production of recombinant proteins with glycoforms that are both compatible with, and bioactive in humans [1]. CHO cells can also be easily manipulated genetically, which has become of great importance more recently [3]. These two characteristics are especially important in the production of biosimilars, where achieving the correct extent of similarity to the reference molecule is a great challenge. The nucleotide sequence of the gene that encodes amino-acid sequence of the desired protein is the same as for the reference molecule. In contrast, post-translational modifications that are the consequence of metabolic pathways can differ between host cell lines, clones, cultivation conditions, medium composition, specific productivity and physiologic state of the cell [4]. Consequently, these need to be fine-tuned during development. In addition to posttranslational modifications, different charge variants can result in heterogeneity in the production of monoclonal antibodies (mAbs). These modifications potentially result in changes to the bioactivity, bioavailability or immunogenicity of the mAbs, and they therefore need to be additionally characterised to ensure the safety, quality and efficacy of the product. Among these, C-terminal amidated structures on the heavy chains of mAbs have recently attracted particular attention [5–8].

C-terminal α-amidation is catalysed by peptidylglycine α-amidating monooxygenase (PAM), and this protein modification is often required to confer full biological activity to peptide hormones [6, 9–11]. Amidation is catalysed starting from a glycine-extended prohormone, by two sequential actions of two enzymes, peptidylglycine α-hydroxylating monooxygenase and peptydilamido-glycolat lyase. In mammals, both of these enzymes are derived from a single gene, which gives rise to the bi-functional PAM protein [12]. PAM thus catalyses the conversion of peptidylglycine substrates into α-amidated products in a two-step reaction, and it is the only enzyme known to catalyse the formation of amidated peptides [13].

Recently, large proteins like immunoglobulins have been reported to be substrates for PAM [6, 7], and the expression of PAM in CHO cells was reported previously [14]. Tsubaki et al. reported that C-terminal α-amidation was detected in 8 of 12 recombinant mAbs, with ratios from 0.3% to 25.9% [6]. During our studies, we have also observed the presence of C-terminal amidated species in recombinant mAbs produced in CHO cells. Prolinamide was detected in up to 14% of all mAb molecules, which was too high to accomplish the desired similarity to the reference molecule.

It was previously shown that the level of mAb amidation in CHO cells can be affected via bioprocesses and medium optimisation, with the addition of copper to the culture medium [7]. On the other hand, metabolic engineering is becoming a powerful tool to manipulate expression hosts for improved product quality, and the introduction of a PAM knocked-down cell line that can produce mAbs with desired comparability to a reference molecule would be a state-of-the-art solution.

In the present study, the expression of PAM, and consequently the C-terminal amidation of recombinant mAbs, was reduced by two approaches: gene manipulation using RNA interference (RNAi) and zinc finger nucleases (ZFN). RNAi has been efficiently used for down-regulation of desired genes, and it can be performed using chemically synthesised small-interfering (si) RNA molecules, or via the endogenous expression of short-hairpin (sh)RNA molecules that are encoded by plasmid vectors [15, 16]. While siRNAs can provide transient knock-down of expression of a target gene, shRNA vectors can induce long-term expression of RNAi silencing in a target cell [15]. The ease and rapidity of these RNAi approaches have made them the method of choice for initial target validation.

However, using RNAi, complete elimination of expression of the target gene is rarely obtained. To achieve permanent gene knock-out, ZFNs can offer a distinct advantage. Moreover, the growing number of reports using ZFNs across different species suggests that ZFN-mediated gene disruption is a robust and general method for targeted gene knock-out. As this approach results in an alteration of the genome itself, these mutations are transmitted stably through all of the subsequent generations of the cell line, as is the case with conventional gene targeting [3, 17]. Rapid gene knock-out can be achieved by using a ZFN approach to create a double-stranded break in the locus of interest, which allows it to be repaired by non-homologous end-joining; this ligates the two broken ends, with the occasional loss of genetic information. ZFNs can therefore be used to introduce small deletions at the site of such a break, an outcome that can be exploited to disrupt a target gene [18, 19]. The high frequencies of gene disruption strongly support the likelihood of achieving a desired genotype, in which each copy of the target gene is functionally knocked out in 1% to 50% of all cells [17, 18].

RESULTS AND DISCUSSION

siRNA and shRNA Experiments

C-terminal amidation is a common posttranslational modification that is seen on therapeutic mAbs [6], and which has also been observed during recombinant mAb development (data not shown). The significant variations between clones during clone selection, and the relatively high content of amidated species in comparison to reference molecules, directed us towards the development of a cell line with reduced production of C-terminal amidated species.

RNAi has already been efficiently used for the improvement of cellular productivity and the quality of recombinant proteins produced in CHO cells. RNAi had thus been used for silencing apoptosis-associated gene expression [20–23], glycosylation-associated gene expression [24–26], and gene expression of lactate dehydrogenase [27] and dihydrofolate reductase [28,29].

Fifteen different siRNAs were designed on the basis of the *C. griseus* *PAM* nucleotide sequence and tested on the CHO*Der2* parental cell line. Up to an 8-fold decrease was observed in PAM mRNA expression levels using siRNAs from Invitrogen, and up to a 5-fold decrease using siRNAs from Ambion (Figure 1). On the basis of these data, and due to shRNA design limitations, siRNAs si5 and si6 (Ambion) were selected for the design of shRNAs, to obtain long-term silencing of *PAM*. siRNA knock-down of the target lasts for 3 to 5 days, and therefore to induce long-term expression of RNAi silencing in the target cells, an shRNA vector has to be used [15]. The shRNA silencing effect was tested on two different CHO parental cell lines, CHO *Der2* and CHO *Der3*, and on two mAb-expressing clones derived from the CHO *Der3* cell line, clone K62 with high (14%), and clone K25 with low (4%), prolinamide contents in the mAb that was produced (Figure 2). After shRNA transfection and antibiotic selection, all of the generated pools were analysed by cation-exchange chromatography (CEX), for evaluation of the prolinamide content (Figure 2). The shRNA designed on the basis of the si6 siRNA was shown to have the most potent silencing effect on all of the transfected cell lines (Figure 2).

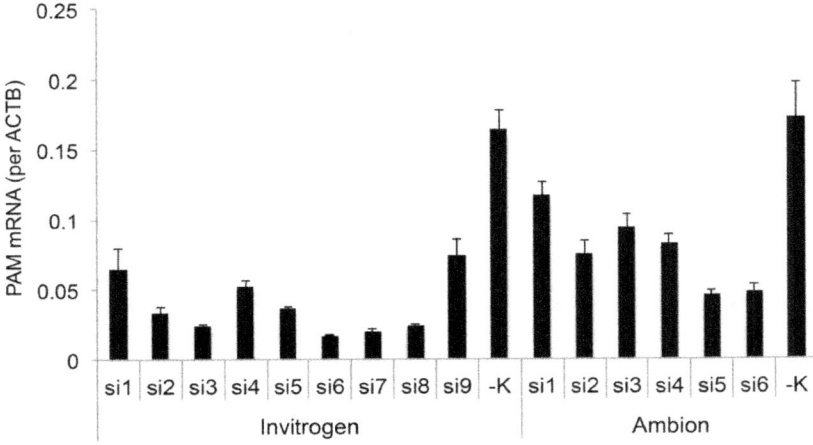

Figure 1: Silencing of *PAM* using siRNA. *PAM* mRNA expression level after transfections using the siRNAs (Invitrogen, Ambion, as indicated) in the CHO *Der2* parental cell line. In comparison to the negative control (−K), there was up to 8-fold reduction in the mRNA expression levels using the si6 Invitrogen siRNA, and up to 5-fold reductions using the si5 and si6 Ambion siRNAs. Overall, better silencing effects were obtained using the Invitrogen siRNAs. The mRNA expression levels were determined using qPCR (calculated per *ACTB* housekeeping gene), and the data are means ± standard deviations of the two biological replicates.

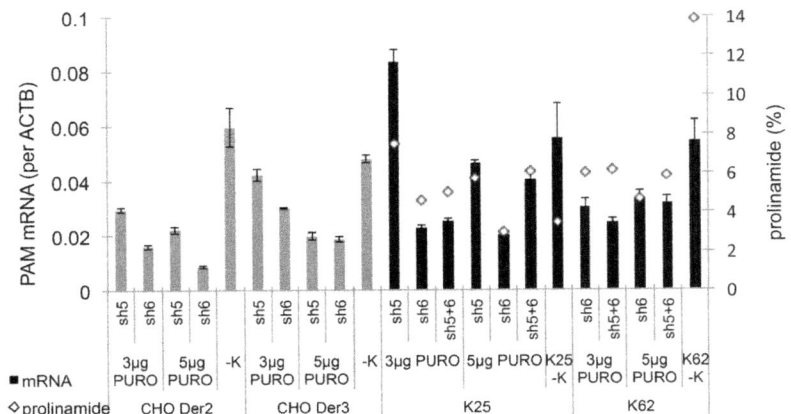

Figure 2: Silencing of *PAM* using shRNA. The mRNA expression levels are shown after the transfections using the Ambion sh5 and sh6 shRNAs on the CHO *Der2* and CHO *Der3* parental cell lines and on the K25 and K62 clones derived from the

CHO *Der3* parental cell line. The data obtained for the parental cell lines are in grey. In comparison to the negative controls (CHO *Der2* -K, and CHO *Der3* -K), the highest silencing effects were achieved with shRNA sh6 and 5 μg/ml antibiotic selection (5 μg puromycin [PURO]). Up to 3.7-fold reduction in the mRNA expression levels was seen for the CHO *Der2* cell line, and up to 2.6-fold reduction for the CHO *Der3* cell line. The data for the mAb-expressing clones K25 and K62 are in black. In this case, the *PAM* expression level and the prolinamide content (%) are presented. In comparison to the negative controls (K25 -K, and K62 -K), the different antibiotic selections did not show any differences in mRNA expression levels and prolinamide content. Up to 3.5-fold reduction in mRNA expression levels was observed for K25, and up to 2.2-fold reduction in K62. Prolinamide was decreased from 3.5% to 3% for K25, and from 14% to 4.6% for K62, which represents a 3-fold decrease. The mRNA expression levels for parental cell lines and the K25 and K62 clones were determined using qPCR (calculated per *ACTB* housekeeping gene), and the data are means ± standard deviations of two biological replicates.

The data presented in Figure 2 show the correlation between *PAM* mRNA expression levels and C-terminal amidation of the recombinant mAb. Up to a 4-fold decrease in mRNA and a 3-fold decrease in prolinamide content were observed. As can be seen from Figure 2, the prolinamide content for clone K62 was decreased to 4.6%, which represents a 3-fold decrease, and for clone K25, where the initial starting point was 3.5% prolinamide content, only a minimal reduction was obtained. Nevertheless, there is an interesting observation here that should be considered. No matter which clone is considered, as one with a previously high (14%) or low (4%) prolinamide content, the reduction in the prolinamide content after shRNA knock-down never decreased below 4%, which is the same as the level of the reference molecule.

ZFN Experiments

The shRNA experiments gave very promising results, as they yielded PAM levels that were comparable to the reference molecule. However, possible toxicity effects on long-term expression and the additional metabolic load on the cells due to the overexpression of these factors during times of stress might also influence cell performance. In addition, shRNA-mediated knock-down relies on the constant expression of repressor molecules, which can be unstable in the knocked-down cells in the long term [17]. RNAi instability was reported by Lim et al. in the silencing of the *BAX* and *BAK* genes, where 1% of the screened clones was silenced for both genes, but there was instability of the generated knock-down clones [20]. This is an undesirable feature during the mAb production process, as it can cause heterogeneity of the product [20]. To avoid such problems, rapid and permanent gene knock-out using ZFNs can offer distinct advantages. Moreover, the growing number of reports using ZFNs across different species have suggested that ZFN-mediated gene disruption is a robust and general method for targeted gene knock-out. ZFN-mediated gene knock-out requires only transient expression of the ZFNs, and it can result in a permanent genetic mutation that is stably transmitted through all of the subsequent generations of the cell line [17].

The first reported example of the use of engineered ZFNs to disrupt an endogenous locus in a mammalian cell was a knock-out of the dihydrofolate reductase gene in CHO cells. The observed bi-allelic mutation rate was 2% to 3%. In comparison to traditional methods, this frequency of gene disruption is very high, and it increases the possibility of achieving the desired genotype in which each copy of the target gene is functionally knocked out [17].

ZFNs have also already been successfully used for the generation of many different knocked-out CHO cell lines. They have been reported as used to create triple knock-out CHO cells, with the disruption of the two selectable marker genes of glutamine synthetase and dihydrofolate reductase, as well as the gene encoding α-1,6-fucosyltransferase (FUT8) [30]. *BAK* and *BAX* deletion was also achieved, to produce apoptosis-resistant CHO cells [31]. Additionally a *FUT8* knock-out CHO cell line has been reported [32, 33].

Thus, ZFNs were used to create a cell line with stably knocked-down *PAM* expression, so as to reduce C-terminal amidated species on the target mAb. These ZFNs were designed on the basis of a sequence that had been shown to generate silencing effects using RNAi. Initially, ZFN plasmid DNA was used for the transfections of the parental CHO *Der2* and CHO*Der3* cell lines. The knock-out effect was evaluated by determination of the *PAM* gene copy number using qPCR. This method was evaluated using the Cell-I Nuclease Mismatch assay (data not shown), which is usually used to evaluate knock-out clones. The qPCR method has been shown to be less laborious and time consuming, and therefore it enables efficient screening of the high numbers of clones generated.

In all, 85 clones were generated from the CHO *Der2* cell line, and 109 clones from the CHO *Der3* cell line. In contrast to the CHO *Der2* cell line, which is diploid, the CHO *Der3* cell line was shown by karyotyping to be a triploid (data not shown), and a knock-out was therefore more difficult to achieve. A partial reduction of the *PAM* copies in the CHO *Der2* and CHO *Der3* cell lines was achieved (Figure 3). Reported knock-out efficiencies for single copy genes have been higher than 1% [17, 30,34], and have reached 5% [33]. Thus, to detect a single clone with a total knock-out that would be reflected in a mAb without functional *PAM*, around 100 clones would need to be generated, and in the case of the triploid CHO *Der3* cell line, this would be higher.

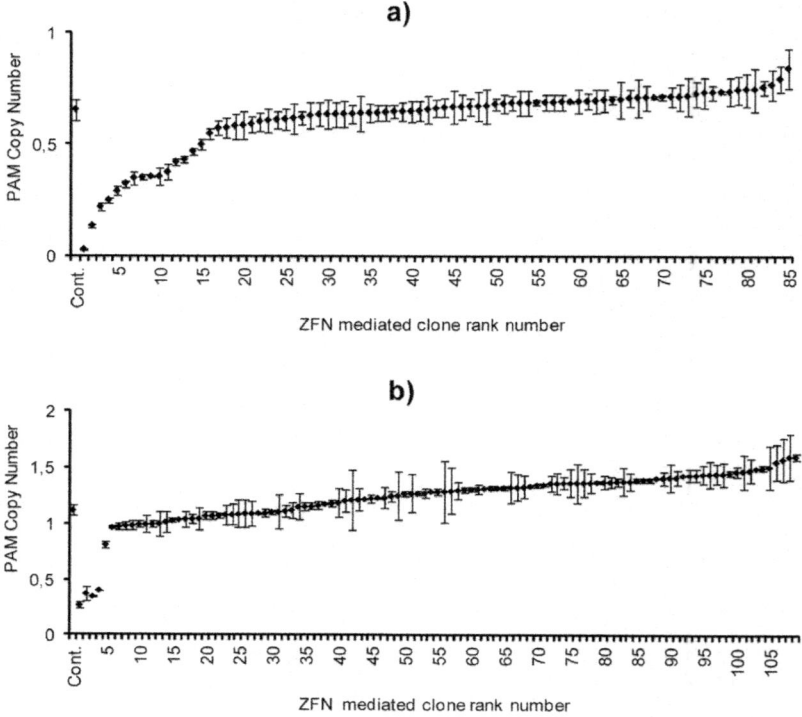

Figure 3: *PAM* gene copy numbers for the clones derived from the CHO *Der2* and CHO *Der3* cell lines transfected using the ZFN plasmid DNA. The PAM gene copy numbers for the 85 ZFN clones derived from the CHO *Der2* parental cells (a) and the 109 ZFN clones derived from the CHO *Der3* parental cells (b), transfected using the ZFN plasmid DNA. The copy numbers were determined using qPCR, and the data are means ± standard deviations of two technical replicates.

The *PAM* mRNA expression was evaluated for the 10 best clones by qPCR, and high correlation was observed with the copy number determination at the genomic DNA level (data not shown).

To determine the mRNA expression levels, the *PAM* mRNA was calculated per housekeeping gene. The choice of a housekeeping gene can often be quite challenging, as expression is usually not the same across different CHO cell lines and clones, and can also vary according to the physiological state of the cells. To avoid these effects, four different housekeeping genes were tested in the present study: *ACTB, GAPDH, G6PD* and *EF1* (Table 1). All four of these housekeeping genes were shown to be suitable for the calculation of the mRNA expression levels (data not shown), with the data below are based on the use of *ACTB* as the housekeeping gene.

Table 1: Nucleotide sequences of the qPCR primers and probes used in the present study

	Forward primer	Reverse primer	Probe
PAM	GGCCGGATCCAAT-GTTTCAGAA	TCCCAAATGACG-CATGTTTAATCTCT	CTGACAC-CAAAGAATTT
ACTB	AGCCACGCTCGGT-CAG	CATCCTGCGTCTG-GACCT	CCGGGACCT-GACAGACT
GAP-DH	CTGATTTTTCTTGC-GTCGAGTTT	GAGTTGTGTTTGTG-GACGAAGTAC	TCCGGTA-AGACCTTTCG
EF1	CCTGGAATGGTGGT-TACCTTTG	CATGGTG-CATTTCAA-CAGACTTTACT	CCAGTCAAC-GTTACAACAGA
G6PD	ACAACATTGCCTGT-GTGATCCT	CCCAAATTCAT-CAAAGTAGCCC	CCACGACCCT-CAGTACCA

PAM, peptidylglycine amidating monooxygenase; ACTB, β-actin; GAPDH, glyceraldehyde-3-phosphate dehydrogenase; EF1, elongation factor 1; G6PD, Glucose-6-phosphate dehydrogenase.

The three best clones were transfected with a plasmid containing the mAb expression cassette. After 4 weeks of antibiotic selection and 14 days of fed-batch cultivation, the mAbs produced from these three clones were evaluated for the presence of prolinamide, using CEX. The data from the *PAM* gene copy numbers, mRNA expression levels, and prolinamide content are presented in Figure 4. The prolinamide levels for both of the CHO *Der2* and CHO *Der3* cell lines after evaluation using CEX were determined to have decreased to 6%. As observed with the shRNA (see above), there was correlation between copy number, mRNA expression, and prolinamide content of the recombinant mAb (Figure 4).

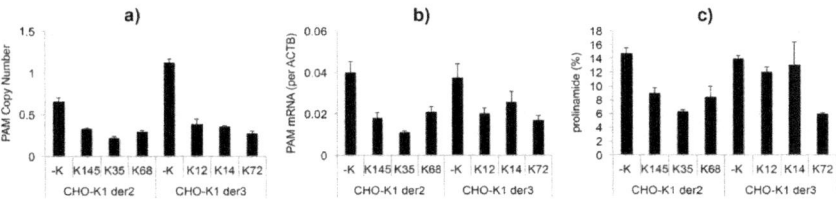

Figure 4: *PAM* gene copy numbers, mRNA expression and C-terminal amidation for clones transfected using ZFN pDNA. The *PAM* gene copy numbers (a), *PAM* mRNA expression levels (b; calculated per *ACTB* housekeeping gene), and prolinamide content on the recombinant mAb (c) for the three best ZFN knock-out clones derived from the CHO *Der2* and CHO *Der3* parental cell lines, where ZFN pDNA was used for the transfection. Prolinamide was reduced to 6%. The copy numbers and mRNA expres-

sion levels were determined using qPCR, and the data are means ± standard deviations of two technical replicates. The prolinamide content was determined using CEX, and these data are means ± standard deviations of two biological replicates.

Transfections using ZFN plasmid DNA result in its random integration into the genome of CHO cells and continuous expression of the ZFN. Genome integration is not desired, due to the continuous ZFN activity, and thus potential off-target effects. The mechanism of ZFN target recognition has been reported to be highly specific, with usually an 18-bp site recognized and the dimerisation of two ZFNs needed to cause a double-stranded break [18, 35]. Despite this, some off-target effects have also been observed. Off-target effects of 5.4% were reported for the disruption of the CCR5 HIV receptor by Klug et al. [19].

To avoid this, the transfections were performed using ZFN mRNA, for transient ZFN expression. The ZFN mRNA was prepared in-house using *in-vitro* transcription from the two ZFN plasmid DNAs. The cells were then transfected with the two ZFN mRNAs; i.e., ZFN1 and ZFN2. Various amounts for the combination of these two mRNAs were tested: total mRNAs of 30 μg to 60 μg per transfection. The 50 μg mRNAs transfection (25 μg ZFN1 mRNA, 25 μg ZFN2 mRNA) was the most promising combination. The concentrations of the mRNAs used for the transfections correlated with the silencing effects, although the higher mRNA concentrations decreased the cell viability to <80% (data not shown). The pools were evaluated by qPCR on day 4 following transfection. A 50% knock-out was achieved for the generated pools, and the cell viability was >80%, which is needed for seeding into the semi-solid medium for clone generation using ClonePix.

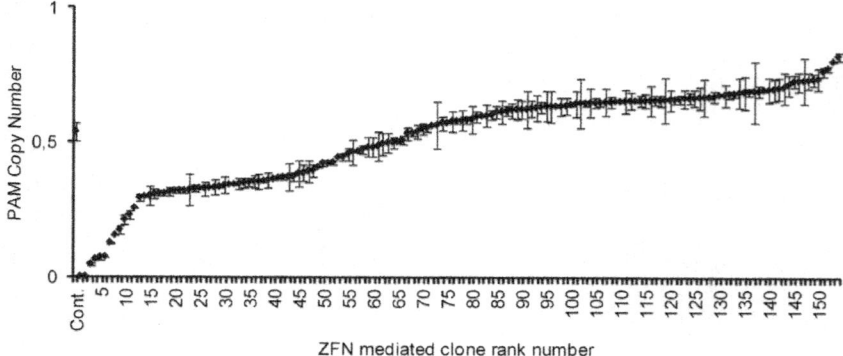

ZFN mediated clone rank number

Figure 5: *PAM* gene copy numbers for 154 ZFN clones from CHO *Der2* parental cell line transfected using ZFN mRNA. The two clones (K62 and K68) identified as knocked-out for all *PAM* gene copies at the genomic DNA level. The copy number was determined by qPCR, and the data are means ± standard deviations of two technical replicates.

These mRNA transfections were used only with the CHO *Der2* cell line. Here, 154 clones were evaluated and two were determined to be completely knocked-out at the genomic DNA level by qPCR (Figure 5), which represented a 1% knock-out efficiency. The 10 best clones were again tested for *PAM* mRNA expression (data not shown), and afterwards the best three of these 10 clones were transfected with a plasmid containing the mAb expression cassette, in duplicates, and evaluated by CEX (using the same procedure as described for the plasmid DNA transfections). These data are presented in Figure 6.

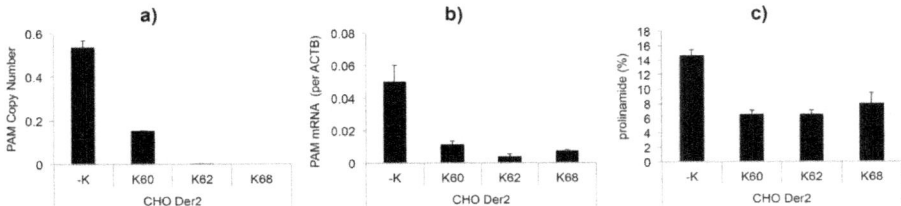

Figure 6: *PAM* gene copy numbers, mRNA expression, and C-terminal amidation for the clones transfected using ZFN mRNA. The *PAM* gene copy numbers (a), *PAM* mRNA expression levels (b; calculated per *ACTB* housekeeping gene), and prolinamide contents of the recombinant mAb determined by CEX (c), for the three best ZFN knock-out clones derived from the CHO *Der2* parental cell line, where ZFN mRNA was used for the transfection. The two clones identified (K62 and K68) were knocked-out for all *PAM* gene copies at the genomic DNA level, with mRNA expression and a prolinamide content of 6% still observed. The copy numbers and the mRNA expression levels were determined using qPCR, and the data are means ± standard deviations of two technical replicates. Prolinamide content was determined by CEX, results are presented as an average of two biological replicates.

The C-terminal amidated structures in all of these tested clones decreased to a limit of 6%, which was the same as observed using ZFN plasmid DNA. mRNA expression and C-terminal amidation were observed for all of these tested clones, even though this was not expected for clones K62 and K68.

Down-regulation of *PAM* did not have any influence on cell growth, in comparison to the non-silenced controls. All of the cells showed similar cell growth, which did not differ from that expected for CHO cells. Additionally, there was no correlation between cell productivity and C-terminal amidation of the produced mAb (data not shown).

The silencing effects that were obtained using the RNAi approach showed more promising results in terms of prolinamide reduction on the recombinant mAb. In this case, prolinamide was reduced to 3%. As indicated above, siRNA effects are only transient [15] and the shRNA approach is also questionable, due to long-term stability problems and the additional metabolic burden on the cells. Using the ZFN approach, the decrease in prolinamide was not as effective as with shRNA, as it was reduced to 6%. On the other hand, this ZFN approach has the advantage of producing a stable genetic modification, which is very important for the production scale. After transfection with the mAb-expressing plasmid, each of these two generated ZFN knocked-down cell lines (CHO *Der2* and CHO *Der3* cells) represented a mixed cell population with a mean prolinamide content of 6%. In comparison to the negative controls, which generated 15% (CHO *Der2*) and 14% (CHO *Der3*) prolinamide content, this represents a decrease in the prolinamide content of >50%. These pools with mean prolinamide content of 6% thus represent good starting points for the isolation of candidate clones that express different amounts of C-terminal amidated species, thus providing the choice for the one that is the most comparable to the reference molecule.

The higher efficiency of the knock-down achieved using shRNA in comparison to ZFN might be because the shRNA targets the degradation of the *PAM* mRNA, which will reduce its expression overall. For the ZFNs, the knock-out effects are at the level of the genomic DNA, and if total knock-out is not achieved in all of the copies, the expression of a functional mRNA will still take place, and thus the translation into the functional PAM protein will not be fully blocked. Disruption of the target gene can occur by causing a shift in the reading frame, the generation of a premature termination codon, or the deletion of critical amino-acid residues. However, the result of non-homologous end joining repair might be only an amino-acid point mutation or a deletion of the protein molecule, which might result in a partially or fully functional enzyme.

However, the goal of the present study was not to generate a completely knocked-out CHO cell line, but to decrease the content of the C-terminal amidated structures to a level that is sufficiently comparable to the reference molecule, and this was successfully achieved using both approaches.

Conclusions

During development, C-terminal amidated species are detected on recombinant mAbs, an observation that was also recently reported in other studies [6]. C-terminal amidation can be reduced by development and optimisation of the grwth medium, and use of cofactors such as copper [7], although these approaches are project specific and time consuming. Due to its extensive development, metabolic engineering has provided a powerful tool to diminish the expression of the *PAM* gene through gene alterations using RNAi and ZFNs. In the present study, both of these approaches were tested. The RNAi was conducted for the initial target validation, for the later ZFN knock-out. Using the shRNA approach, decreases in C-terminal amidation were achieved for two mAb-producing clones: the K25 with low prolinamide content (4%), and the K62 clone with high (14%) prolinamide content, which were decreased to 3.5% and 5%, respectively. The prolinamide levels of the reference molecule were 4%, and therefore a comparability level was successfully achieved. In this case, the gene silencing was performed on two clones that were already expressing the desired mAb.

For recombinant mAb production, a parental cell line with stably reduced *PAM* expression is desired. To achieve this, genome alterations using ZFNs are the more appropriate tool, as breaks are created at the genomic DNA level that are stably transferred through the subsequent generations of the cell line [3, 17]. Two CHO cell lines were generated, as CHO*Der2* and CHO *Der3* cells, and they were evaluated for *PAM* at the genomic DNA and mRNA expression levels. After the selection of the three most promising clones, these were transfected with a mAb-producing plasmid and evaluated for the C-terminal amidated species. The C-terminal amidation was decreased to 6% in both of these cell lines, which represents a 50% lower content in comparison to the ZFN non-treated controls. The two generated cell lines now represent two pools from which candidate clones with the highest comparability to the reference molecule can be selected.

Thus, this use of genetically manipulated cell lines now gives us the opportunity to go even further towards providing high-quality and safe therapeutics.

METHODS

Design of siRNAs, shRNAs and ZFNs

The Chinese hamster *(Cricetulus griseus) PAM* nucleotide sequence was derived from a public database (Table 2). The siRNAs, shRNAs and ZFNs were constructed on the basis of this sequence.

Table 2: The *Cricetulus griseus PAM* gene sequence used for the RNAi and ZFN construction

PAM gene nucleotide sequence	GGGAGTGCTCCTAAGCCAGGCCAGTTCAGTGTTCCTCACAGTTTGGCCCTTGTGCCTCATTTGGACCAGTTGTGTGTGGCAGACAGGGAA AATGGCCGGATCCAATGTTTCAGAACTGACACCAAAGAATTTGTGAGAGAGATTAAACATGCGTCATTTGGGAGAAATGTATTCGCAATT TCATATATATCAGGTTTGCTCTTTGCAGTAAATGGGAAGCCTTACTTTGGAGACCATGAACCTGTGCAAGGCTTTGTGATGAACTTTTCCAG TGGGGAAATTATAGATGTCTTCAAGCCAGTACGGCAAGCACTTTGACATGCCTCACGATGTGGTTGCCTCTGACGATGGGAATGTGTACATT GGAGACGCACACACGAACACGGTGTGGAAGTTCACCCTGACTGAAAAAATGGAGCATCGATCGGTTAAAAAGGCAGGCATTGAGGCTC AGGAAATCAAAGAAACCGAGGCAGTTGTTGAATCCAAAATGGAGAACAAACCCACCTCCTCAGAATTGCAGAAGATGCAAGAGAAACA GAAACTGATCAAAGAGCCAGGTTCGGGAGTGCCCGTGGTTCTCATTACAACCCTTCTGGTTATTCCTGTGGTTGTCCTGCTGGCCATTGTC ATGTTTATTCGGTGGAAAAAATCAAGGGCCTTTGGAGGAAAA

The online design tools of Invitrogen and Ambion were used to design the siRNA sequences (nine according to Invitrogen, six according to Ambion;

Table 3). After the siRNA evaluation, shRNAs were designed using the Ambion online design tool. Two complementary oligonucleotides were synthesised for each shRNA by Methabion (Table 4), and then these were annealed in-house, to generate the double-stranded oligonucleotides. Subsequently, these annealed oligonucleotides were cloned into the *pSilencer* 2.1-U6 puro vector (Ambion). DNA sequencing was performed to verify the sequences of the oligonucleotide inserts.

Table 3: Nucleotide sequences of the siRNAs used in the present study

Name	siRNA oligonucleotide	Supplier
siRNA_si1	CAGUUGUGUGUGGCAGACAGGGAAA	Invitrogen
siRNA_si2	CGGAUCCAAUGUUUCAGAACUGACA	Invitrogen
siRNA_si3	CCAAUGUUUCAGAACUGACACCAAA	Invitrogen
siRNA_si4	GAGAGAGAUUAAACAUGCGUCAUUU	Invitrogen
siRNA_si5	CAUGCGUCAUUUGGGAGAAAUGUAU	Invitrogen
siRNA_si6	UGGGAGAAAUGUAUUCGCAAUUUCA	Invitrogen
siRNA_si7	GGGAGAAAUGUAUUCGCAAUUUCAU	Invitrogen
siRNA_si8	CACACGAACACGGUGUGGAAGUUCA	Invitrogen
siRNA_si9	CAAAGAAACCGAGGCAGUUGUUGAA	Invitrogen
siRNA_si1	CAGUAAAUGGGAAGCCUUAUT	Ambion
siRNA_si2	AGGCAGUUGUUGAAUCCAAUT	Ambion
siRNA_si3	AACAGAAACUGAUCAAAGAUT	Ambion
siRNA_si4	CAAGAGAAACAGAAACUGAUT	Ambion
siRNA_si5	GAACUGACACCAAAGAAUUTT	Ambion
siRNA_si6	UUUCAGAACUGACACCAAAUT	Ambion

Table 4: Nucleotide sequences of the shRNAs used in the present study

Name	shRNA top strand oligonucleotide	shRNA bottom strand oligonucleotide
shRNA_sh5	GATCCGAACTGACACCAAAGAATTCTCAAGAGAAAT TCTTTGGTGTCAGTTCTGTTTTTTGGAAA	AGCTTTTCCAAAAAACAGAACTGACACCAAAGAATTTCTCTT GAGAATTCTTTGGTGTCAGTTCG
shRNA_sh6	GATCCGTTTCAGAACTGACACCAAACTCAAGAGATTT GGTGTCAGTTCTGAAACATTTTTTGGAAA	AGCTTTTCCAAAAAATGTTTCAGAACTGACACCAAATCTCTT GAGTTTGGTGTCAGTTCTGAAACG

For the ZFN experiments, two ZFN plasmids that targeted the *PAM* gene (pZFN PAM1, pZFN PAM2) were designed by Sigma (CompoZr™ Custom Zinc Finger Nuclease, PN. SAFCZFN-1kt). These two ZFN plasmids were later used as templates for the preparation of PAM1 and PAM2 mRNA, by *in-vitro* transcription using MessageMax T7 ARCA-Capped Transcription Kit (Cellscript). The transcripts were poly-A tailed using A-Plus Poly(A)

Polymerase Tailing Kit (Cellscript), and purified using RNeasy Micro Kit (Qiagen). The RNA concentrations were determined by the NanoDrop, and the transcript sizes were evaluated using a Bioanalyser.

CELL CULTURE AND TRANSFECTION

siRNA and shRNA Experiments

Two different CHO parental cell lines (CHO *Der2* and CHO *Der3*) and two different mAb-producing clones that were derived from the CHO *Der3* cell line (clone K25 and clone K62) were used in this study. Clones K25 and K62 were prepared by nucleofection of the parental CHO *Der3* cell line using a plasmid vector containing the mAb expression cassette. The cells were cultivated at 37°C, under 10% CO_2, and with agitation at 90–110 rpm, in animal-component-free growth medium. The C-terminal amidation of the mAb in clone K25 was determined as 4%, and in clone K62, as 14%, using CEX. It was shown that 4% of all of the reference molecules contained C-terminal amidated structures.

For the siRNA experiments, CHO *Der2* cells were transfected with siRNAs, in duplicate, using nucleofection, and cultivated for 4 days. On day 4, the cell pellets were collected for analysis of gene expression by qPCR.

For the shRNA experiments, all four cell lines (CHO *Der2*, CHO *Der3*, clone K25, and clone K62) were transfected with shRNAs, in duplicate, using nucleofection. Antibiotic selection of all of the transfected pools was performed using 3 µg/ml puromycin. The puromycin was added to the cell culture 2 days after the transfection, when the cell viability exceeded 60%. The cells were split on a 2-2-3-day schedule at 2-3E5 cells per ml, in the appropriate pre-warmed medium to maintain exponential cell growth. After reaching the appropriate cell density and viability, the cell pellets were collected for gene expression analysis by qPCR, and a 10-day simple batch containing 3 µg/ml puromycin was inoculated (after 10 days, the supernatants were collected, purified on protein A columns, and evaluated for prolinamide content using CEX). The cells were further cultivated in medium containing 5 µg/ml puromycin. After reaching the appropriate cell density and viability, this procedure was repeated.

ZFN experiments

For the ZFN experiments, the CHO *Der2* and CHO *Der3* parental cell lines were transfected using 3 µg of each ZFN plasmid (PAM1, PAM2), by nucleofection. After 4 days, the cells were seeded into the semi-solid medium, and on day 10 the colonies were transferred into 96-well plates by ClonePix FL. After colony formation in the 96-well plates, the individual colonies were transferred into

24-well plates, and later into 6-well plates. Finally, the clones were transferred into 125-ml shaking flasks, and after achieving the appropriate cell density and viability, the cell pellets were collected for the determination of gene copy number and gene expression by qPCR. The same procedure was repeated on the CHO *Der2* cell line using 25 µg of each ZFN mRNA (PAM1, PAM2), which was prepared in-house from the PAM1 and PAM2 ZFN plasmids.

After the selection of the three clones with the highest knock-down efficiency generated using ZFN plasmid DNA and mRNA, the transfections were performed using 3 µg mAb-expressing vector, in duplicate. After 4 weeks of antibiotic selection with geneticin, a 14-day fed-batch was inoculated. After 14 days, the supernatants were collected and purified using protein A columns. The prolinamide contents were evaluated using CEX.

DNA/RNA Purification and qPCR Template Preparation

The genomic DNA and the total RNA were isolated from pellets containing 5E6 cells on an automated QIAcube workstation. The genomic DNA was isolated using Blood & Cell Culture DNA Mini Kit (Qiagen), and the total RNA using RNeasy Mini Kit (Qiagen). After isolation, the nucleic acid concentrations were measured using the NanoDrop. After the determination of the total RNA concentrations, DNase I (Ambion) was added to 5 µg total RNA and incubated according to the manufacturer instructions. After this DNase treatment, the RNA was transcribed into cDNA using SuperScript VILO Kit (Invitrogen).

The genomic DNA and cDNA were used later as templates for the qPCR. The cDNA was used for evaluation of the siRNA and shRNA silencing effects, while the genomic DNA was used for evaluation of the knock-out effects of clones produced using the ZFNs.

QPCR EXPERIMENTS

siRNA and shRNA Experiments

The silencing effects of the siRNAs and shRNAs were examined by qPCR, using TaqMan chemistry and absolute quantification. The standard curve was constructed from CHO *Der1* genomic DNA, which was amplified using the *PAM* and *ACTB* primers (Table 1). The *PAM* and *ACTB* mRNA copy numbers for each sample were extrapolated from the standard curve, and the *PAM* mRNA expression levels were expressed as the number of *PAM* mRNA transcripts per reference *ACTB* gene. The standard curves were constructed from five dilution points, each as three parallel determinations. The *PAM* mRNA expression levels were determined as four dilution points, each as two parallel determinations for each sample.

ZFN experiments

The efficiencies of the ZFNs were determined by qPCR, using TaqMan chemistry and absolute quantification. The TaqMan probe was designed on the ZFN cutting site, and therefore no amplification of the *PAM* gene was expected in completely knocked-out clones. The standard curve was constructed from the CHO *Der1* genomic DNA amplified using the *GLUC* and*PAM* primers (Table 5). The copy numbers for both of these genes were extrapolated from the standard curve for each sample, and the ratio between them was calculated. *GLUC* is a single-copy gene in CHO cells, and was used as the reference gene for the *PAM* gene copy number determination. Standard curves were constructed from five dilution points, each as three parallel determinations. The *PAM* copy numbers were determined as two dilution points, each as three parallel determinations for each sample.

Table 5: Nucleotide sequences of the qPCR primers and probes used in the present study

	Forward primer	Reverse primer	Probe
GLUC	ATTGCCAAAC-GCCACGAT	CCAAGCAAT-GAATTCCTTTGC	CT-GAAGGGACCTT-TACCA
PAM	GCCCCAGCTCG-GACATG	CCTTGCAGGGC-GAGCA	CCGGCTGCTGCT-GCT

GLUC, glucagon; PAM, peptidylglycine amidating monooxygenase.

After the selection of the 10 clones with the highest knock-out efficiency, the *PAM* expression was determined as described above, using four different housekeeping genes: *GAPDH*, *ACTB*, *G6PD* and *EF1* (Table 1).

Cation-Exchange Chromatography

The mAb concentrations were determined using the ForteBio system. The mAbs were purified using protein A columns and analysed using CEX, with an analytical HPLC chromatographic system. Using this method, Lys and prolinamide were eluted in the same peak. The amount of prolinamide was then determined by C-terminus treatment with carboxypeptidase. Following the further analysis using the same CEX system, the remaining peak defined the prolinamide levels. The prolinamide levels are given as percentages of the

C-terminal amidated species of the total mAb content. After carboxipeptidase treatment, in addition to prolinamide, the basic peak can include multiple species, like N-terminal cyclised species, succinimide species, hybrid glycan-containing species, and others. Therefore the CEX analysis was always performed in parallel with the non-silenced and non-knockout controls, represented by samples from the same cell lines that had undergone blank transfections and were grown under the same conditions. The effects of RNAi and the ZFNs on the treated samples are highly specific, and the probability of other post-translational modifications other than reductions in prolinamide is very low. The reductions in the C-terminal amidation were determined relative to the controls, and therefore the reduction in the basic peak defined the reduction in C-terminal amidation.

ACKNOWLEDGMENTS

The authors express their sincere thanks to Holger Laux for help with the siRNA and shRNA design, and to Uroš Jamnikar for helpful information and discussions. They would also like to thank Petja Šušteršič for handling the cells during the RNAi experiments, and Irena Filipovič for performing part of the qPCR analysis.

COMPETING INTERESTS

This study was performed at Sandoz Biopharmaceuticals, Mengeš, a Member of the Novartis group, which also financed the article-processing charges. The patent application entitled "Reduction of formation of amidated amino acids in cell lines for protein expression" (No. 12164264.9 - 1212) was filed in 2012. The authors confirm that there are no known conflicts of interest associated with this publication, and there has been no significant financial support for this study that might have influenced its outcome. The authors confirm that they have given due consideration to the protection of the intellectual property associated with this study, and that there are no impediments to its publication, including the timing of the publication, with respect to the intellectual property. In so doing, the authors confirm that they have followed the regulations of their institutions concerning intellectual property. The authors also confirm that the manuscript has been read and approved by all of the named authors, and that there are no other persons who satisfy the criteria for authorship who are not listed. All of the authors declare they have no competing interests of financial or non-financial natures.

AUTHORS' CONTRIBUTIONS

MS and DP carried out all of the experiments and MS drafted the manuscript. DG participated in the planning the design of the study. DG, MK and RZ critically revised the manuscript and all authors approved the final version for publication.

REFERENCES

1.   Kim JY, Kim YG, Lee GM: CHO cells in biotechnology for production of recombinant proteins: current state and further potential. Appl Microbiol Biotechnol. 2012, 93: 917-930.

2.   Butler M: Animal cell cultures: recent achievements and perspectives in the production of biopharmaceuticals. Appl Microbiol Biotechnol. 2005, 68: 283-291.

3.   Warner TG: Enhancing therapeutic glycoprotein production in Chinese hamster ovary cells by metabolic engineering endogenous gene control with antisense DNA and gene targeting. Glycobiology. 1999, 9: 841-850.

4.   Werner RG, Kopp K, Schlueter M: Glycosylation of therapeutic proteins in different production systems. Acta Paediatr Suppl. 2007, 96: 17-22.

5.   Harris RJ: Processing of C-terminal lysine and arginine residues of proteins isolated from mammalian cell culture. J Chromatogr A. 1995, 705: 129-134.

6.   Tsubaki M, Terashima I, Kamata K, Koga A: C-terminal modification of monoclonal antibody drugs: amidated species as a general product-related substance. Int J Biol Macromol. 2013, 52: 139-147.

7.   Kaschak T, Boyd D, Lu F, Derfus G, Kluck B, Nogal B, Emery C, Summers C, Zheng K, Bayer R, Amanullah A, Yan B: Characterization of the basic charge variants of a human IgG1: effect of copper concentration in cell culture media. MAbs. 2011, 3: 577-583.

8.   Johnson KA, Paisley-Flango K, Tangarone BS, Porter TJ, Rouse JC: Cation exchange-HPLC and mass spectrometry reveal C-terminal amidation of an IgG1 heavy chain. Anal Biochem. 2007, 360: 75-83.

9.   Cuttitta F: Peptide amidation: signature of bioactivity. Anat Rec. 1993, 87-93. 172–173; discussion 193–175

10. Merkler DJ: C-terminal amidated peptides: production by the in vitro enzymatic amidation of glycine-extended peptides and the importance of the amide to bioactivity. Enzyme Microb Technol. 1994, 16: 450-456.

11. In Y, Fujii M, Sasada Y, Ishida T: Structural studies on C-amidated amino acids and peptides: structures of hydrochloride salts of C-amidated

Ile, Val, Thr, Ser, Met, Trp, Gln and Arg, and comparison with their C-unamidated counterparts. Acta Crystallogr B. 2001, 57: 72-81.

12.   Bolkenius FN, Ganzhorn AJ: Peptidylglycine alpha-amidating mono-oxygenase: Neuropeptide amidation as a target for drug design. Gen Pharmacol. 1998, 31: 655-659.

13.   Prigge ST, Mains RE, Eipper BA, Amzel LM: New insights into copper monooxygenases and peptide amidation: structure, mechanism and function. Cell Mol Life Sci. 2000, 57: 1236-1259.

14.   Hayashi N, Kayo T, Sugano K, Takeuchi T: Production of bioactive gastrin from the non-endocrine cell lines CHO and COS-7. FEBS Lett. 1994, 337: 27-32.

15.   Wu SC: RNA interference technology to improve recombinant protein production in Chinese hamster ovary cells. Biotechnol Adv. 2009, 27: 417-422.

16.   Amarzguioui M, Rossi JJ, Kim D: Approaches for chemically synthesized siRNA and vector-mediated RNAi. FEBS Lett. 2005, 579: 5974-5981.

17.   Santiago Y, Chan E, Liu PQ, Orlando S, Zhang L, Urnov FD, Holmes MC, Guschin D, Waite A, Miller JC, Rebar EJ, Gregory PD, Klug A, Collingwood TN: Targeted gene knockout in mammalian cells by using engineered zinc-finger nucleases. Proc Natl Acad Sci U S A. 2008, 105: 5809-5814.

18.   Urnov FD, Rebar EJ, Holmes MC, Zhang HS, Gregory PD: Genome editing with engineered zinc finger nucleases. Nat Rev Genet. 2010, 11: 636-646.

19.   Klug A: The discovery of zinc fingers and their applications in gene regulation and genome manipulation. Annu Rev Biochem. 2010, 79: 213-231.

20.   Lim SF, Chuan KH, Liu S, Loh SO, Chung BY, Ong CC, Song Z: RNAi suppression of Bax and Bak enhances viability in fed-batch cultures of CHO cells. Metab Eng. 2006, 8: 509-522.

21.   Sung YH, Hwang SJ, Lee GM: Influence of down-regulation of caspase-3 by siRNAs on sodium-butyrate-induced apoptotic cell death of Chinese hamster ovary cells producing thrombopoietin. Metab Eng. 2005, 7: 457-466.

22.   Sung YH, Lee JS, Park SH, Koo J, Lee GM: Influence of co-down-regulation of caspase-3 and caspase-7 by siRNAs on sodium butyrate-induced apoptotic cell death of Chinese hamster ovary cells producing thrombopoietin. Metab Eng. 2007, 9: 452-464.

23.   Wong DC, Wong KT, Nissom PM, Heng CK, Yap MG: Targeting early

apoptotic genes in batch and fed-batch CHO cell cultures. Biotechnol Bioeng. 2006, 95: 350-361.

24. Ngantung FA, Miller PG, Brushett FR, Tang GL, Wang DI: RNA interference of sialidase improves glycoprotein sialic acid content consistency. Biotechnol Bioeng. 2006, 95: 106-119.

25. Mori K, Kuni-Kamochi R, Yamane-Ohnuki N, Wakitani M, Yamano K, Imai H, Kanda Y, Niwa R, Iida S, Uchida K, Shitara K, Satoh M: Engineering Chinese hamster ovary cells to maximize effector function of produced antibodies using FUT8 siRNA. Biotechnol Bioeng. 2004, 88: 901-908.

26. Imai-Nishiya H, Mori K, Inoue M, Wakitani M, Iida S, Shitara K, Satoh M: Double knockdown of alpha1,6-fucosyltransferase (FUT8) and GDP-mannose 4,6-dehydratase (GMD) in antibody-producing cells: a new strategy for generating fully non-fucosylated therapeutic antibodies with enhanced ADCC. BMC Biotechnol. 2007, 7: 84-

27. Kim SH, Lee GM: Down-regulation of lactate dehydrogenase-A by siRNAs for reduced lactic acid formation of Chinese hamster ovary cells producing thrombopoietin. Appl Microbiol Biotechnol. 2007, 74: 152-159.

28. Hong WW, Wu SC: A novel RNA silencing vector to improve antigen expression and stability in Chinese hamster ovary cells. Vaccine. 2007, 25: 4103-4111.

29. Wu SC, Hong WW, Liu JH: Short hairpin RNA targeted to dihydrofolate reductase enhances the immunoglobulin G expression in gene-amplified stable Chinese hamster ovary cells. Vaccine. 2008, 26: 4969-4974.

30. Liu PQ, Chan EM, Cost GJ, Zhang L, Wang J, Miller JC, Guschin DY, Reik A, Holmes MC, Mott JE, Collingwood TN, Gregory PD: Generation of a Triple-Gene Knockout Mammalian Cell Line Using Engineered Zinc-Finger Nucleases. Biotechnol Bioeng. 2010, 106: 97-105.

31. Cost GJ, Freyvert Y, Vafiadis A, Santiago Y, Miller JC, Rebar E, Collingwood TN, Snowden A, Gregory PD: BAK and BAX Deletion Using Zinc-Finger Nucleases Yields Apoptosis-Resistant CHO Cells. Biotechnol Bioeng. 2010, 105: 330-340.

32. Yamane-Ohnuki N, Kinoshita S, Inoue-Urakubo M, Kusunoki M, Iida S, Nakano R, Wakitani M, Niwa R, Sakurada M, Uchida K, Shitara K, Satoh M: Establishment of FUT8 knockout Chinese hamster ovary cells: an ideal host cell line for producing completely defucosylated antibodies with enhanced antibody-dependent cellular cytotoxicity. Biotechnol Bioeng. 2004, 87: 614-622.

33. Malphettes L, Freyvert Y, Chang J, Liu PQ, Chan E, Miller JC, Zhou Z, Nguyen T, Tsai C, Snowden AW, Collingwood TN, Gregory PD, Cost GJ: Highly Efficient Deletion of FUT8 in CHO Cell Lines Using Zinc-Finger Nucleases Yields Cells That Produce Completely Nonfucosylated Antibodies. Biotechnol Bioeng. 2010, 106: 774-783.

34. Klug A: The discovery of zinc fingers and their development for practical applications in gene regulation and genome manipulation. Q Rev Biophys. 2010, 43: 1-21.

35. Durai S, Mani M, Kandavelou K, Wu J, Porteus MH, Chandrasegaran S: Zinc finger nucleases: custom-designed molecular scissors for genome engineering of plant and mammalian cells. Nucleic Acids Res. 2005, 33: 5978-5990.

Chapter 6

THE HYPERACTIVE SLEEPING BEAUTY TRANSPOSASE SB100X IMPROVES THE GENETIC MODIFICATION OF T CELLS TO EXPRESS A CHIMERIC ANTIGEN RECEPTOR

Z Jin[1], S Maiti[1], H Huls[1], H Singh[1], S Olivares[1], L Mátés[2], Z Izsvák[2,3], Z Ivics[2,3], D A Lee[1], R E Champlin[4] and L J N Cooper[1]

[1]Division of Pediatrics, Children's Cancer Hospital, The University of Texas Graduate School of Biomedical Sciences, The University of Texas MD Anderson Cancer Center, Houston, TX, USA

[2]Max Delbrück Center for Molecular Medicine, Berlin, Germany

[3]University of Debrecen, Debrecen, Hungary

[4]Stem Cell Transplantation and Cellular Therapy, University of Texas MD Anderson Cancer Center, Houston, TX, USA

ABSTRACT

Sleeping Beauty (SB[3]) transposon and transposase constitute a DNA plasmid system used for therapeutic human cell genetic engineering. Here we report a comparison of SB100X, a newly developed hyperactive SB transposase, to a previous generation SB11 transposase to achieve stable expression of a CD19-specific chimeric antigen receptor (CAR[3]) in primary human T cells. The electro-transfer of SB100X expressed from a DNA plasmid or as an introduced mRNA species had superior transposase activity in T cells based on the measurement of excision circles released after transposition and emergence of CAR expression on T cells selectively propagated upon CD19[+] artificial antigen-presenting cells. Given that T cells modified with SB100X and SB11 integrate on average one copy of the CAR transposon in each T-cell genome, the improved transposition mediated by SB100X apparently leads to an augmented founder effect of electroporated T cells with durable integration of CAR. In aggregate, SB100X improves SB transposition in primary human T cells and can be titrated with an SB transposon plasmid to improve the generation of CD19-specific CAR[+] T cells.

INTRODUCTION

To overcome immune tolerance, T cells can be genetically modified to express chimeric antigen receptors (CARs) to redirect specificity to tumor-associated antigens, such as CD19 (ref. 1). These transgenes can be introduced into T cells *ex vivo* using virus-based vectors and non-viral systems. Viral-based vectors are widely used in research and clinical trials as they provide stable transduction of target cells.[2, 3] However, retroviruses' non-random patterns of integration, at least in hematopoietic stem cells, could potentially activate/inactivate oncogenes/tumor suppressor genes leading to deleterious autonomous T-cell proliferation.[4] Non-viral gene transfer systems based on transposable elements are an alternative approach to transduction to introduce desired transgenes into the genome.[5, 6] The *Sleeping Beauty* (SB) transposon system, which integrates at TA dinucleotides apparently randomly across the genome, can be adapted for genetic engineering of T cells.[7, 8, 9] This is a result of the stable and efficient integration of an electro-transferred SB transposon by the enzymatic activity of an SB transposase, which is typically introduced as a separate DNA plasmid *in trans* from the DNA plasmid expressing the transposon. SB11 is a hyperactive SB transposase reported to achieve about 100-fold higher integration rates than those achieved by DNA plasmids without transposase activity that use illegitimate recombination to achieve integration.[10] On the basis of the SB11 transposase, we are undertaking gene therapy clinical trials (INDs no. 14193 and no. 14577) infusing CD19-specific T cells that have been electroporated to introduce SB transposon and transposase to generate CAR$^+$ T cells, which can be selectively propagated on CD19$^+$ artificial antigen-presenting cells (aAPC^3).[11, 12] Using this approach, clinically sufficient numbers of T cells can be obtained within a few weeks after electroporation of the SB DNA plasmids.

Improvements to the efficiency of transposition may augment our ability to generate CAR$^+$ T cells. Therefore, we investigated the integration efficiency of a CD19-specific CAR transposon, designated CD19RCD28, using a new mutant of SB transposase termed SB100X,[13] which had been systematically engineered to have increased enzymatic activity in mammalian cells. Follow-up studies have validated the superiority of SB100X transposase activity in mouse embryonic stem cells and human hematopoietic stem cells.[5, 14] In preparation for a next-generation trial using the SB system, we compare for the first time the ability of SB11 and SB100X to generate CAR$^+$ T cells from human peripheral blood mononuclear cells (PBMC). Our data reveal that SB100X results in 10 to 100 times more transposition events than SB11, as determined by the excision of transposon from DNA plasmid, which resulted in three to four times more efficient outgrowth of CD19-specific CAR$^+$ T cells within 28 days after electroporation, and with approximately one copy of CAR transposon per T-cell genome. This apparent increase in enzymatic activity of SB100X is highlighted by our ability to achieve superior outgrowth of CAR⁺ T cells using just one-tenth the amount of DNA plasmid coding for SB100X compared with SB11.

RESULTS

Measuring SB Transposition by Quantifying Excision Circles

It has been reported that SB100X results in improved transposition in mouse and human cells.[14, 18] Therefore, we determined the relative ability of SB100X versus SB11 to mediate a transposition event by quantitative-polymerase chain reaction (Q-PCR) adapted to measure DNA fragments (excision circle)[19] that are the expected by-product produced when CAR transgene (transposon) integrates into the T-cell genome, as schematically shown in Figure 1a. The DNA plasmids used to introduce SB transposase in this study are shown in Figures 1b and c.

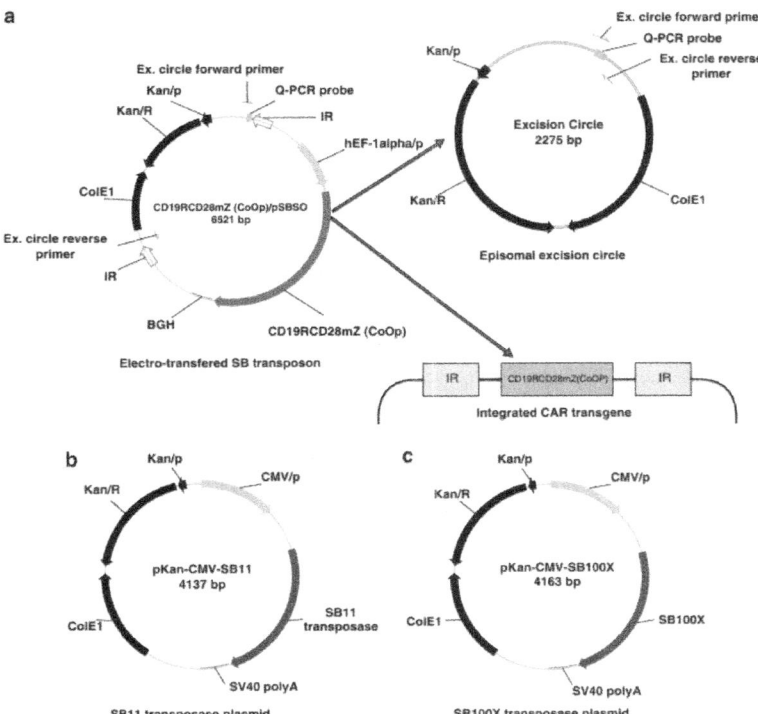

Figure 1: Schematic showing the formation of excision circles and integration of

transgene using DNA plasmids from SB system. (**a**) The transgene (transposon) to be integrated is flanked by two IR and mobilized from the CD19RCD28mZ(CoOp)/pSBSO plasmid by SB transposase. Upon SB transposition, the CAR transposon (CD19RCD28) is inserted into the T-cell genome, whereas the non-integrated DNA forms an episomal excision circle. The PCR to detect a released excision circle reveals a 77 base pair band, whereas the same primers bound to sites in the CD19RCD28mZ(CoOp)/pSBSO plasmid are 4298 base pairs apart. (**b**) Schematic of DNA plasmids expressing SB11 transposase and (**c**) SB100X transposase. BGH, bovine growth hormone polyadenylation signal sequence; CDS, coding sequence of gene; CMV/p, cytomegalovirus promoter; ColE1, colicin E1 (origin of replication); hEF-1α/p, human elongation factor-1α hybrid promoter; IR, inverted repeats; Kan/R, kanamycin resistance gene; Kan/P, kanamycin resistance gene promoter; SV40 poly A, Simian virus 40 polyadenylation signal sequence.

SB100X Transposase Improves Frequency of Transposition

To assess directly the ability of SB100X to improve the frequency of transposition compared with SB11 transposase, we serially measured the formation of non-integrated excision circles (products) by real-time Q-PCR after electroporation of T cells from PBMC. The transient accumulation of excision circles represents the enzymatic activities of the two SB transposases, whereas measurement of the electro-transferred SB transposon and subsequent expression of CD19RCD28 CAR indicates the overall integration efficiency of the non-viral gene transfer process. To account for variations in electro-transfer efficiency, we normalized the excision circle data to the amount of DNA transposon recovered after electroporation (excision circle to CD19RCD28 ratio). This revealed the individual SB transposase's enzymatic activities and adjusted for possible difference in electroporation efficacy between different samples. Measurement of RPPH1, a subunit of RNase P,[20] which is present at one copy on chromosome 14q11.2, was used as internal control in real-time Q-PCR for both quantification of excision circles and transposon DNA species. By varying the relative amounts of DNA plasmids coding for the transposases, SB100X reached peak activity at a concentration of 5 μg per electroporation (Figure 2b). However, at 10 μg per electroporation, SB100X resulted in significant loss of PBMC viability (data not shown). This decrease in excision products when 10 μg of SB100X DNA plasmid was used per electroporation could be due to DNA toxicity, although this was not observed with SB11, but is more likely due to over-production inhibition. SB11 also showed maximal activity at 5 μg per electroporation (Figure 2a), but the amount of excision circles produced was significantly less than that achieved with SB100X. Overall, SB100X was apparently 10 to 100 times more active compared with SB11. As expected, the episomal excision circles are lost to detection as the T cells propagate on the aAPC. To avoid the possibility that the SB transposases could integrate, we assessed whether improved transposition could also be achieved by mRNA coding for SB100X. The two transposases were electro-transferred as mRNA species along with a fixed amount of SB DNA transposon into T cells pre-activated by crosslinking CD3 with the monoclonal antibody OKT3. We observed that the introduction of SB100X mRNA at 0.1 μg per electroporation was as active as 10 × the amount of SB11. When SB100X mRNA was used at 0.25 μg per electroporation, there was a higher transposition activity compared with SB11 at 1 μg per electroporation (Figure 2c). These data indicate the superior activity of SB100X in primary T cells whether this transposase is expressed from electro-transferred DNA or mRNA.

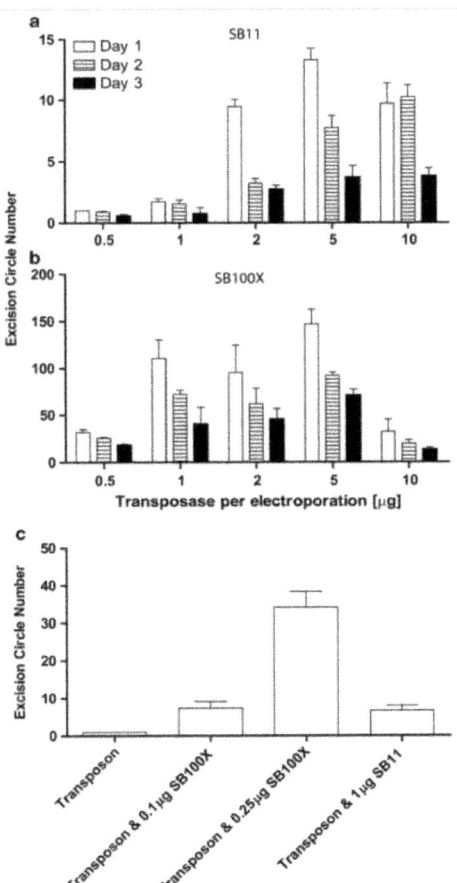

Figure 2: Evaluation of transposase activity by detecting excision-circle formation

after SB transposition event. Excision circle to CD19RCD28 ratio is based on the amount of (**a**) SB11 and (**b**) SB100X transposase as detected by Q-PCR at days 1 to 3 after electroporation. The DNA transposon plasmid (CD19RCD28mZ(CoOp)/pS-BSO) to express CD19RCD28 CAR was used at 15 μg per electroporation. (**c**) Excision circle to CD19RCD28 ratio after electro-transfer of mRNA coding for SB100X and SB11, combined with SB DNA transposon plasmid, on day 1 after electro-transfer.

Generation of CAR$^+$ T Cells by SB Transposition

Primary human T cells from PBMCs were electro-transferred on day 0 of cell culture with SB transposon (CD19RCD28mZ(codon optimized (CoOp))/pSBSO) and one of the two DNA plasmids, pKAN-CMV-SB11 and pKAN-CMV-SB100X, coding for SB11 or SB100X, respectively. The T cells were subsequently propagated for up to 28 days on γ-irradiated CD19$^+$ aAPC, added every 7 days in the presence of soluble recombinant interleukin-2 (IL-2) cytokine (Figure 3a). Our approach to generating clinical-grade CAR$^+$ T cells uses 5 μg of pKAN-CMV-SB11 along with 15 μg of CD19RCD28mZ(CoOp)/pSBSO in the electroporation of 2×10^7 PBMC per cuvette; therefore, this was used as a starting point to assess the ability of SB100X to improve the rate of transposition and subsequent outgrowth of CAR$^+$ T cells. However, when we electroporated T cells with the DNA plasmid coding for SB100X at 5 μg per electroporation with 15 μg per electroporation of the DNA plasmid coding for CAR transposon, this accentuated cell death the day after electro-transfer, as shown by Trypan blue staining and failure to propagate CAR$^+$T cells. However, the electro-transfer of DNA plasmid coding for SB100X at decreased amounts did not compromise cell viability as the genetically modified T cells could be readily propagated on aAPC.

Indeed, upon reducing the concentration of the DNA plasmid coding for this transposase to 0.1 and 0.5 μg per electroporation, SB100X successfully integrated the transposon to support the outgrowth of CAR$^+$ T cells. When using 10-fold less (0.5 μg per electroporation) than the input concentration of SB11 (5 μg per electroporation), we calculate that SB100X was about 3.6 times more efficient than SB11 in generating CAR$^+$ cells as assessed at day 28 of co-culture with CD19$^+$ aAPC (based on dividing the number of CAR$^+$ T cells associated with SB100X with the number of CAR$^+$ T cells associated SB11, in two independent experiments). Indeed, even 0.1 μg per electroporation of the DNA plasmid coding for SB100X was almost as efficient as 5 μg per electroporation of SB11 when the number of CAR$^+$ T cells were counted at 28 days of tissue culture (Figures 3b and c). Expression of the CAR on electroporated and propagated T cells was documented by flow cytometry (Figure 3d). Thus, transposition mediated by SB100X results in improved outgrowth of CAR$^+$ T cells.

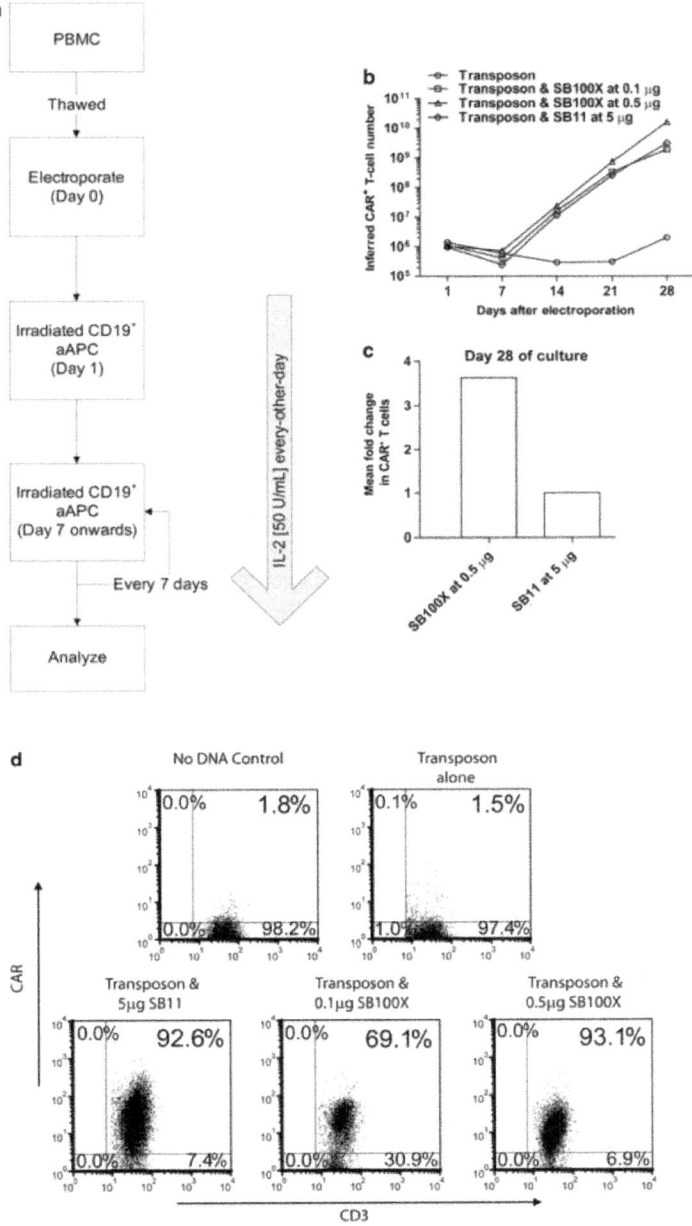

Figure 3: Selective outgrowth of CD19-specifc CAR⁺ T cells after transposition with SB100X versus SB11 transposases. (**a**) Schematic outlining co-culture process to generate CD19-specifc CAR⁺ T cells. A total of 10⁵CAR⁺ T cells from PBMC were stimu-

lated with γ-irradiated CD19⁺ aAPC (clone no. 4) every 7 days at a 1:2 (CAR⁺ aAPC) ratio in the presence of soluble IL-2. (**b**) Kinetics of CAR⁺ T-cell numeric expansion by repetitive co-culture with CD19⁺ aAPC. T cells were electroporated on day 0 with SB DNA plasmid transposon expressing CD19RCD28 and graded doses of DNA plasmids expressing SB100X or SB11. (**c**) Fold change in the number of CAR⁺ T cells sampled at day 28 of co-culture on aAPC from two independent experiments. (**d**) Expression of CAR (CD19RCD28) on CD3⁺ T cells by flow cytometry at day 28 of culture.

CAR⁺ T Cells can be Generated Using SB100X mRNA

As SB transposase coded by mRNA species was capable of accomplishing SB transposition, we determined if CAR⁺ T cells could be selectively propagated on aAPC after electro-transfer of DNA plasmid coding for CD19RCD28 and mRNA coding for SB100X or SB11. We adapted our propagation method to generate CAR⁺ T cells (Figure 4a), so that T cells were pre-activated with OKT3 to improve uptake of and expression from mRNA. As shown in Figure 4b, 0.25 μg per electroporation of SB100X and 1 μg per electroporation of SB11 successfully produced CAR⁺ T cells that could be propagated on CD19⁺ aAPC. The superiority of the SB100X transposase to support the outgrowth of CAR⁺ T cells was apparently not due to differences in integrity of the mRNA (Figure 4c).

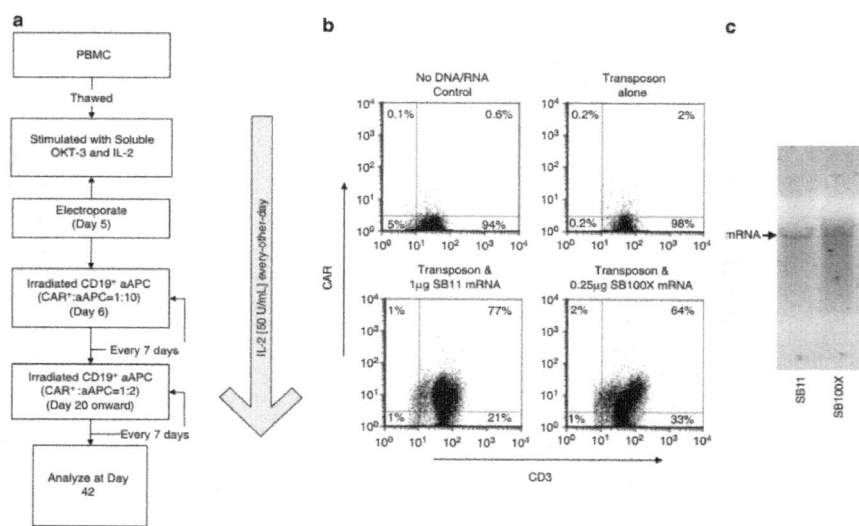

Figure 4: CAR⁺ T cells generated by electro-transfer of mRNA coding for SB transposases and DNA plasmid coding for CD19RCD28 CAR transposon. (**a**) Schematic used to generate CAR⁺ T cells. (**b**) Expression of CAR on numerically expanded CD3⁺ T cells at day 42 of co-culture with aAPC. The SB100X mRNA was used at 0.25 μg per electroporation, whereas the SB11 mRNA concentration was at 1 μg per

electroporation. The DNA plasmid expressing CD19RCD28 was used at 10 μg per electroporation. (**c**) Integrity of mRNA species after *in vitro* transcription used to express SB11 and SB100X.

Transposition using SB100X and SB11 Result in Comparable Number of Integration Events per T Cell

Given that SB100X gives rise to a greater number of transposition events compared with SB11, we investigated whether this enzyme resulted in multiple integration events. To evaluate the number of integration events per electroporated and propagated T cell, we measured the copy number of CD19RCD28 transpose relative to the copy number of RNase P gene by real-time Q-PCR. The calculated transgene copy per T cell is 0.95±0.068 gene (mean±s.e.m.) for 0.5 μg per electroporation SB100X and 0.80±0.033 (mean±s.e.m.) for 5 μg per electroporation SB11, respectively (Figure 5a). This difference is not statistically significant (P=0.185) when measured at a time point when sufficient (clinical grade) CAR$^+$ T cells are available that can be harvested from cultures for adoptive immunotherapy. These data revealed that the number of integrated copies of CAR is approximately 1 per T-cell genome upon transposition with both SB100X and SB11.

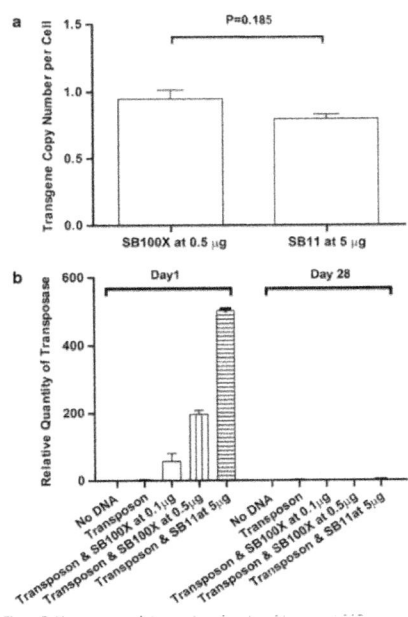

Figure 5: Measurement of the number of copies of integrated CAR transgene and detection of SB transposases. (**a**) Copy number of CAR transgene, normalized to

RNase P, at day 28 of co-culture on CD19⁺ aAPC after SB transposition with SB100X or SB11. Q-PCR using CAR-specific primers revealed CD19RCD28 copy numbers at 0.95 and 0.80 transgene per T-cell genome after electro-transfer of 0.5 μg DNA plasmid coding for SB100X, and 5 μg DNA plasmid coding for SB11, respectively. There was no statistical difference between the copy number of integrated CAR transgenes. (**b**) Measurement by Q-PCR using transposase-specific primers at day 1 (day after electroporation) and at day 28 of co-culture of genetically modified T cells on CD19⁺ aAPC. Transposon was added at 15 μg per electroporation along with graded doses of DNA plasmids coding for SB100X and SB11 transposases.

SB100X DNA cannot be Detected in Electroporated and Propagated T Cells

To establish that SB100X or SB11 DNA is not present in propagated T cells, we developed a Q-PCR assay to reveal integrated transposase plasmid. Our data show that after 28 days of *in vitro* culture, the SB100X transposase, as well as that of SB11, are absent in the cultured cells (Figure 5b). The standard curves associated with this assay are in the Supplementary Data.

CAR⁺ T Cells Generated by SB Transposition with SB100X Exhibit Redirected Killing for CD19⁺ Tumor Cells

We have previously shown that CAR⁺ T cells genetically modified with SB11 transposase specifically lyse B-cell tumor cells.[8] The electroporated and propagated T cells generated using SB100X transposase (using 0.1 μg per electroporation) were evaluated for their ability to be activated for effector functioning in a CAR-dependent manner.

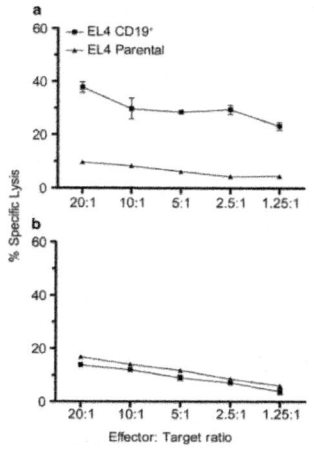

Figure 6: Killing of tumor cells by genetically modified T cells. Lysis by chromium

release assay of CD19⁺ EL4 tumor cells compared with CD19⁻ parental EL4 cells by **(a)** T cells genetically modified with SB100X and propagated on aAPC for 28 days. **(b)** CAR^neg T cells that were not genetically modified were numerically expanded on OKT3-loaded aAPC and these cells failed to lyse the EL4 targets.

We showed by chromium release assay that CAR⁺ T cells exhibited redirected specific lysis of a genetically modified EL4 target expressing 92% human CD19 (Figure 6a). T cells not genetically modified, but propagated by crosslinking CD3 using γ-irradiated aAPC loaded with OKT3 did not appreciably lyse CD19⁻ EL4 cells (Figure 6b).

DISCUSSION

We and others have used transposon and transposase systems to improve integration efficiency of DNA plasmids expressing immunoreceptors.[8, 9, 17, 21, 22]Building upon these data, we have adapted the SB system for human application[11] to use the SB11 transposase to integrate CAR into TA dinucleotide repeats across the genome of populations of human T cells.

The development of SB100X raised the possibility that we could use this new SB transposase to improve the integration frequency of CAR transposon in T cells. We were able to show that SB100X had up to 100 times higher rates of transposition compared with SB11 based on the release of excision circles. This is attributed to improved enzymatic activity, but the improved transposition efficiency may also be due to different translational efficiencies as well as post-translational modifications. The efficiency of transposition was dependent on the amount of electro-transferred transposase, but an elevated input concentration of SB100X led to cell death. The reasons for this toxicity are not known, however, it is apparent that overexpression of transposase can lead to inhibition of SB transposition.[6, 16] This highlights the need to titrate the input DNA plasmid concentrations between the transposon and the transposase. By reducing the amount of SB100X relative to SB11, we were able to show its superior activity.

The ability of SB100X to augment the selective propagation of CAR⁺ T cells on CD19⁺ aAPC raised the possibility that this transposase led to the insertion of multiple copies of the transposon per cell. Indeed, multiple integrations have been observed after SB100X-mediated transposition in other cells.[6, 13] However, this was disproved when we found that on average, there was approximately one copy of the CAR transgene per T cell, which was similar to the integration efficiency associated with SB11. Our lower number of integrants per genome compared with the published reports of SB100X activity may be due to an intrinsic property of the T-cell genome, genotoxicity leading to loss of viability of T cells carrying multiple copies of the transposon, a reduced amount of SB100X DNA plasmid electro-transferred, and/or a reflection of the selective pressure provided by the aAPC to selectively propagate CAR⁺ T cells bearing just one copy of the transposon.

That SB100X has intrinsically improved transposase activity is revealed by the release and detection of more excision circles when SB100X was used compared with SB11. Presumably, the release of the excision circles from a population of electroporated T cells is correlated with the stable integration of the CAR transgene, which is measured per T cell. Thus, it appears that the superior enzymatic activity of SB100X results in an improved efficiency of transposon integration and a beneficial founder effect leading to the subsequent improved outgrowth of CAR⁺ T cells upon *in vitro* propagation on aAPC. This finding pertains to the safety of SB100X for, as with SB11, it does not lead to multiple insertions of the CAR transgene when T cells are electroporated and co-cultured on aAPC.

The increased enzymatic activity of SB100X was also evident when we reduced the amount of transposase DNA plasmid needed to accomplish the integration of CAR into T cells. Indeed, we could achieve superior numeric expansion of CAR⁺ T cells on CD19⁺ aAPC using 0.5 μg per electroporation of SB100X compared with 10 times as much DNA plasmid coding for SB11. This also has implications for improved safety of SB system in human trials as a decrease in the amount of SB transposase delivered by DNA plasmid presumably decreases the chance of inadvertent integration of the transposase into the genome and the potential for re-mobilization of the inserted transposon. When we evaluated for the presence of integrated plasmid expressing SB100X, we did not detect a signal by Q-PCR, which curtails the possibility of a re-hopping event after SB-mediated transposition. However, it is possible that DNA for SB transposase was present in the genome at a level below the limit of detection by this assay. To exclude the possibility that SB100X can integrate into the T-cell genome, we showed that electro-transfer of mRNA species coding for this integrase could mediate transposition, and further that the efficiency of integration was again higher than SB11. Previously, it has been shown that SB11 transposase coding by mRNA can mediate transposition.[23] Furthermore, viral vectors have been used to deliver SB transposase to improve the pattern of integration of an integrase-deficient lentivirus to achieve a more random pattern of integration than can be achieved with lentiviral integrase.[24, 25]

We are currently undertaking the first clinical applications of the SB system, which has successfully received Investigational New Drug Applications from the Food and Drug Administration, to electro-transfer the SB11 transposase to express CD19RCD28 transposon in autologous and allogeneic T cells for infusion in patients with B-lineage lymphoma. With the development of

SB100X, our data show that this transposase may be a desirable alternative transposase to the use of SB11 for use in future clinical trials to adoptively transfer CAR⁺ T cells.

MATERIALS AND METHODS

Plasmids

The SB transposon contains the CoOp second-generation CD19RCD28 CAR, specific for human CD19, flanked by the SB inverted repeats. This CAR has been described previously.[15] In brief, the ampicillin resistance gene (*AmpR*) and origin of replication from the plasmid CoOpCD19RCD28/pT-MNDU3 was replaced with a DNA fragment encoding the kanamycin resistance gene (*KanR*) and origin of replication (ColE1) from the pEK Vector.[8] The human elongation factor-1αpromoter from pVitro4 vector (InvivoGen, San Diego, CA, USA) was swapped with MNDU3 promoter to generate the DAN plasmid CD19RCD28mZ(CoOp)/pSBSO (Figure 1a). The pKan-CMV-SB11 DNA plasmid (Figure 1b), coding for the SB11 transposase, was constructed by digesting pCMV-SB11 (kindly provided by Dr Perry Hackett, University of Minnesota, Minneapolis, MN, USA)[16] with *Pvu*II, harvesting the fragment coding for SB11 transposase and ligating to the *Ase*I and*Pac*I fragment, which contained the *KanR* and *ColE1* origin of replication from pEK vector. pKan-CMV-SB100X (Figure 1c), coding for SB100X transposase, was built by digesting both the pKan-CMV-SB11 plasmid and pCMV(CAT)T7-SB100X plasmid[13] with *Ava*I and *Psi*I, and then annealing the fragment coding for SB100X with the backbone of pKan-CMV-SB11 plasmid. All DNA plasmids were purified using Qiagen (Valencia, CA, USA) endotoxin-free reagents. The integrity of the DNA plasmids coding SB100X and SB11 transposase was assessed by Experion automatic electrophoresis station (Bio-Rad, Hercules, CA, USA).

Cells

After obtaining consent, PBMC from healthy donors were isolated by density gradient centrifugation over Ficoll-Paque-Plus (Pharmacia Biotech, Piscataway, NJ, USA) and stored in liquid nitrogen. K562 were transduced with lentivirus to express CD64, CD86, CD137L and membrane-bound IL-15 and were cloned (clone no. 4, kindly provided by Dr Carl June, University of Pennsylvania, Philadelphia, PA, USA). The construction of CD19⁺ K562 aAPC was reported previously.[17] The parental murine immortalized EL4 T-cell line (catalog no. TIB-39; ATCC, Manassas, VA, USA) and human CD19⁺ EL4 transfectants were cultured in RPMI 1640 medium (Hyclone, Logan, UT,

USA), supplemented with 2 mM Glutamax-1 (Gibco-Invitrogen, Carlsbad, CA, USA) and 10% heat-inactivated fetal calf serum (FCS).

Generation of CAR⁺ T Cells by Transposition with SB DNA Plasmids Coding For CAR and Transposases

On day 0, PBMCs were thawed at 37 °C, washed once with phenol-free RPMI 1640 and rested for 2 h at 37 °C. A total of 2×10^7 PBMC/cuvette were resuspended in Human T Cell Nucleofector buffer (Lonza Inc., Basel, Switzerland) along with 15 µg CD19RCD28mZ(CoOp)/pSBSO plasmid coding for CD19RCD28 CAR and graded doses of DNA plasmids coding for SB100X or SB11. After electroporation with Nucleofector II (Lonza Inc.) using program U14, the cells were washed once with phenol-free RPMI 1640 and resuspended in 1 ml of phenol-free RPMI 1640 (without FCS) and cultured in a 12-well plate at 37 °C for 4 h.

Then, 1 ml of phenol-free RPMI 1640 with 20% of FCS was added. The next day, 3×10^5 cells were harvested for DNA extraction and immunophenotyping. The remaining cells were stimulated by 1:2 (CAR⁺ T cells:aAPC) weekly additions of γ-irradiated (100 Gy) CD19⁺ aAPC (clone no. 4) for 28 days of continuous co-culture in RPMI 1640 supplemented with 2 mM Glutamax-1 and 10% heat-inactivated FCS. Recombinant human interleukin 2 (50 IU ml⁻¹; Novartis Pharmaceuticals Corporation, East Hanover, NJ, USA) was added every other day beginning at day 1.

In vitro Transcription to Generate mRNA Coding for SB100X and SB11

Twenty micrograms of the SB transposase plasmids were digested with SpeI and purified with Qiaquick gel extraction kit (Qiagen) and the DNA concentration was measured. Ten micrograms of the linearized plasmid DNA was used to synthesize mRNA with T7 RiboMAX Express Large Scale RNA Production System (Promega, Madison, WI, USA). A PolyA tail was added to the newly synthesized mRNA molecule with Poly(A) Tailing Kit (Ambion, Austin, TX, USA). After quantification, the mRNA was analyzed by gel electrophoresis, aliquoted and stored in Nalgene cryogenic vials (Thermo Fisher Scientific, Waltham, MA, USA) at −80 °C for future use.

Generation of CAR⁺ T Cells by Transposition with SB DNA Plasmid Coding CAR and mRNA Coding for Transposases

PBMC were resuspended in RPMI 1640 with 5% heat-inactivated AB human serum (Invitrogen) along with 50 ng ml⁻¹ of OKT3 monoclonal antibody (eBioscience, San Diego, CA, USA) and $50\,U\,ml^{-1}$ of IL-2, which was re-added to the culture every other day. After 5 days, 10^7 cells per cuvette were electroporated using $10\,\mu g$ DNA plasmid (CD19RCD28mZ(CoOp)/pSBSO) coding for CD19RCD28 and graded amounts of SB transposase mRNA. The propagation of the T cells was achieved using 1:10 (CAR$^+$ T cells to aAPC) ratio of γ-irradiated CD19$^+$ aAPC for first 2 weeks and 1:2 ratio thereafter. aAPC clone no. 4 were added every 7 days and IL-2 every other day beginning at day 1.

Flow Cytometry

Cells were first incubated with 1:200 diluted APC-labeled goat anti-human immunoglobulin G Fc (category (cat.) no. 12-0569-42; Jackson ImmunoResearch Laboratories Inc., West Grove, PA, USA), then washed once, and finally incubated with 1:50 anti-CD3-FITC (cat. no. 349201; BD Biosciences, Los Angeles, CA, USA) and 1:50 anti-CD56-PE (cat. no. 12-0569; eBiosciences). After staining, all cells were resuspended in $100\,\mu l$ fluorescence-activated cell Sorting buffer (2% FCS and 0.1% sodium azide in phosphate-buffered saline) and live/dead cells were differentiated upon the addition of propidium iodide (catalog no. P4864; Sigma-Aldrich, St Louis, MO, USA), and then analyzed with flow cytometry (FACSCalibur; BD Biosciences).

Real-Time Q-PCR

All of the primers, probes and TaqMan Gene Expression Master Mix were purchased from Applied Biosystem (Foster City, CA, USA). The Q-PCR reactions were performed in a Steponeplus Real-time PCR system (Applied Biosystem) with TaqMan real-time Q-PCR technique as recommended. To measure excision circles, the forward primer (5'-TCCCAGTCACGACGTTGTAAAA-3') and probe (5'-CCAGTGAATTCGAGCTC-3') bound 5' of the CAR cassette and the reverse primer (5'-CGTTGGCCGATTCATTATCG-3') bound 3' of the CAR cassette in the SB transposon DNA plasmid CD19RCD28mZ(CoOp)/pSBSO as illustrated in Figure 1a. A positive PCR reaction reveals a 77 base pair band that is generated only after the CAR transposon was excised. To measure CD19RCD28 CAR, the primer sequences are as follows: forward primer, 5'-CAGCGACGGCAGCTTCTT-3'; reverse primer, 5'-TGCATCACGGAGCTAAA-3'; and probe, 5'-AGAGCCGGTGGCAGG-3'. To measure SB transposases, a common primer and probe set were designed to target the plasmid backbone shared by pKan-CMV-SB100X and pKan-CMV-SB11. These sequences are as follows: forward primer, 5'-AAGGCCAGGAACCGTAAAAAG-3'; reverse primer,

5'-GGCGGAGCCTATGGAAAAA-3'; and probe, 5'-CCGCGTTGCTGGC-3'. RNase P primer and probe set of TaqMan RNase P Control Reagents Kit (cat. no. 4316844; Applied Biosystem) were used as Q-PCR internal control.

Analysis of Q-PCR Results

An RNase P C_T versus cell number standard curve (Curve A, Supplementary Figure 1s 1a) was achieved by a serial dilution (200, 20 and 2 ng) of genomic DNA from a genetically modified Jurkat T-cell clone bearing one copy of CAR transgene based on Southern blotting (Maiti *et al.*, in preparation). As one cell has about 7 pg of DNA, 200 ng total DNA is equivalent to approximately 28570 cells. C_T represents the threshold circle where the Q-PCR was deemed positive. With Curve A, we calculated the cell number from corresponding RNase P C_T value. In parallel, a CD19RCD28 transgene C_T versus transgene number curve (Curve B, Supplementary Figure 1s 1b) was generated to compute the number of integrated transgenes. Twenty nanograms of sample genomic DNA from genetically modified and propagated primary T cells could then be analyzed by Q-PCR. The RNase P C_T was used to calculate cell numbers with Curve A and the integrated transgene copy number was deduced from transgene C_T with Curve B. The copy numbers per cell were calculated with the following formula:

$$\text{Copy number per cell} = \frac{\text{Tg}N}{\text{Cell}N}$$

$\text{Tg}N$ is the number of integrated transgenes and $\text{Cell}N$ represents the cell number deduced from RNase P C_T. Quantification of excision circle was achieved with comparative C_T method provided by the Steponeplus real-time PCR system (Applied Biosystem).

Chromium Release Assay

The redirected specificity of the genetically modified T cells was determined by chromium release assay using ^{51}Cr-labeled EL4 cells as targets. The T cells (effectors) were harvested 28 days following stimulation with aAPC, washed and plated in V-bottom microtiter plates (Costar, Cambridge, MA, USA) in triplicate at 10^5, 5×10^4, 2.5×10^4, 1.25×10^4 cells per well and with 5×10^3 target cells. After incubation at $37\,°C$ for 4 h followed by centrifugation, 50 µl aliquots of cell-free supernatant were harvested and counted with Topcount NXT (PerkinElmer, Waltham, MA, USA). The percent of specific cytolysis was calculated from the release of ^{51}Cr as follows:

$$\frac{(\text{Experimental}\,^{51}\text{Cr}) - (\text{Control}\,^{51}\text{Cr})}{(\text{Maximal}\,^{51}\text{Cr}) - (\text{Control}\,^{51}\text{Cr})} \times 100$$

Control wells contained target cells incubated in media. The maximal ^{51}Cr was determined by measuring the ^{51}Cr content released by target cells lysed with 1%Triton X-100.

Statistical Analysis

The Student's t-test was employed to determine statistical significance and$P<0.05$ was considered significant. Where applicable, the data are reported as an average and s.d. Electroporations were performed on two different donors while Q-PCR reactions were carried out in triplicate.

REFERENCES

1.   Jena B, Dotti G, Cooper LJ. Redirecting T-cell specificity by introducing a tumor-specific chimeric antigen receptor. *Blood* 2010; **116**: 1035–1044.

2.   Tan PH, Tan PL, George AJ, Chan CL. Gene therapy for transplantation with viral vectors—how much of the promise has been realized? *Expert Opin Biol Ther* 2006; **6**: 759–772.

3.   Ciuffi A. Mechanisms governing lentivirus integration site selection. *Curr Gene Ther* 2008; **8**: 419–429.

4.   Hacein-Bey-Abina S, Von Kalle C, Schmidt M, McCormack MP, Wulffraat N, Leboulch P *et al.* LMO2-associated clonal T cell proliferation in two patients after gene therapy for SCID-X1. *Science* 2003; **302**: 415–419.

5.   Liang Q, Kong J, Stalker J, Bradley A. Chromosomal mobilization and reintegration of *Sleeping Beauty* and *PiggyBac* transposons. *Genesis* 2009;**47**: 404–408.

6.   Grabundzija I, Irgang M, Mátés L, Belay E, Matrai J, Gogol-Döring A *et al.* Comparative analysis of transposable element vector systems in human cells. *Mol Ther* 2010; **18**: 1200–1209.

7.   Huang X, Wilber AC, Bao L, Tuong D, Tolar J, Orchard PJ *et al.* Stable gene transfer and expression in human primary T cells by the *Sleeping Beauty*transposon system. *Blood* 2006; **107**: 483–491.

8.   Singh H, Manuri PR, Olivares S, Dara N, Dawson MJ, Huls H *et al.* Redirecting specificity of T-cell populations for CD19 using the *Sleeping Beauty* system. *Cancer Res* 2008; **68**: 2961–2971.

9.   Peng PD, Cohen CJ, Yang S, Hsu C, Jones S, Zhao Y *et al.* Efficient nonviral*Sleeping Beauty* transposon-based TCR gene transfer to

peripheral blood lymphocytes confers antigen-specific antitumor reactivity. *Gene Therapy*2009; **16**: 1042–1049.

10. Geurts AM, Yang Y, Clark KJ, Liu G, Cui Z, Dupuy AJ *et al.* Gene transfer into genomes of human cells by the *Sleeping Beauty* transposon system.*Mol Ther* 2003; **8**: 108–117.

11. Williams DA. *Sleeping Beauty* vector system moves toward human trials in the United States. *Mol Ther* 2008; **16**: 1515–1516.

12. Hackett PB, Largaespada DA, Cooper LJ. A transposon and transposase system for human application. *Mol Ther* 2010; **18**: 674–683.

13. Mátés L, Chuah MK, Belay E, Jerchow B, Manoj N, Acosta-Sanchez A *et al.* Molecular evolution of a novel hyperactive *Sleeping Beauty* transposase enables robust stable gene transfer in vertebrates. *Nat Genet* 2009; **41**: 753–761.

14. Xue X, Huang X, Nodland SE, Mátés L, Ma L, Izsvák Z *et al.* Stable gene transfer and expression in cord blood-derived CD34+ hematopoietic stem and progenitor cells by a hyperactive *Sleeping Beauty* transposon system. *Blood* 2009; **114**: 1319–1330.

15. Kowolik CM, Topp MS, Gonzalez S, Pfeiffer T, Olivares S, Gonzalez N *et al.* CD28 costimulation provided through a CD19-specific chimeric antigen receptor enhances *in vivo* persistence and antitumor efficacy of adoptively transferred T cells. *Cancer Res* 2006; **66**: 10995–11004.

16. Geurts AM, Yang Y, Clark KJ, Liu G, Cui Z, Dupuy AJ *et al.* Gene transfer into genomes of human cells by the *Sleeping Beauty* transposon system.*Mol Ther* 2003; **8**: 108–117.

17. Manuri PV, Wilson MH, Maiti SN, Mi T, Singh H, Olivares S *et al.* piggyBactransposon/transposase system to generate CD19-specific T cells for treatment of B-lineage malignancies. *Hum Gene Ther* 2010; **21**: 427–437.

18. Izsvák Z, Chuah MK, Vandendriessche T, Ivics Z. Efficient stable gene transfer into human cells by the *Sleeping Beauty* transposon vectors. *Methods* 2009; **49**: 287–297.

19. Liu G, Aronovich EL, Cui Z, Whitley CB, Hackett PB. Excision of *Sleeping Beauty* transposons: parameters and applications to gene therapy. *J Gene Med* 2004; **6**: 574–583.

20. Szilagyi A, Blasko B, Szilassy D, Fust G, Sasvari-Szekely M, Ronai Z. Real-time PCR quantification of human complement C4A and C4B genes. *BMC Genet* 2006; **7**:

21. Nakazawa Y, Huye LE, Dotti G, Foster AE, Vera JF, Manuri PR *et al*. Optimization of the *PiggyBac* transposon system for the sustained genetic modification of human T lymphocytes. *J Immunother* 2009; **32**: 826–836.

22. Huang X, Guo H, Kang J, Choi S, Zhou TC, Tammana S *et al*. *Sleeping Beauty* transposon-mediated engineering of human primary T cells for therapy of CD19+ lymphoid malignancies. *Mol Ther* 2008; **16**: 580–589.

23. Wilber A, Frandsen JL, Geurts JL, Largaespada DA, Hackett PB, McIvor RS. RNA as a source of transposase for *Sleeping Beauty*-mediated gene insertion and expression in somatic cells and tissues. *Mol Ther* 2006; **13**: 625–630.

24. Staunstrup NH, Moldt B, Mátés L, Villesen P, Jakobsen M, Ivics Z *et al*. Hybrid lentivirus-transposon vectors with a random integration profile in human cells. *Mol Ther* 2009; **17**: 1205–1214.

25. Vink CA, Gaspar HB, Gabriel R, Schmidt M, McIvor RS, Thrasher AJ *et al*. Sleeping beauty transposition from nonintegrating lentivirus. *Mol Ther* 2009; **17**: 1197–1204.

Chapter 7

GENETIC INFORMATION AND TESTING IN INSURANCE AND EMPLOYMENT: TECHNICAL, SOCIAL AND ETHICAL ISSUES

Béatrice Godard[1,6], Sandy Raeburn[2], Marcus Pembrey[3], Martin Bobrow[4], Peter Farndon[5] and Ségolène Aymé[1]

[1]INSERM SC11, Paris, France

[2]Department of Medical Genetics, Nottingham, UK

[3]Institute of Child Health, London, UK

[4]Department of Medical Genetics, University of Cambridge, UK

[5]Department of Medical Genetics, Birmingham, UK

[6]Current address: Faculty of Law and Department of Social and Preventive Medicine, University of Montreal, Montreal, Canada.

ABSTRACT

The present paper examines the professional and scientific views on the social, ethical and legal issues that impact on genetic information and testing in insurance and employment in Europe. For this purpose, many aspects have been considered, such as the concerns of medical geneticists, of the insurers and employers, of the public, as well as the regulatory frameworks and unresolved issues. The method used was primarily the review of the technical, social, economical and ethical aspects of advances in genetics and the concerns of parties who are involved, that is, the insurers, the employers and the public. The existing guidelines and legislation on this topic were also reported. Then, the method was to examine the issues debated by these parties in Europe, as well as by 47 experts from 14 European countries invited to an international workshop organized by the European Society of Human Genetics Public and Professional Policy Committee in Manchester, UK, 25–27 February 2000. The result of this was that the most important issues raised by genetic information and testing in insurance and employment in Europe include a need for clear definitions of terms used in genetics, declaring the grounds on which genetic information is or is not used, and promoting confidence between the public and the insurance industry. There is currently very little use of genetic information in relation to employment, but the situation should be kept under review.

INTRODUCTION

Genetic information or genetic test results can be used to prevent the onset of diseases, or to assure early detection and treatment, or to make reproductive decisions. This information can also be used for nonmedical purposes, such as insurance and employment purposes. Insurers might wish to use a genetic test result for underwriting, just as other medical or family history data. Employers might wish to ensure that an individual does not have a genetic risk which might affect his ability to work or which might lead to problems of safety to the individual or to others. Applicants might wish to voluntarily disclose their genetic status in order to pay cheaper premiums; or applicants who are prone to disease might wish to seek out the companies with the best benefits. The impact of the use of genetic information for nonmedical purposes justify special attention. The issues which could arise need to be very carefully assessed. Being denied insurance or charged higher premiums on the basis of genetic traits could have serious consequences and could affect individuals, families, or groups who may be already disadvantaged. The choices of the present may affect future generations.

A number of international and national committees have developed recommendations for policy-makers to protect individuals against genetic discrimination. The UNESCO Universal Declaration on the Human Genome and Human Rights (1997) states that 'no one shall be subjected to discrimination based on genetic characteristics that is intended to infringe or has the effect of infringing human rights, fundamental freedoms and human dignity'. The 1997 Council of Europe's Convention for the Protection of Human Rights and Dignity of the Human Being with Regard to the Applications of Biology and Medicine specifies in Article 11: 'Any form of discrimination against a person on grounds of his or her genetic heritage is prohibited'. At the national level, the approaches used vary greatly.[a] In respect to insurance, three solutions are usually proposed: (1) prohibition of any use of genetic information by insurers outright; (2) legislation prohibiting this below a certain amount of coverage; and (3) moratoria; the adoption of moratoria on the use of genetic information has been a widespread response of the insurance industry throughout Europe. Among the countries where there is no regulation, bills have been presented or states that have ratified the 1997 Council of Europe's Convention are bound by it.

Despite the desire to protect individuals against genetic discrimination and consequently to restrict the use of genetic information for nonmedical purposes, it seems necessary to find a balance between the interests of insurers, those of applicants, as well as those of other policy holders. This appears especially relevant and sensitive under health-care systems and welfare sectors with increasing budgetary restrictions.

The present paper aims to examine the professional and scientific views on the social, ethical, and legal issues that impact on Genetic Information and Testing in Insurance and Employment in Europe. For this purpose, many aspects have been considered, such as the concerns of medical geneticists (II), of the insurers (III), of the employers (IV) and of the public (V), as well as the regulatory frameworks (VI) and unresolved issues (VII).

METHODS

The method used for analyzing the professional and scientific views on the social, ethical, and legal issues that impact on genetic information and testing in insurance and employment in Europe was primarily the review of the technical, social, economical, and ethical aspects of advances in genetics and the concerns of parties who are involved, that is, the insurers, the employers, and the public. The existing guidelines and legislation on this topic were also reported. Then, the method was to examine the issues debated by these parties in Europe, as well as the results of discussions held during an international workshop. This workshop was organized by the European Society of Human Genetics Public and Professional Policy Committee in Manchester, United Kingdom, February 25–27, 2000.

The purpose of the workshop was to identify, from a professional viewpoint, the most important/pressing/burning ethical issues raised by genetic information and testing in insurance and employment in Europe. The formal workshop presentations covered the following themes: the fundamentals of genetics, of insurance, family histories, actuarial relevance and genetic testing and employment issues. Small multidisciplinary groups were convened to take these discussions further, in particular to consider the specific issues involved in employment, life insurance, private medical insurance, long-term care and critical illness insurance, and total permanent disability and income replacement insurance. Their initial task was to explore the insurance needs and rights in the countries represented and to consider the extent to which these needs were currently being met. Following the small group sessions, conclusions were fed back to the whole group where there were opportunities for further discussion.

A group of 47 experts from 14 European countries was invited. These experts were representatives of the seven following sectors:

- Medical Genetics
- Human Genetics Societies
- Ethical, Legal and Social Issues
- Support Groups
- Biotechnology/Pharmaceutics
- Insurance/Employment
- European Union Institutions

A first background document was discussed during the workshop. A second document, including discussions of the workshop, was sent for comments to representatives of the human genetic societies and European experts in the fields of insurance and genetics, as well as to all ESHG members. This document was also put on the ESHG website (www.eshg.org) for public consultation and discussion. The final document was approved by the ESHG board.

CONCERNS OF MEDICAL GENETICISTS

Complexity of Genetic Tests

Genetic tests are available for two forms of genetic diseases: monogenic and multifactorial diseases. Monogenic disorders are rare but highly penetrant; the genetic test will indicate whether a person has or will get the disease. Presymptomatic testing identifies healthy individuals who may have inherited a gene for a late-onset disease and if so will develop the disease if they live long enough. Multifactorial diseases are frequent and most likely triggered by specific combinations of functional DNA polymorphisms interacting with the environment in ways that are subject to behavioral changes. Susceptibility testing identifies healthy individuals who may have inherited a gene that puts them at increased risk of developing a multifactorial disease, although these individuals may never develop the disease in question. In these situations, the most that the genetic test can do is to show a propensity to a disease.[1,2]

Genetic testing classifies people into those who have the mutant gene and those who do not have it. Now, a mutant gene is not a disease. Genetic disorders show different degrees of severity and diverge with respect to the age of onset. Some genetic disorders affect people with near-certainty but others not. Predictions are therefore complicated by these phenomena.

Finally, our ability to identify individuals at risk for genetic diseases often exceeds our ability to prevent or treat the diseases.[3,4] This has been described as the 'therapeutic gap' and as a reason for tension between policy-makers and health professionals.[5] The use of computerized medical data banks within large companies could exacerbate this problem, genetic information becoming not only a medical fact but also a disease.[6,7]

We are forced to note that genetic tests present some limits, including the possibility of uninformative results, the inability to predict the exact age of onset or the severity of symptoms and, in the case of multifactorial diseases, the inability to predict if the individuals will develop the disease in question. In fact, genetic tests cannot account adequately for the external factors, which can be as important as inborn characteristics.[3,6,8] Tests using genetic markers linked to a disease gene (as opposed to testing directly for disease-causing mutations) are not totally reliable since they provide only statistical probabilities based on the presumption that people have inherited genes with the identified markers. In other respects, a clear distinction must be drawn between genetic tests carried out in a research setting (aimed at establishing new genetic tests or developing quality control of tests) and those carried out in clinical practice. Research projects can be experimental and the results of the tests can be uncertain.

Calling Ethical Principles into Question

Different arguments suggest that there is something special about genetics, and yet ethical principles in medical genetics are the same in medicine, even if this has been questioned. These principles are: respect for the autonomy of persons, beneficence, non-maleficence, and justice. At present, in regard to medical genetics, these principles are not applied with equal force around Europe. The principle of nonmaleficence which aims at avoiding and preventing harm to persons, is called into question if genetic information is used for discrimination or favoritism in insurance and employment. The principle of justice, which may consist in distributing benefits and burdens fairly and with equity, varies depending on whether healthcare is founded on the principle of social solidarity or on the basis of mutuality. Although the market for private health insurance in Europe is small and in some countries nonexistent, the possible use of genetic information in insurance and employment has gradually generated debate and increasingly causes concern.[9,10,11,12,13,14,15]

Furthermore, medical geneticists' concerns extend beyond the traditional ethical guidelines in medicine. For instance, genetic testing for insurance and employment purposes could disturb family relations. Family cooperation is often necessary to detect genetic problems, but genetic information may affect an entire family rather than only one individual, and the choices of the present may affect future generations. Genetic information links the members not only of families but also of whole communities. Genetic disorders are often over-represented in ethnic groups and intensive genetic research on some populations could exaggerate the presence of problems.[6,16,17]

CONCERNS OF THE INSURERS

Genetic information through family history was already used by some insurance companies before anyone considered genetic testing, and individuals were covered or denied coverage or charged higher premiums according to family history. Nevertheless, the progress made in predicting diseases alters the information available with regard to the risk of disease. Genetic information contains more certainty than information traditionally gathered by insurers to investigate the existence of diseases running in the family.[6,18] This may have important consequences for insurance industry.

Goal of Insurers

The insurers' goal is to maximize their profits. This is usually reached with an increased number of people under coverage. In this regard, developments in medical science have resulted in an increase of life insurance sales.[19] In other respects, everyone carries some potentially abnormal genes and insurers will not wish to deny coverage to a significant segment of the population. However, the insurance industry would like to use genetic information as just part of the (predictive) information that they should be able to use less for deciding to accept a private, voluntary application, than for setting the premium level according to the individuals' risk, and for avoiding the possibility of adverse selection.

Underwriting

Underwriting is the method used to classify people according to their risk. Insurers classify the risk by asking questions and through medical investigation. The questions sometimes cover the medical histories of family members. Depending on the case and the amount of coverage involved, medical questions might be followed by medical tests or complete medical examinations.[20,21]

In the underwriting process, the expectations of individuals in relation to longevity and health are quantified and expressed as statistical probabilities. Insurers can predict that the overall mortality rate of a specific group of people, classified in the same substandard risk category, will be higher than the mortality rate in the general population.[7] Usually, underwriting leads to classification in three groups: standard, substandard, and uninsurable. Individuals in the first group have few problems getting insurance. Individuals in the second group must pay higher than average premiums, based on the risk they represent. Individuals in the third group are excluded because the cost of their coverage is unquantifiable or would exceed any reasonable premium. Experience shows that the assessment of substandard risks due to genetic information is proved fair since the observed mortality is very close to what had been expected. Requesting genetic tests from insurance applicants could then constitute another source of information for insurers. This would permit to classify individuals more accurately in various categories of risk, or to assess risk premiums more accurately. Genetic testing would enhance equity by allowing a precise calculation of which people are really in the same situation and which are not.[6,18] The concept of equity in insurance means that people who have similar health or similar life expectancies should pay equal premiums and those who have worse health or lower life expectancies should pay more.

To date, insurers do not require applicants to submit to genetic testing. In some countries, this is due to legal barriers which prohibit insurers from asking for genetic tests. This is also due to the lack of information on the predictive value of certain tests and on the costs of diseases.[1,6,22] But that does not mean that insurers are not using genetic information. Insurers can currently make genetic inferences from routine and well-accepted questions on family history. Insurers can use genetic information available in medical files; the registered information in medical files is usually more accurate and complete than what is known by the insurance applicants.[15] Since genetics is integrated in medical practice, insurers will have access more and more to genetic information. This will allow insurers, among other things, to know whether applicants have neglected to mention that they are carriers of genetic disorders or that these run in the family.

Adverse Selection

Adverse selection occurs when people have undergone testing and conceal positive test results from insurers.[16] If the insured person does not disclose information which the insurer needs to know, then this disrupts the equilibrium of the relationship and the possibility of adverse selection arises. Insurers require symmetry of information. If insurers are prohibited from having access to pertinent information at the time of underwriting or when the policy is renewed, the applicants could use genetic information to abuse the insurance system, taking advantage of private knowledge of the risks they are submitting for coverage.[19] The consequences of a lack of symmetry in information between insurers and applicants or insured persons could force insurers to adjust premiums. In this way, in the Netherlands, after the Medical Examination Act has been in force (1998), insurers have taken measures to prevent the risk of adverse selection by implementing premium increases in advance, by prescribing a maximization of the pension pay-out or basing payments on a maximum salary, or by including an option to increase the premium in the policy.[23] Dutch insurers have also introduced waiting times for existing illnesses when issuing the insurance. This means that if, within a term stipulated in the waiting time, the insured becomes disabled or dies as a result of an illness that he had when he took out the insurance, no payment will be made. This measure does not apply for life insurance. Sweden (1999) has the same policy. In the United Kingdom (2000), the Genetics and Insurance Committee stated that the reliability and relevance of the genetic test for Huntington's Disease was sufficient for insurance companies to use the result when assessing applications for life insurance. But in October 2001 the UK government reached an agreement with the Association of British Insurers (ABI) to institute a 5-year moratorium on the use of genetic tests results up to a certain value.

CONCERNS OF THE EMPLOYERS

Concerns of employers and of insurers are similar. The main difference between life insurers and employers is that for employers, sickness represents a greater financial risk than death, while for health insurers, the opposite is usually the case.

Goal of Employers

It is in employers' interests to have a healthy workforce. Some employers provide facilities to encourage the staff to achieve a good health, like regular medical check-ups and sport. It has been argued that if it could be demonstrated that genetic screening would encourage more healthy lifestyles, it would be possible to envisage that employers would fund such screening for their staff.[23,24]

Employers are particularly interested in the health of the employees for jobs where there is a substantial investment in training or for very senior positions. Different sources of information can be used to assess whether an individual has a risk of either sickness or death: medical examination, medical history, family history, age, lifestyle. Genetic testing might confirm the risk of developing a genetic disease, for which some jobs could make the person unacceptable.[23,25] What would also change is that some employees would move from 50 to 0% chance and they would have opportunities which are currently denied them.

For most jobs, employers do not insist on intensive health testing of prospective employees, because the extent of the employers' investment in new employees is not great enough to warrant such expense.[23] The prospective employees are simply asked to make a declaration about their state of health.

Constraints Imposed on Employers

The costs of any health investigation by employers are significant: if employers investigate every prospective employee, they will have to pay the investigation costs for all of them, but in only a few will the investigation show anything at all. The decision for employers, where there is a known health risk, is whether the value that employees will give to the firm justifies the risk.[21,23]

Many employers provide a range of health insurance coverage for their employees: sick pay, permanent health insurance, spouse's pension, retirement pensions, health-care benefits. Most employees are covered without having to provide any information about their health. But in recent years there has been some trend towards flexible remuneration packages, under which employees get some measure of choice as to which employees benefits they take. Where employees have a choice, some measure of individual underwriting is required.[21,23]

Although the use of genetic information might conceivably be of some benefit for employers, it runs counter to the fundamental rights of workers to nondiscrimination for health reasons and those relating to protection of privacy. For instance in France, such rights which have been reinforced by the laws on bioethics in 1994, are proclaimed in several articles in the labor and penal codes. In those countries that do not have specific regulations prohibiting or limiting employer access to, and use of, genetic information, existing antidiscrimination and privacy legislation may provide individuals with some protection.

CONCERNS OF THE PUBLIC

Right to Underwrite

People are becoming aware that they are exposed to global risks, such as rising unemployment, collapse of pension funds, funding problems of welfare programs, and are therefore vulnerable. In this context of cost-shifting, public funding for insurance may be threatened, while community rating in commercial insurance may happen, as for instance with private medical insurance **cover** in Ireland.

Private insurance is based on mutuality and consequently discriminates in setting premiums. Mutual insurance refers to the notion of forming a risk pool in which each of the members participate according to the risk they represent to the pool. The cost of the insured risk is distributed between the members of the pool, each paying its own part.[26] Individuals assessed as representing a higher perceived risk may pay more, and some may be denied cover, although the great majority are treated as standard risks.[13,25,27,28]

Duty of Disclosure

The duty of disclosure, which is established by legislation, states that the insurance applicants must declare everything relevant to their risk's appreciation and their classification.[6] If the applicants have neglected to mention that they are carriers of genetic disorders or that these run in the family, this could be invoked to prove that the applicants have made a false declaration and that the contract is invalid.

The duty of disclosure raises many questions: (1) Are genetic test results always relevant for insurers? Applicants who test positive for genetic mutations in a context of research might not have health problems that are relevant for insurance purposes; (2) How relevant is it, when people neglect to inform their insurer about medical problems or conceal health information from them, if their death has nothing to do with the missing information? (3) Insurers may have access to confidential information that applicants do not want to know, thus infringing on their right not to know. (4) The duty of disclosure may also generate social pressure on a would-be applicant to have a genetic test and disclose a negative result to show that their family history does not put them at increase risk.

Fear of Discrimination

The fear of genetic discrimination by insurers or employers may tip the scales against somebody seeking testing to obtain improved medical management and reassurance.[29] This fear has been observed among people with a family history of Huntington disease who requested presymptomatic gene identification: people attempted to avoid insurance or employment discrimination by withholding the decision to seek testing from their primary care providers.[29] People may also be encouraged not to share the result with their general practitioner for fear of disclosure to insurance companies.[30] Genetic testing could then cause insurance applicants and their relatives to be rated up or denied insurance and lead to social exclusion, especially since genetic information would not only be used for insurance purposes but also employment purposes. The practice of some clinicians to advise people to buy insurance before having a predictive DNA-test highlights the current perception that people at high genetic risk of late-onset disease face the additional social disadvantage of higher premiums or application rejection.[31]

REGULATORY FRAMEWORKS

In regard to the above, two principles govern the use of genetic information and testing in insurance and employment; firstly, no one should be subjected to discrimination based on genetic characteristics; secondly, the disclosure of information to a third party or accessibility to personal genetic data should be allowed only with the individual's informed consent. These principles can be found in all international and regional texts. There is a general consensus that applicants should not undergo genetic testing as a condition of obtaining insurance.

On the contrary, national texts (legislation and recommendations) vary greatly. Three solutions are usually proposed: (1) Prohibition of any use of genetic information by insurers outright; such as Austria, Belgium, Denmark, Estonia, France, Luxembourg, and Norway. In Belgium, a notable feature of the legislation is that it prohibits the use of genetic information even in circumstances where it is to the benefit of the applicant. The rationale is to protect privacy. (2) Legislation prohibiting this below a certain amount of coverage, like in Sweden, the Netherlands, and the United Kingdom. In the United Kingdom, the government also set up the Genetics and Insurance Committee (1998) whose role is to assess the actuarial validity of genetic tests that insurance companies would like to be able to take into account in setting insurance premiums. And 3) Moratoria; Moratoria are either indefinite (Finland, Germany), or for a limited number of years (France, Switzerland), or still limited to insurance policies which do not surpass a certain value (Sweden, The Netherlands, the United Kingdom).

Among the countries where there is no regulation, bills have been presented, like in Iceland and Switzerland, or states that have ratified the European Convention for the Protection of Human Rights and Dignity of the Human Being with Regard to the Application of Biology and Medicine are bound by it. The Council of Europe's Convention on Biomedicine upholds that the rights and dignity of humans should be respected with regard to the application of biology and medicine and have primary over the goals of science or society.

The ceiling system (ie 'no questions' asked below a certain amount) of regulation is a policy response that mitigates the problems associated with genetics and insurance. This approach protects the insurance industry against the dangers of adverse selection and, for the applicants, it permits the acquisition of social goods such as healthcare or housing. It is assumed that the risk of adverse selection only truly comes into play with large amounts of capital. This is the case in the Netherlands where insurance companies are prohibited by the Medical Examinations Act from seeking disclosure of the results of any genetic test where the amount being sought is less than 300.000 00 guilders.

Although it is not a legislative decision, in the United Kingdom, the Association of British Insurers (2001) announced that its members would no longer request results of genetic tests in respect of applications for any type of insurance up to a certain value.

A system of regulation combined with a pragmatic board of examination of ongoing scientifically validated tests, like in the United Kingdom, is also another policy option. The advantage of this approach is that in an area of rapidly developing technology, a responsive system of procedural regulation can react to changing circumstances.[11] Until the moratorium, GAIC had approved only tests for Huntington's disease in respect of life insurance, but a small number of additional tests was under consideration.

Table 1: shows the current responses to the use of genetic information by insurers in Europe (dated the 1st January 2003)

Country	Legislation	Moratorium	No regulation
Austria	+		
Belgium	+		
Denmark	+		Has ratified the Oviedo Convention
Estonia	+		
Finland		+	
France	+	+	
Germany		+	
Greece		+	Has ratified the Oviedo Convention
Iceland			Bill
Ireland			
Italy			
Luxembourg	+		
Norway	+		
Portugal			
Spain			Has ratified the Oviedo Convention
Sweden		+	
Switzerland		+	Bill
The Netherlands	+	+	
United Kingdom		+	

The Oviedo convention refers to the Convention of the Council of Europe: 'Convention for the protection of Human Rights and Dignity of the Human Being with Regard to the Application of Biology and Medicine', April 1997.

A further policy option is the use of moratoria. The adoption of moratoria on the use of genetic information has been a widespread response of the insurance industry throughout Europe. The rationale is that the consequences of the use of genetic information and testing on health and medical research can be studied. This practice affords the insurance industry time to formulate an alternative policy strategy. However, since moratoria are voluntary, they may only survive for as long as there are no commercial advantages to be gained in using genetic information. For instance, in October 2001 the UK government reached an agreement with ABI to institute a 5-year moratorium on the use of genetic tests results in assessing applications for life insurance policies up to a value of £500 000, and for critical illness, long-term care and income protection policies up to a value of £300 000. For an amount over those limits ABI will be able to use genetic tests results if they have been approved by GAIC. These limits will be reviewed after 3 years.

In the context of employment, there has been some public anxiety that employers may use personal genetic information to discriminate improperly against employees who are seen to be at risk of a particular illness or condition. Yet, it should not be forgotten that employers are bound to protect the health and welfare of their employees. Some countries (Austria, Estonia, France) have adopted legislation which prohibits genetic testing by employers. In other countries, such as Switzerland and the Netherlands, genetic tests can only be used by employers where there is an unambiguous health requirement for the job, or where the protection of the employee›s health in the workplace calls for such a test. In the United Kingdom, where there is no legislative prohibition on the use of genetic information in employment, discrimination on the basis of an existing disability of genetic origin would be prevented by the *Disability Discrimination Act* 1995, but there is currently no specific legislation to prevent discrimination against asymptomatic employees.

Finally, in both insurance and employment unless genetic information becomes increasingly normalized (eg blood pressure, cholesterol, etc), the ‹consent› of the individual to access to the medical record will limit participation in genetic testing and research, if such ‹consented-to› access has more negative economic consequences than access to other medical information. However, the Association of British Insurers (2001) in a joint statement with the British Society for Human Genetics and the UK Forum for Genetics and Insurance, announced that the results of a genetic test taken as part of a research project need not be declared to insurers.

ISSUES

The following discussion has been largely inspired by the workshop organized by the European Society of Human Genetics Public and Professional Policy Committee in February 2000 in Manchester (UK) (see Methods). This discussion has then allowed the ESHG: PPPC to issue recommendations on genetic information and testing in insurance and employment (www.eshg.org).

Need for Definition

Insurance contracts laws state that the contract must be written up in utmost good faith otherwise the contract may be void. This means that the applicant is under an obligation to reply honestly, without withholding information. But if the definition of what can be considered genetic information is not clear, how can an applicant reply honestly and how can an insurer ask specific questions which are relevant to risk assessment? There is a need then for clear definitions of terms used in genetics, insurance and employment, so that different professions and their clients have a common understanding of the issues. A genetic test is a test of anything that is, or potentially can be, inherited according to mendelian laws. This covers not only DNA, RNA, and chromosome analysis, but also protein truncation test and clinical examination of a patient for a mendelian condition that is diagnosable in that way.[30] But does the test result have predictive value for the subject or family members? If the answer is no, there are no special features. If it is predictive for the subject but not the family, it is ethically similar to several other medical tests. Only if there are also implications for the family is there a special case. It is also important to distinguish between research and clinical genetic tests. A lot of people's worries concern tests for disease susceptibility, and these are almost always part of research, but only clinically validated tests should be considered for insurance purposes. Legislation without a precise definition of these terms may confuse insurers and applicants when underwriting or renewing an insurance policy.

Risk Pooling and Underwriting

A common objection is that classifying policyholders according to risk is an objectionable practice because it amounts to discrimination.[32] The insurance industry argues that it is not engaged in discrimination but in differentiation. It differentiates between risk categories rather than between individuals. To date, legislators have reacted to the claim that since an increased use of genetic information will mean that some people will be refused insurance premiums, this amounts to an unjustified form of discrimination. This could then lead to the conclusion that legislation should be passed to limit the use of genetic information by insurers.

Another objection has to do with the distinction which arise between an industry based on equity and one that is based on equality. The insurance industry does not claim to be based on equality – as a social insurance system would – but rather the principles of equity, mutuality, and actuarial fairness produce a system whereby the individual consumer pays a premium which seeks to reflect the risk which she/he brings to that mutuality or risk pool. However, genetic testing skews the fairness principle because (1) some will be aware of their risk status whereas others will not and (2) because the risks associated with particular genotypes are not voluntarily assumed by individuals, but are rather the result of the luck of the draw. It has been argued that if the principle of equity in insurance is replaced by the principle of equality, this could signal the end of the involvement of the insurance industry in certain sectors of the market, notably life insurance and perhaps medical expenditure insurance. If legislators decide to intervene in this area and alter the balance of the insurance industry, efforts might be directed at finding an alternative method of producing the social benefits currently provided by the insurance industry.[33]

Adverse Selection

More and more, people might be able to undergo confidential genetic testing and hide their results.[6,32] Genetic testing will be readily available in doctors' offices and free-standing commercial laboratories. Those who will know that they are at high risk might start buying substantial amounts of insurance and insurance companies would be overwhelmed by claims. Insurers are concerned that many individuals could attempt to use genetic test results to create an estate when none would have existed prior to testing and for many people, the temptation to buy insurance under these circumstances could be irresistible.[34] Those opposed to sharing genetic information with insurers argue that antiselection will be a rare event. A recent study assessing the potential for adverse selection in the life insurance market when tested individuals know their genetic test results but insurers do not, shows that women who test positive for the BRCA1 gene mutation do not capitalize on their informational advantage by purchasing more life insurance than those women who have not undergone genetic testing.[35]

There is an element of speculation involved in the possibility of adverse selection due to the information provided by genetic tests. Only about 5% of diseases are caused by a single gene. Most are caused by interactions between many genes that are subtle and difficult to determine. Consequently, only individuals with mutations for late onset untreatable diseases will be able to deceive the industry. The number of such diseases will probably also be reduced as treatments will be available for these diseases.

There is controversy about the ruling by insurance companies that genetic test results have to be disclosed by people seeking new policies. For monogenic disorders, the effect of any anti-selection by an individual exploiting knowledge of his undisclosed genotype depends largely on the size of the sum for which his life is insured[7] while for multifactorial diseases, it is difficult to establish genotype-specific predictive empirical risk figures. Therefore, there should be no role for genetic data concerning multifactorial diseases in underwriting decisions.[33] Although insurance companies may vary in the stringency with which they scrutinize medical record or use research data to determine insurability, one denial may have far-reaching effects on the individual's opportunities from other insurers. Critics fear that people could be deterred from taking tests whose outcome may be vital for determining their need for prophylactic treatment, or from participating in research involving genetic testing. They are also concerned that the duty of disclosure may infringe on individuals' privacy and violate their right not to know.[12,36]

Predictive Genetic Information in Insurance

There is an issue surrounding the boundary between predictive genetic information and other health-relevant data. A person's sex is genetic information predictive of health outcomes, but being overt and covered by its own anti-discrimination legislation would not normally be included within 'genetic information' for insurance purposes.

Family history is also predictive genetic information, although it is recognized that the self-reported family history may be inaccurate. There is a need to resolve inconsistencies in current attitudes and policies on use of family history in relation to the use of genetic test results. If the ceiling for life insurance cover, without use of genetic information is intended to allow all healthy people to obtain this basic cover without disclosure of their genetic risk of late onset disease, then it is illogical to still take family history into account.[7] Most of the high risks relevant to life insurance that are contained in genetic test results are revealed by an accurate family history. Thus it would be of little benefit to the genetically disadvantaged if a company agreed to forego the use of genetic test results, but would still require family history information.

In other respects, it is difficult to predict the extent to which genetic tests might become relevant for health prediction in multifactorial diseases, and even more difficult to predict the extent of their influence and timing of such advances in knowledge. A predictive DNA test cannot be regarded as a diagnostic test but rather as a prognostic factor because, at the time the genetic test is performed, there is not yet a disease established. However, one can compare how many tested persons have a positive result and how many a negative result and how many will develop a certain disease during a period. The final result is a likelihood of disease and not a predictive value which can then be expressed as a relative risk. Therefore, as with all genetic information used in an insurance context, sound knowledge of the real predictive value of the information needs to be accrued and validated before being put into practice. It is important that insurance applicants should be clearly aware of the limits of genetic information that is required and utilized by the insurance industry in relation to these complex diseases.

Types of Insurance Product Affected by Predictive Medicine

Some argue that genetic test results will not affect all types of insurance product in the same manner. Genetic test results may almost exclusively affect individuals insurance as members of groups usually not individually underwritten.[19] In the same way, genetic test results may only concern individuals applying for high levels of coverage, for which a medical examination is usually requested,[20] or for personally purchased insurance in general.[37]

It is recommended that insurance product which should be exposed to restrictive legislation are those which are perceived as necessary to guarantee a service considered as a basic need, such as health and social insurance.[19,38] There is a clear case for a solidarity-based system for basic needs, with optional extras being provided through a system based on mutuality. Insurance with respect to basic needs has often been compulsory (state or private). There is an issue as to how much 'solidarity for basic needs' can be incorporated into private, voluntary insurance without serious threat to the industry (through adverse selection for example). If it is considered that a substantial solidarity element can be provided by the private sector, the questions arise as to who should finance this solidarity and whether guidelines or legislation are required to regulate this insurance.

Requesting Genetic Tests

In the future, could applicants be asked to undergo a genetic test in order to obtain any type of insurance? Insurers may therefore have access to information that applicants do not want to know. Furthermore, requesting genetic tests from insurance applicants could create problems if counseling services are missing and if social pressure increases on those affected by genetic disorders.

Some published works indicate that despite the significant scientific progress, there are currently not sufficient grounds for requiring individuals to undergo genetic testing and to disclose genetic test results to insurers. This is because the current state of knowledge about patterns of genetic test results does not generally support good predictions of the incidence, timing and severity of disease or of time of death.[3,31] Further research is needed in order to yield useful information. Well-described conditions such as Huntington disease have yielded such information, but this has been gathered over periods of several years. Nevertheless, in the United Kingdom for instance, there had been public disquiet following the ABI report and personal experience had shown increased anxiety regarding testing in Huntington disease clinics.[12] Some of the public found that negative genetic test results could be used to their advantage in lowering already high premiums.[12]

In Employment

At least two types of employment discrimination based on genetic testing have been identified.[26,39] First, an employer may not hire someone who is likely to develop a genetic disease. An at-risk individual may be viewed as someone who would frequently be absent from work, would be less productive than others, or might require more healthcare services. Second, an employer may not permit an individual to work in an area in which he would be exposed to a toxic chemical if that individual is known to have a susceptibility to its toxic effects. It might be proved easier to test for genetic susceptibility than to remove whatever environmental health hazards there are in the work place. Genetic testing in this situation may increase productivity by reducing absenteeism caused by illness linked to susceptibilities to occupational hazards. However, it has been argued that, at least for multifactorial diseases, there is no scientific evidence yet to link unexpressed genetic factors and the ability to perform a job function.[40] The Human Genetics Advisory Committee (1999) said that individuals should not have to disclose the results of previous genetic tests without clear evidence that the information was needed to assess whether they could do the job safely. Also, genetic tests are unlikely to identify susceptibility to disease with any precision as it might be aggravated by the workplace environment.[27] Now even if genetic testing in the workplace may lead to individuals with an increased susceptibility to the effects of workplace toxins being banned from working in these areas,[26,41] prevention of most genetically determined defects that may lead to illness and disability seems an unattainable goal. Consequently, it has been argued that genetic testing by employers should be limited to screening individuals at-risk for developing diseases that may result from certain exposures that exist in the workplace; employment decisions should not be based on genetic factors.[42,43,44] However, there is a range of ethical issues with which the occupational health professional may be confronted as genetic technology advances.[44,45] Genetic testing could be used to improve preventive medicine but also to reduce the costs of sickness in the workplace.[23]

Finally, as for insurance, the fear of employment discrimination through employers access to medical files might discourage at-risk individuals from undergoing medically indicated genetic testing.[2,46]

Education

The fear of genetic discrimination by insurers and employers has spread throughout society.[4,33,45] It is likely that many people who might benefit from such testing will be reluctant to be tested unless laws are in place to protect them. However, a law is not enough to provide a comprehensive solution to genetic discrimination in insurance. One cannot be certain in the present economic context, that pressure might not be put upon applicants for an insurance contract in order to obtain genetic information about them. Nor can one exclude the possibility that the candidates themselves might wish to produce the information spontaneously if it were in their favor. Education is needed. Insurance decisions are sometimes made by inexperienced people, or because of a lack of knowledge about particular genetic conditions. Educational programs on the basic principles of genetics and insurance will have to be developed to improve the insurance coverage. This is important especially since the funding problems of most welfare programs lead many governments to shift a portion of the State's financial burden onto private insurers, particularly in relation to medical costs and the costs of long term care.

CONCLUSION

Insurers and employers are told that unreliable genetic tests must be ignored. Ultimately, objections to the use of genetic information will be subsumed by economic and scientific realities: individually underwritten insurance cannot be sold without risk classification, and some of the medical information needed to classify risks will be genetic.[36] It will become increasingly difficult to distinguish genetic from nongenetic diseases, genetic information from nongenetic information, or to talk of medical and genetic tests as separate categories.

However, in attempting to develop practice fair to both insured and insurer, it is widely accepted that there is a need for clarification of the best means for determining the extent of increased genetic risk of late onset disease, so that there is demonstrable evidence of validity and consistency in the use of any genetic information in underwriting. It is accepted that in time when more reliable actuarial data are available for single gene disorders, genetic test results may be used but it is felt strongly that for multifactorial diseases the results should not be used. Most susceptibility genes are already shared by many people currently insured at standard rates. The unfolding of such results would stratify society in an unacceptable way.

Clear definitions of the terms used in genetics and insurance are revised for the transparency of the process by which genetic information is incorporated into insurance decisions, and for ensuring that genetic information is not used to the detriment of other family members. There is a broad consensus that insurance or employment considerations should not unduly influence the uptake of appropriate clinical care, which may increasingly involve genetic tests. There is also a broad consensus that applicants should not be asked to undergo genetic tests, in order to obtain insurance or employment.

At present, the fear of genetic discrimination remains intense; perhaps because there are very little data to support or refute that discrimination is actually taking place. How to reassure people and protect them? Can a law provide a solution to the problems of insurance, employment and genetics? There are diverging approaches among the various states which have sought to establish binding norms. The legislative activities in several countries show a growing consensus on the need to define the use of genetic information for insurance purposes. Some restrictions on the use of genetic information may be found and be compatible with the continued existence of the insurance industry, such as a ceiling below which no genetic information (genetic test results or family history) has to be disclosed. A valid explanation for selecting a particular ceiling also needs to be provided and should relate to the point where basic economic security (basic house purchase, necessary provision for dependants, and protection for the self-employed) gives way to personal investment. As to genetic testing by employers, it should stay limited to screening individuals at risk for developing diseases that may result from certain exposures that exist in the workplace.

NOTES

[a] This paper addresses the issues raised by the use of genetic information by insurers in Europe. Comparisons between the European and American norms ought to be undertaken with caution given the health-care funding differences between the two. Despite the range of developments in relation to genetic information and insurance in both Europe and the United States, the question of the appropriate policy response remains an open one. In Europe, the debate on genetic and insurance has centered upon potential restrictions on the availability of life insurance and related products which are closely linked to the acquisition of primary modern socioeconomics goods (eg homes, cars, loans, etc). In the United States, much of the legislation has developed within a privacy or discrimination paradigm due to the absence of universal healthcare.

APPENDICES

Appendix A

National and international regulatory frameworks (dated the 1st January 2003)

European Institutions

European Union, Resolution on the Ethical and Legal Problems of Genetic Engineering of the European Parliament (March 16, 1989, n. R89, 2, n. R89, 14) (http://europa.eu.int)

Two principles refer to insurance: Principle 19: 'Insurance companies have no right to demand that genetic testing be carried out before or after the conclusion of an insurance contract nor to demand to be informed of the results of any such tests which have already been carried out and that genetic analysis should not be made a requirement for the conclusion of an insurance contract'. Principle 20: 'The insurer has no right to be notified by the policyholder of all the genetic data known to the latter'

The Resolution has no legal authority; it sensitizes people to the arisen problems of the developments in genetics.

Council of Europe, Recommendation on genetic testing and screening for health-care purposes of the European Committee of Ministers (1992, n. R92, 3) (http://www.coe.fr/cm/ta/rec/1992/92r3.htm)

Principle 7 refers to insurance: 'Insurers should not have the right to require genetic testing or to inquire about results of previously performed tests, as a pre-condition for the conclusion or modification of an insurance contract'.

All members of the Council of Europe adopted this Recommendation, except the Netherlands.

European Union, The Data Protection Directive, 1995 (http://www.privacy.org/pi/intl_orgs/ec/eudp.html)

In 1995 the Council and Parliament of the European Union adopted the Directive 95/46/EC in order to harmonize the protection of data privacy in the EU. The Directive was implemented in national laws and regulations by October 24, 1998. The Directive was designed to establish minimum standards for the processing and use of personal data throughout the EU, for two reasons: (1) to ensure that the Member States protect the 'fundamental right' to privacy with respect to the processing of personal data, and (2) to prevent Member States from restricting the 'free flow of personal data' among Member States on grounds of privacy protection.

Council of Europe, Convention for the Protection of Human Rights and Dignity of the Human Being with Regard to the Application of Biology and Medicine (April 1997, DIR/JUR 96, 14) (http://www.coe.fr/fr/txtjur/16 4fr.htm)

Three articles refer to insurance. Article 11: 'Any form of discrimination against a person on grounds of his or her genetic heritage is prohibited'. Article 12: 'Tests which are predictive of genetic diseases or which serve either to identify the subject as a carrier of a gene responsible for a disease or to detect a genetic predisposition or susceptibility to a disease may be performed only for health purposes or for scientific research linked to health purposes, and subject to appropriate genetic counseling'. Article 26: 'No restrictions shall be placed on the exercise of the rights and protective provisions contained in this Convention other than such as are prescribed by law and are necessary in a democratic society in the interest of public safety, for the prevention of crime, for the protection of public health or for the protection of the rights and freedoms of others'. These restrictions may not be placed on Articles 11, 13, 14, 16, 17, 19, 20 and 21.

Council of Europe, Recommendation on the Protection of Medical Data of the European Committee of Ministers (1997, n. 97, 5) (www.coe.fr/cm/ta/ rec/1997/97r5.html)

Article 4.7 states that 'Genetic data collected and processed for preventive treatment, diagnosis or treatment of the data subject or for scientific research should only be used for these purposes or to allow the data subject to take a free and informed decision on these matters'. Article 4.9 stipulates that for purposes other than those provided for in Principles 4.7, 'the collection and processing of genetic data should, in principle, only be permitted for health reasons and in particular to avoid any serious prejudice to the health of the data subject or third parties. However, the collection and processing of genetic data in order to predict illness may be allowed for in cases of overriding interest and subject to appropriate safeguards defined by law'.

European Union, The Data Protection Act of the European Committee of Ministers, 1998 (europa.eu.int/comm/dg03/publicat/)

The Data Protection Act 1998 implements the EU Data Protection Directive and provides a system of general protection and security for personal data which covers, amongst other things, medical data.

European Union, Charter of Fundamental Rights of the European Union (December 18, 2000) (http://www.europarl.eu.int/charter/default_en.htm)

Article 21 of the Charter apply to insurance and states that 'any discrimination based on any ground such as sex, race, color, ethnic or social origin, genetic features, language, religion or belief, political or any other opinion, membership of a national minority, property, birth, disability, age or sexual orientation shall be prohibited'.

EUROPEAN COUNTRIES

Austria

The Gene Technology Act (1994) (http://www.gentechnik.gv.at/gentechnik/ B1_orientierung/gen_10084.html)

This Act regulates work with genetically modified organisms, the release and marketing of genetically modified organisms, and the use of genetic testing and gene therapy in humans. Section 67 stipulates that it is forbidden for insurers and employers including their representatives and collaborators to obtain, request, accept or in any other way make use of the results of genetic analyses on their employees, candidates, policyholders, or insurance applicants. In practice, the state insurance system does not refuse cover to any applicant, but the private insurance companies are able to refuse to grant cover or only grant it at the cost of an increased premium.

Belgium

Law on terrestrial insurance contracts, 1992

Article 95 prohibits the use of genetic testing that enables to predict the future state of health while Article 5 states that 'genetic data may not be declared'. Applicants are prohibited from subletting the results of genetic testing to insurers, whether these results are positive or negative.

Denmark

Danish Council of Ethics, Protection of Sensitive Personal Information – A Report, Copenhagen, 1992 (www.etiskraad.dk/english/publi cations.htm)

The Council recommends very strict control on the use of medical records and medico-biological banks. The Council recommends that the individual is given full control of the gathering and use of 'person-sensitive' data and biological material. The Council also recommends legislation to secure individual autonomy, integrity and the right to know about and control the use of person sensitive data.

Danish Council of Ethics, Genetic Testing in Appointments. Copenhagen, 1993

Law No. 286 of 24 April 1996 on the Use of Health Information on the Labor Market (Intl. Dig. Hlth. Legis., 47, 1996:371-72)

This Act strictly limits employers' rights to ask potential employees for health information including information based on genetic testing.

Act No. 413 of 10 July 1997, Act to Amend the Insurance Agreement Act and Act on the Supervision of Company Pension Funds (http://www. forsikringenshus.dk/htmm/eng/annualr.htm)

The way insurers used to get health information when the law was passed is not prohibited and this means that the insurer is allowed to ask for information on blood samples e.g. HIV test. Insurers may only ask for HIV test and family history when the sum insured is high and over a certain level.

Estonia

Estonian Parliament, Human Gene Research Act, 2001 (http://www.genomics. ee/genome/act1312.html)

This Act has been enacted to protect persons from misuse of genetic data and from discrimination based on interpretation of the structure of their DNA and the genetic risks arising therefrom.

Article 25 on Prohibition on discrimination states that (1) 'It is prohibited to restrict the rights and opportunities of a person or to confer advantages on a person on the basis of the structure of the person's DNA and the genetic risks resulting therefrom; (2) It is prohibited to discriminate against a person on the basis of the person being or not being a gene donor'.

Article 26 is devoted to discrimination in employment relationships: (1) 'Employers are prohibited from collecting genetic data on employees or job applicants and from requiring employees or job applicants to provide tissue samples or descriptions of DNA; (2) Employers are prohibited from imposing discriminatory working and wages conditions for people with different genetic risks'.

Article 27 is devoted to discrimination in insurance relationships: (1) 'Insurers are prohibited from collecting genetic data on insured persons or persons applying for insurance cover and from requiring insured persons or persons applying for insurance cover to provide tissues samples or descriptions of DNA; (2) Insurers are prohibited from establishing different insurance conditions for people with genetic risks and from establishing preferential tarif rates and determining insured events restrictively'.

Under Article 31, the Criminal Code is amended as follows: 'unlawful restriction of the rights of a person or conferral of unlawful preferences on a person based on the genetic risks of the person is punishable by a fine, detention or up to one year imprisonment'.

Finland

By law, policyholders are obliged to give correct and complete answers to questions posed by insurance companies before policies are approved. In principles, such questions include those about genetic tests. However, the Finnish Insurance Companies have adopted a policy of not asking questions about genetic tests in connection with their risk assessment. Nor do they make use of such information if they obtain the results of genetic tests undergone by their customers. Nor, in their risk assessment, do they pose questions or use information on the state of health of applicants' relatives (Federation of Finnish Insurance Companies, 1999). About the occupational aspects, there is a law on the privacy in occupational life that is under preparation. The proposal states that genetic tests in occupational settings can be used only with a permission of The National Board of Medical Legal Affairs and that permission could be attained only if the test is for protecting the individuals health.

France

Law of December 1989 related to the protection of persons against discrimination on the basis of their state of health or of their handicap, J. O. of January 3, 1990

Law n. 94-653 of July 29, 1994 on respect for the human body (Article 16-10 of the Civil Code) (http://www.cnrs.fr/SDV/loirespectcorps.html)

According to the Article 16-10, 'the genetic study of an individual's characteristics can only be carried out for medical purposes or scientific research'.

Article 226-26 of the *Code pénal* states that 'the use of information about an individual which has been obtained by studying his genetic characteristics other than for medical purposes for scientific research is punishable with one year's imprisonment and a fine of FRF 100 000'.

Law n. 94-654 of July 29th regarding the donation and use of elements and products of the human body, medically assisted procreation and prenatal diagnostic (http://www.cnrs.fr/SDV/loirespectcorps.html)

Article L1131-1 states that the Genetic characteristics of a person or his iden-tification through the use of genetic prints, when not performed for a judiciary procedure can only be done for medical or scientific goals after having ob-tained that person's consent.

In *1994, the French Federation of Insurance Companies* (http://www.ffsa.com/pub/pub.htm) announced that for a period of 5 years, which coincided with the 5-years period upon expiry of which the law n. 94-653 of July 29, 1994 was to be revised, its members would not use genetic information to determine applicants' insurability. This moratorium has been extended for another period of 5 years (2004).

National Consultative Ethics Committee, Opinion and Recommendations on Genetics and Medicine: from Prediction to Prevention, Reports, Paris, 1995 (http://www.ccne-ethique.org/english/avis/)

The report recommends prohibiting insurers from using genetic information, even if that information is voluntarily provided by applicants.

Decree n. 2000-548 date June 15, 2000 Predictive Medicine, Genetic Identification and Genetic Research (http://www.legifrance.gouv.fr)

This decree states that the examination of genetic characteristics of a person, when not done for a judiciary procedure can only be performed for medical purposes or scientific research after having obtained the consent of that person.

Germany

Contractual liberty allows insurers to ask applicants to undergo tests that are relevant for the determination of risks. According to the medical committee within the German insurance federation, paragraph 16 of the German insurance contract law states that an insured is already bound to give information regarding all particulars known to him which could be important for the acceptance of a risk. This includes the results of a genetic test. However, a moratorium exists since 1988, according to which insurers neither make genetic tests a prerequisite for insurance contracts nor do they ask for the results of genetic tests performed in the past. This moratorium has been renewed in 1999 by the German insurers' association (Lauth & Schmidtke 1999). Regarding genetic testing in the workplace, there is a requirement to obtain genetic knowledge for certain occupations at pre-employment stage. This consists of traditional questions, such as those about family history. Genetic testing designed to analyze genes in relation to employment is not undertaken. Because of the dynamic character of molecular genetics and the fact that future developments can hardly be predicted there is general agreement that legal regulations are not suitable for the regulation of genetic testing (Karlic & Horak 1998).

Enquette-Kommission des Deutschen Bundestages, Chancen und risiken der Gentechnologie, Dokumentation des Berichts an den Deutschen Bundestag, Frankfurt, 1987

This document recommended a new criminal offence where an employer discriminates against an employee on the basis of the results of his genetic test. In most instances the report did not recommend that legislation be enacted but rather that these matters be supervised by authoritative professional bodies (McGleenan 1999).

The German insurers' roof organization, Moratorium on genetic tests, 1988

This moratorium states that insurers neither make genetic tests a prerequisite for insurance contracts nor do they ask for the results of genetic tests performed in the past.

Greece

To date, there is no legislation concerning practice in genetics. Insurance companies have agreed to a voluntary code of conduct and do not ask for genetic testing prior to insuring patients.

Iceland

There is no legislation dealing specifically with the issue of genetic discrimination in life insurance and employment. However, discrimination based on genetic characteristics might be prevented by the following regulation and Act.

Ministry of Health and Social Security, Government Regulation No 32/2000 on a Health Sector Database (2000)

According to Article 14, 'providing information on individuals from the Health Sector Database is prohibited. Only statistical information involving groups of individual may be provided'.

Ministry of Health and Social Security, Act on the Rights of Patients No 74/1997 (1997)

Article 1 stipulates that 'It is prohibited to discriminate against patients on grounds of gender, religion, beliefs, nationality, race, skin color, financial status, family relation or status in other respect'.

Ireland

Irish Insurance Federation, Code of practice on genetic testing, 2001

1. Applicants must not be required to undergo a genetic test in order to obtain insurance
2. Disclosure of the result of a genetic test will not be required in new

applications for life cover unless the sum assured on the new application exceeds £300 000 or the total of the sum assured on the new application and other policies, if any, taken out with any insurer between 1st April 2001 and 31st December 2005 exceeds £300 000.

Italy

Law n. 675, 31 December 1996, D.P.R. n. 318, 28 July 1999, on Medical Information Privacy

There is no specific legislation on the use of genetic information by insurers and employers in Italy, but the Law n. 675 states the privacy of all medical information.

The Italian Committee on Bioethics, Orientamenti bioetici per i test genetici, 19 November 1999 (http://www.palazzochigi.it/bioetica/orientamenti%20 biomedici.htm)

These recommendations state that genetic information must be treated as the general medical information and therefore it is forbidden to give this information to insurers or employers without consent.

Luxembourg

Insurance Contracts Act of 27th July 1997

This law stipulates that the prohibition on the use of genetic test by insurers is of public matter and cannot be bypassed, even with the consent of the insurance applicant.

Norway

Act Relating to the Application of Biotechnology in Medicine, Law n. 56 of 5 August 1994 (http://www.helsetilsynet.no/htil/avd2/bio_act.htm)

Chapter 6 states that genetic testing can only be performed for medical diagnosis and/or therapeutic purposes. (...) It is forbidden to request, receive, possess or use information resulting from a genetic test on any person. It is also prohibited to ask whether such a test has been carried out previously.

Chapter 8 stipulates that anyone violating this law will be punished with an economic fine or will be sentenced to prison for three months.

Norwegian Biotechnology Advisory Board, Genetic Testing: When & Why? Oslo, March 1996.

Norwegian Biotechnology Advisory Board, The Use of Genetic Information

about Healthy People by Insurance Companies. Oslo, April 1997.

Portugal

The Ratification of the 'Convention for the Protection of Human Rights and Dignity of the Human Being and the additional protocol on the prohibition of cloning human beings' was published in January 2001.

Act No 10/95 related to the Protection of Personal Information

Spain

The Spanish Constitution of 1978

The Spanish Constitution forbids any kind of discrimination on grounds of any personal or social circumstance or condition. This prohibition should be concerned for employers as well as for insurers, if they try to refuse to contract with some applicants being carriers of genetic susceptibility for certain diseases (Karlic & Horak 1998).

The Organic Law regulating the automated processing of personal data of 29 October 1992

This law provides special measures of protection for personal health data (articles 7.3 and 8).

Labor Risk Preventive Act of 8 November 1995

Article 25 'Protection of the specially sensitive workers to determined risks' stipulates that employers will guarantee the protection of the workers who will be specially sensitive to the risk derived from work. This article does not refer to the situations of susceptibility to known genetic predisposition or to future monogenic illnesses also known without any type of symptom at the moment of entering the work post (Karlic & Horak 1998). There is no provision for applicants to a job.

The Organic Law regulating the automated processing and protection of personal data of 13 December 1999

This law includes automated data and any type of personal data.

Sweden

Law 114 of March 1991 on the Use of Certain Gene Technologies within the Context of General Medical Examinations (1993)

This law examines the use of certain genetic technology in medical screening. There must be a permission from the National Board of Health and

Welfare. Authorization from this body is required before DNA testing can be carried out. This requirement extends to the use of genetic screening techniques for diagnostic purposes.

The use of information about an individual which has been obtained by studying his genetic characteristics other than for medical purposes is prohibited.

Genetic discrimination can be subject to penalties in the form of fines or prison sentences up to a maximum of 6 months.

The Agreement between the Swedish State and the Swedish Insurance Federation concerning genetic testing, 1999

According to this agreement, insurance companies have undertaken not to start requiring insurance applicants to undergo genetic investigations, nor – as a condition of individual life and health policies up to an inflation-indexed once-only lump sum – to ask them to submit the findings of previous genetic tests, if any. The state is entitled to cancel the agreement with immediate effect if any insurance company disregards what the Insurance Federation has undertaken. This agreement is valid to the year 2002.

Switzerland

The Federal Code of Obligations

The federal Code of Obligations stipulates the nullity of any contract against the law or against common morality (art. 20). Read in connection with article 27 II of the Civil Code which protects the individual against excessive commitments, this article speaks for the nullity of a contractual clause in an insurance contract which would release the applicant physician altogether from his obligation of confidentiality. Article 321 of the federal Criminal Code punishes the professionals who reveal confidential information. Article 328b of the Code of Obligations stipulates that employers may only use data regarding the employee if they concern the employment relationship or if they are necessary to carry out the employment contract. This rule concerns existing or imminent diseases, thus excluding presymptomatic investigations (Karlic & Horak 1998).

The Swiss Federal Constitution, 1992

Article 119 (introduced in 1992 as article 24novies, old numbering) paragraph 2 states that the genetic heritage of an individual may be analyzed, registered or divulged only with his consent or on the basis of a legal prescription.

The Swiss Academy of Medical Sciences, Medical-ethical Guidelines for Genetic Investigations in Humans, Approved by the Senate of the Swiss Academy of Medical Sciences on 3rd June 1993 (http://www.samw.ch/e/richtlini en/richtlinien_fs.html)

Paragraph 3.7 states that 'medical doctors may make the medico-genetic findings available to third parties only with the consent of the person investigated or of his legal representative, and only after the implications of such disclosure of information have been explained to them'.

Paragraph 3.8 states that 'genetic investigations must not be carried out for the purpose of assessing the suitability of a person for certain activities or work, unless the investigation is performed in order to detect factors which, if present, would render a particular activity a considerable risk to the health of the individual or for other persons'.

Paragraph 3.9 recommends 'particular reservations when the results of a requested genetic investigation are to be used in connection with the taking out or the revision of an insurance policy. The results are to be communicated exclusively to the person investigated or his legal representative, after the implications of the passing on of such information to third parties have been explained to them'.

The Swiss Academy of Medical Sciences guidelines about genetic investigations in humans have been included into the Code of Deontology of the Swiss Medical Association and apply directly to all the physicians who are members of the Association. These guidelines are not legally binding, unless cantonal legislation gives them binding force.

Bill regarding Genetic Investigations in Humans, 1998 (http://www. admin.ch/cp/d/384b8f91.0@fwsrvg.bfi.admin.ch.html)

This bill has not yet been debated in Parliament. Section 3 stipulates that when establishing an employment relationship, or during employment, the industrial doctor may order a presymptomatic investigation only if all of the following conditions are met (Art. 19 § 1): The workplace represents a risk for an industrial disease or a serious damage to the environment or an extraordinary risk of accidents or health hazard for third parties. Safety measures according to the law are not sufficient to eliminate this risk. The workplace is put under the regulation of preventive industrial medicine by order of the competent authority or by law. The specific risk for the employee or the imminent and serious risk for third parties or the environment cannot be evaluated in another way. A federal panel for genetic investigations has pronounced the method safe and reliable on detecting a risk. The employee agrees to the investigation. The employee shall inform the industrial doctor, on the latter's initiative, of the results of former presymptomatic investigations relevant to the ability to perform the specific work (Art. 19 § 2).

Section 4 stipulates that insurers are not allowed to demand a presymptomatic or prenatal investigation as a condition of insurance (Art. 22 § 1). As for the results of former investigations, the Bill differentiates: As a rule, insurers are not allowed to ask for or use the results of former presymptomatic or prenatal investigations or investigations for family planning (Art. 22 § 2). The competent federal authority, however, can make an exception in the case of non-compulsory insurance (Art. 23 § 2). The applicant is obliged to answer the medical examiner's questions on the results of a former presymptomatic investigation, if this investigation is reliable and if the scientific value of the result for calculating the premium is shown (Art. 23 § 2). The applicant may inform the insurer of the results of former presymptomatic or prenatal investigations in order to demonstrate that he has wrongly been classified in a high-risk group (Art. 23 § 1). The competence to specify which genetic information can be requested by insurers must rest in the hands of a federal authority (Art. 24 § 1). The questions must be relevant to evaluating the insured risk (Art. 24 § 2).

The Netherlands

Verzekeraars verlengen moratorium erfelijkheidsonderzoek, December 1990 (1995)

The moratorium, originally for 5 years, became indefinite in 1995. Insurers must abstain from using existing genetic test results for life applications up to NLG 300 000 and for disability applications up to NLG 60 000. Insurers must abstain from requesting genetic tests for all applications.

Medical Examination Act, 1 January 1998

The basic principle of the Act is that individuals must have unimpeded access to socially important facilities such as work and certain insurances; employers and insurers may not discriminate people with some blemish. The legislature was of the opinion that in a number of cases this principle could only be achieved by a prohibition of the medical examination. The Medical Examination Act prohibits employers and insurers from requiring medical tests that could indicate that the applicant may be suffering from a severe incurable disease. Regarding genetic testing, when carrying out a medical examination for taking out or changing insurance, insurers may not ask an insured whether the prospective insured has any hereditary, serious, untreatable disease, unless the illness has already manifested itself in the prospective insured. Insurers may not ask whether any blood relatives have any hereditary, serious, untreatable diseases, not even if the illness has already manifested itself or the blood relative has died from it. Finally, insurers may not ask about the results of previous genetic tests among blood relatives or the prospective insured himself. However, these prohibitions apply only for life policies below NLG 300 000 and for disability policies below NLG 60 000 (Goedvolk 1999).

United Kingdom

In May 2001, the Human Genetics Commission (HGC) recommended a three-year moratorium on the use of genetic information by insurers, except in respect of policies over £500 000 in value. In the case of these high-value policies, the HGC says insurers should be permitted to use only the results of tests approved by the Genetics and Insurance Committee (GAIC). GAIC has so far approved only tests for Huntington's disease in respect of life insurance, but a small number of additional tests is currently under consideration. The HGC recommends that the moratorium on the use of genetic information in insurance should be enforced by legislation. The HGC recommendation follows a report by the House of Commons Select Committee on Science and Technology on Genetics and Insurance (April 2001), which found that the current system of self-regulation by the insurance industry was not satisfactory. On the same day that the HGC released its recommendations, the Association of British Insurers announced that its members would no longer request results of genetic tests in respect of applications for any type of insurance up to a value of £300 000. For policies above this value, only the results of tests approved by GAIC would be used. The ABI also announced in a joint statement with the UK Forum for Genetics and Insurance and the British Society for Human Genetics, that the results of a genetic test taken as part of a research project, rather than in the context of a clinical consultation, need not be declared to insurers.

In October 2001 the UK government reached an agreement with ABI to institute a 5-year moratorium on the use of genetic tests results in assessing applications for life insurance policies up to a value of £500 000, and for critical illness, long-term care and income protection policies up to a value of £300 000. For an amount over those limits ABI will be able to use genetic tests results if they have been approved by GAIC. These limits will be reviewed after 3 years.

House of Commons Select Committee on Science and Technology, Human Genetics: the science and its consequences, 3rd report, HMSO, London, 1995 (http://www.parliament.the-stationery-office.co.uk/pa/cm199899/cmselect/cmsctech/489/48902.htm)

The House of Commons Select Committee in its report on human genetics recommended that the insurance industry should find ways to avoid a conflict between their interests and the medical interests in genetic testing. The Association of British Insurers subsequently issued a Code of Practice on Genetic Testing (see below) and the Government appointed the HGAC, who took on insurance as one of their first projects.

Government Response to the Third Report of the House of Commons Select Committee on Science and Technology, Human Genetics: The science and its consequences, Department of Trade and Industry, 1996 (http://www. parliament.the-stationery-office.co.uk/pa/cm/cmsctech.htm)

Association of British Insurers, Code of Practice on Genetic Testing, November 1997 (revised August 1999) (http://www.abi.org.uk)

In its Code of Practice, the ABI undertakes not to require applicants to take any genetic test. In addition, genetic test results are disregarded when setting premiums for life insurance policies up to a value of £100 000 that are linked to new mortage applications. Insurers may not ask for the results of tests taken by other family members, nor offer individuals lower-than-standard premiums on the basis of genetic test results, nor disclose test results to any other party without the individual's consent. In the interim before further applications are put to GAIC, ABI member companies may continue to require disclosure of the results of certain tests that had been identified by its Genetics Adviser as at November 1998. These are tests for myotonic dystrophy, multiple endocrine neoplasia, hereditary motor and sensory neuropathy, familial Alzheimer's disease, familial adenomatous polyposis, and BRCA1/2-associated familial breast cancer. If any of these tests are subsequently rejected by GAIC, the insurance companies will refund any extra premiums paid by applicants on the basis of their results, or contact them to offer them insurance if it had been refused.

Human Genetics Advisory Committee, The implications of genetic testing for insurance, November 1997 (www.dti.gov.uk/hgac/papers/papers_b.htm)

The HGAC report made a number of recommendations of which the three most important were that insurers should not be allowed to use any genetic tests results unless they had satisfied an independent body that there was a good factual actuarial basis for using these results; that there should be a transparent, open and independent appeals process; and that there should be a moratorium on all testing for 2 years while these arrangements were being put in place.

Department of Trade and Industry, Genetic Testing and Insurance, Government formal response to the HGAC report, 5 November 1998 (www. hgc.gov.uk/about_regulatory.htm)

The British Government accepted all of the HGAC recommendations apart from the moratorium. While not agreeing to the moratorium, they suggested that the insurance industry should immediately stop using test results, until the Genetics and Insurance Committee (GAIC), had validated them. In November 1998, the British Government set up the Genetics and Insurance Committee, a nonstatutory, advisory body whose role is to assess the actuarial validity of genetic tests that insurance companies would like to be able to take into account in setting insurance premiums.

British Society for Human Genetics, Statement on Genetics and Life Insurance, 1998 (http://www.bshg.org.uk/insuranc.htm)

This statement recognizes that insurers need to protect themselves against an unacceptable degree of anti-selection. It endorses the recognition of the ABI stating that applicants must not be asked to undergo a genetic test in order to obtain any type of insurance. Genotypes present in more than 5% of the population should not be disclosed or considered for any life insurance. Cover up to an agreed sum should be available for all life insurance purposes without any genotype disclosure. If an insurer requires disclosure of any genetic test results, that requirement should be restricted to results where published and actuarially validated data allow evidence-based underwriting. Finally, insurers should recognize and counter the fear of undue discrimination (BSHG 1998).

The BSHG statement will be reviewed not later than summer 2003.

Human Genetics Advisory Committee, The implications of genetic testing for employment, June 1999 (www.dti.gov.uk/hgac/papers/papers_f/f_03.htm)

The HGAC report does not recommend a total ban on the testing of employees for genes that might predispose them to various conditions. The report suggests that employers could be allowed to ask for tests to detect a potentially dangerous illness, in the way that pilots are currently tested. In effect, testing should only be for the employees' benefit and not for the benefit of shareholders. The Commission also said that genetic tests should not play a part in recruitment. The Commission concluded that an individual's right not to know his or her genetic pre-dispositions should be upheld. Individuals should not have to disclose the results of previous genetic tests without clear evidence that the information was needed to assess whether they could do the job safely. Finally, the report recommends that testing be covered by the principles of data protection.

The Government's response to the HGAC report has now been published. This accepts all the main findings of the HGAC report and agrees that this issue should be kept under review. It asks the Human Genetics Commission to include this issue in the Commission's wider study of the uses of genetic information and to provide advice to Ministers in due course.

Genetic tests and future need for long-term care in the UK, A report of a *work group of the Continuing Care Conference Genetic Tests and Long-term Care Study Group, July 1999 (with update published January 2000)*http://www.medinfo.cam.ac.uk/phgu/info_database/Policy/cccreport.asp)

The group's report concentrates mainly on Alzheimer's disease, for which an actuarial model is presented to predict the costs of long-term care depending on levels of risk as predicted by apoE genotype. The report also contains information about the genetic and environmental basis of other adult-onset conditions including cancers, diabetes, ischaemic heart disease and stroke, osteoarthritis, rheumatoid arthritis and some psychiatric conditions, and preventive options.

*The Genetics and Insurance Committee, Decision of the Genetics and Insurance Committee Concerning the Application for Approval to Use Genetic Test Results For Life Insurance Risk Assessment in Huntington's Disease, October 2000 (*http://www.doh.gov.uk/genetics/gaichuntington.htm)

The Genetics and Insurance Committee (GAIC) was asked to examine the actuarial evidence for using individual genetic tests. The insurance industry, through the main trade body the Association of British Insurers, has agreed to abide by GAIC decisions. If GAIC decides that the evidence on the reliability and relevance of a particular test is insufficient to justify its use, the Association have agreed to stop using them and retrospectively reassess affected individual insurance premiums. The broader social and ethical issues surrounding the use of genetic tests in insurance and employment have been referred to the new Human Genetics Commission. An application for approval of two genetic tests for Huntington's Disease was submitted to GAIC by the Association of British Insurers (ABI) in July 2000. The application was sent to a clinical geneticist and an independent actuary for expert review and also to support groups for Huntington's Disease and to the Genetic Interest Group (GIG) for their comments. At their meeting in September, GAIC considered the application, in the presence of observers from the ABI, GIG and Huntington's Disease Association.

The committee recognizes that this complex subject is an important issue to the public, industry and government alike. GAIC will work closely with the new Human Genetics Commission when they begin their inquiry into the use of genetic data including in insurance and employment.

*House of Commons Select Committee on Science and Technology, 5th report, Genetics and Insurance, HMSO, London, 2001 (*http://www.publications.parliament.uk/pa/cm200001/cmselect/cmsctech/174/17404.htm)

The House of Commons Select Committee in its report on genetics and insurance recommends a two-year moratorium on the use of positive genetic test results by insurers, to allow time for further research on the actuarial relevance of test results.

UK Forum on Genetics and Insurance, Association of British Insurers, British Society for Human Genetics, Joint statement on Genetics and Insurance, 24 April 2001 (http://www.ukfgi.org.uk/joint%20statement%20 abi,%20bshg,%20ukfgi%2024%2004%2001.htm)

The UK Forum on Genetics and Insurance said that 'it would continue to work to ensure the use of genetic information is handled appropriately by all parties'. The Association of British Insurers said that 'results from genetic testing arising from research projects will not be used for underwriting policies. Also, if someone already has an insurance policy it will not be affected by the policyholder participating in a research project concerned with genetic testing'. The British Society for Human Genetics 'welcomes the ABI's confirmation that research genetic tests will not affect any insurance proposal and do not need to be declared in any insurance application. This removes one source of anxiety for people asked to take part in genetic research, and should help avoid the risk that research will be hampered because of people's worries about insurance'.

Human Genetics Commission, The use of genetic information in insurance: Interim recommendations of the Human Genetics Commission, May 2001 (http://www.hgc.gov.uk/business_publications_statement_01may.htm)

In the HGC's view the moratorium should embrace the following features: 'No insurance company should require disclosure of adverse results of any genetic tests, or use such results in determining the availability or terms of all classes of insurance. The moratorium should last for a period of not less than three years. This will allow time for a full review of regulatory options and afford the opportunity to collect data which is not currently available. The moratorium should continue if the issues have not been resolved satisfactorily within this period. (...) An exception should be made for policies greater than £500 000. This will address concerns about adverse selection, the process by which persons having a known risk set out to acquire substantial insurance cover. (...) We recommend this upper financial limit on the basis of the industry's own tables and information as a protection from significant financial loss. Only genetic tests approved by the Genetics and Insurance Committee (GAIC) should be taken into account for these high-value policies'.

Association of British Insurers, Insurers Confirm Decision To Extend Moratorium On Use Of Genetic Test Results, 1 May 2001 (http://www.abi.org. uk/HOTTOPIC/nr415.asp)

'The (industry's) existing Code includes a moratorium on the use of test results in respect of life insurance linked to a mortgage of up to £100 000.

Following very careful consideration within the industry, we propose to extend this moratorium to cover all classes of insurance up to £300 000. This will have the effect of excluding genetic test results from underwriting other than for a very small number of high value policies. The House of Commons Select Committee on Science and Technology called for a 2-year moratorium. The advantage of this will be to provide a period of stability while new and more permanent arrangements can be put in place. We continue to be keen to work with you and the Government to bring this about'.

Government Response to the Report from the House of Commons Science and Technology Committee: Genetics and Insurance, (October 2001) (http://www.doh.gov.uk/genetics/gaicgovrespoct2001.pdf)

International Organizations

The World Medical Association, World Medical Association Declaration on the Human Genome Project (September 1992, doc. 17.S/1) (www.wma.net/e/policy/17-s-1_e.html)

In its Declaration of the Human Genome Project, the World Medical Association considers that 'Medical secrecy should be kept and information should not be passed on to a third party without consent. (...) The disclosure of information to a third party or the accessibility to personal genetic data should be allowed only with the patient's informed consent'.

UNESCO, The Universal Declaration on the Human Genome and Human Rights,   (November1997) (http://www.unesco.org/ibc/uk/genome/project/index.html)

Article 6 states that 'No one shall be subjected to discrimination based on genetic characteristics that is intended to infringe or has the effect of infringing human rights, fundamental freedoms and human dignity'. And according to article 7, 'genetic data associated with an identifiable person and stored or processed for the purposes of research or any other purpose must be held confidential in the conditions foreseen set by law'.

The World Medical Association, Proposed international guidelines on ethical issues in medical genetics and genetics (1998) (http://wwwlive.who.ch/ncd/hgn/hgnethic.htm) 'Genetic data should not be given out to insurance companies, employers, schools or governments, other than after the full informed consent of the person tested. In some countries it may be possible or necessary to protect both confidentiality and non-discrimination through legal means'

HUGO, Statement on the DNA Sampling Control and Access (February 1998) (http://www.gene.ucl.ac.uk/hugo/conduct.htm)

Unless authorized by law, there should be no disclosure to institutional third parties – such as employers, insurers, schools, and government agencies because of possible discrimination – of participation in research, nor of research results identifying individuals or families. Like other medical information, there should be no disclosure of genetic information without appropriate consent.

The World Health Organization, Cloning in Human Health, (1ˢᵗApril 1999) (http://www.who.int/gb/EB_WHA/PDF/WHA53/ea15.pdf)

Article **8** stipulates that 'Genetic information should not be used as the basis for refusing employment or insurance. Exceptions would have to be legally defined'.

APPENDIX B

Contributions

This document was reviewed by the ESHG Public and Professional Policy Committee (PPPC). Members of the PPPC are:

- Ségolène Aymé (Paris, France) Chair
- Martin Bobrow (Cambridge, UK)
- Jean-Jacques Cassiman (Leuven, Belgium)
- Domenico Coviello (Modena, Italy)
- Gerry Evers-Kiebooms (Leuven, Belgium)
- Peter Farndon (Birmingham, UK)
- Helena Kääriäinen (Helsinki, Finland)
- Ulf Kristoffersson (Lund, Sweden)
- Marcus Pembrey (London, UK)
- Sandy Raeburn (Nottingham, UK)
- Joerg Schmidtke (Hannover, Germany)
- Leo ten Kate (Amsterdam, The Netherlands)
- Lisbeth Tranebjaerg (Tromso, Norway) Secretary

The first draft of this document was sent out to a wide range of people and organizations for consultation. It was reviewed critically by the following consultants:

Australia

Agnes Bankier, Institute Royal Children's Hospital, Parkville Victoria

Austria

Gertrud Hauser, Institute of Histology & Embryology, University Vienna, Vienna

Belgium

Herman Nys, Center for Bio-medical Ethics and Law, Catholic University Leuven

Myriam Welkenhuysen, Psychogenetics Unit, Centre for Human Genetics, Leuven

Cyprus

Kyproula Christodoulou, The Cyprus Institute of Neurology and Genetics, Nicosia

Denmark

Jane Emke, Danish Insurance Association, Copenhagen

Lene Koch, Institut of Public Health, University of Copenhagen, Copenhagen

Eastern Countries

Vladislav Baranov, Institute of Obstetrics and Gyneaecology, Russian Academy of Medical Sciences, St. Petersburg, Russia

Georges Kosztolanyi, National Institute of Hygiene, Department of human Genetics and Teratology, Budapest, Hungary

Maria Kucerova, Genetics Department, Thomayer University Hospital, Prague, Czech Republic

Astrida Krumina, Department of Medical Biology and Genetics, Medical Academy of Latvia, Riga, Latvia

Finland

Kaija Holli, Tampere University Central Hospital, Tampere

Pekka Koivisto, Retro Life Assurance Company Ltd, Helsinki

Veikko Launis, Department of Philosophy, University of Turku, Turku

Nina Meincke, Medical Law Project, University of Helsinki, Helsinki

Minna Poyhonen, The Family Federation of Finland, Department of Medical Genetics, Helsinki

Riitta Salonen, Department of obstetrics and Gynecology, Helsinki University Central Hospital, Helsinki

France

Yves-Jean Bignon, Department of Oncology and Genetics, Jean Perrin Centre, Clermont-Ferrand

François Ewald, French Federation of Insurance Companies, Paris

François Eisenger, Institut Paoli Calmettes, Marseille

Josué Feingold, Laboratoire d'Anthropologie Biologique - Paris XII University, Paris

André Klarsfeld, Centre National de la Recherche Scientifique, Institut Alfred Fessard, Gif-sur-Yvette

Dominique Stoppa-Lyonnet, Department of Oncology and Genetics, Curie Institute, Paris

Germany

Achim Regenauer, Munich Reinsurance Company, Munich

Ireland

David Barton, National Centre for Medical Genetics, Ous Lady's Hospital for Sick Children, Dublin

Tony McGleenan, Faculty of Law, Queen's University, Belfast

Peter Whittaker, Biology, National University of Ireland, Maynooth

Israel

Ariella Oppenheim, Hebrew University-Hadassah Medical School, Jerusalem

Italy

Bruno Dallapiccola, Medicina Sperireniato, La Spapienza University, Roma

Giovanni Neri, Instituto di Genetica Umana, Universita Cattolica, Roma

Norway

Kare Berg, Institute of Medical Genetics, University of Oslo, Oslo

Erik Rosaeg, Scandinavian Institute of Maritime Law, University of Oslo, Oslo

Portugal

Maria Cristina Rosamond Pinto, Department of Human Genetics, Faculdty of Medicine, Lisboa

Heloisa Santos, Department of Medical Genetics, De Santa Maria Hospital, Lisboa

Yorge Sequeiros, UmiGENE, IBMC, University of Porto, Porto

Spain

Maria Ramos, Department of Genetics, Virgen Del Camino Hospital, Pamplona

Sweden

Goran Flood, Swedish Skandia Life, Stockholm

Jan Wahlström, Department of clinical Genetics, Sahlgrenska University Hospital/East, Göteborg

Switzerland

Suzanne Braga, FMH medizinische Genetik, Bern

André Chuffart, Swiss Re Life & Health, Zurich

Christian Kind, Ostschweizer Kinderspital, St Gallen

Olivier Guillot, Institut de droit de la santé, Neuchatel University, Neuchatel

United Kingdom

Peter Brett, Peridontology, Eastman Dental Institute, London

Jerry Brown, Swiss Re Life & Health Limited, London

Ruth Chadwick, Center for Professional Ethics, University of Central Lancashire, Preston

David Cook, The Whitfield Institute, Oxford

Stephen Diacon, Centre for Risk and Insurance Studies, University of Nottingham, Nottingham

Robert Dingwall, Genetics and Society Unit, University of Nottingham, Nottingham

Bethalee Jones, Association of British Insurance, London

Beryl Keeley, Genetics Science Policy, Department of Health, London

Alastair Kent, Genetic Interest Group, London

Doris Littlejohn, Central Office of Industrial Tribunals, Glasgow

Angus MacDonald, Department of Actuarial Mathematics & Statistics, Heriot-Watt university, Edinburgh

David Muiry, Swiss Re Life & Health Ltd, International Financial Center, London

D. Paul, British United Provident Association Ltd, BUPA House, London

Andrew Read, Department of Medical Genetics, St Marys Hospital, Manchester

Virginia Warren, Group Medical, UPA, London

Pamela Watson, Genetics & Society Unit, University of Nottingham, Nottingham

United States of America

Dorothy Wertz, The Eunice Kennedy Shiver Center for Mental Retardation, Waltham, MA

This document forms part of a BIOTECH program financed by the Commission of the European Communities (CEE BIO4–CT98–0550).

REFERENCES

1. Evans JP, Skrzynia C, Burke W: The complexities of predictive genetic testing. *BMJ* 2001; **322**: 1052–1056. Article PubMed ChemPort |

2. Holtzman NA: Are we ready to screen for inherited susceptibility to cancer?*Oncology* 1996; **10**: 57–64.

3. Holtzman NA: Putting the search for genes in perspective. *Int J Health Serv*2001; **31**: 445–461.

4. Reilly PR: Genetic Risk Assessment and Insurance. *Genet Test* 1998; **2**: 1–2.

5. Kumar S, Gantley M: Tensions between policy makers and general practitioners in implementing new genetics: grounded theory interview study. *BMJ* 1999; **319**: 1410–1413. PubMed ChemPort |

6. Lemmens T, Bahamin P: Genetics in life, disability and additional health insurance in canada: a comparative legal and ethical analysis. In: Knoppers BM (ed): *Socio-Ethical Issues in Human Genetics*. Montreal:

Les Editions Yvon Blais Inc: 1998, pp 107–276.

7. MacDonald AS: How will improved forecasts of individual lifetimes affect underwriting? *Philos Trans R Soc London* 1997; **352**: 1067–1075.

8. Murthy A, Dixon A, Mossialos E: Genetic testing and insurance. *J R Soc Med* 2001; **94**: 57–60.

9. Chadwick R, ten Have H, Hoedemaekers R *et al:*: Euroscreen 2: towards community policy on insurance, commercialization and public awareness. *J Med Philos* 2001; **26**: 63–72.

10. Harmon C, Nolan B: Health insurance and health services utilization in Ireland. *Health Econ* 2001; **10**: 135–145.

11. McGleenan T, Wiesing U, Ewald F (eds): *Genetics and Insurance.* Oxford: BIOS Scientific Publishers Ltd; 1999.

12. Morrison PJ: Genetic testing and insurance in the United Kingdom. *Clin Genet* 1998; **54**: 375–379. PubMed ISI ChemPort |

13. Morrison PJ, Steel CM, Vasen HF *et al:*: Insurance implications for individuals with a high risk of breast and ovarian cancer in Europe. *Dis Markers* 1999; **15**: 159–165. PubMed ChemPort |

14. Nys H, Nederveen-Van de kragt CJM, Roscam-Abbing HDC *et al: Predictive Genetic Information and Life Insurance: Legal Aspects. Towards European Policy?.* University of Limburg, Department of Health Law, Maastricht; 1993.

15. Rosén E: Genetic information and genetic discrimination how medical records vitiate legal protection. *Scand J Public Health* 1999; **27**: 166–172.

16. Sandberg P: Genetic information and life insurance: a proposal for an ethical European policy. *Soc Sci Med* 1995; **40**: 1549–1559.

17. Weijer C, Emanuel EJ: Ethics. Protecting communities in biomedical research. *Science* 2000; **289**: 1142–1144. Article PubMed ISI ChemPort |

18. Association of British Insurers: Insurers will Use Genetic Test Result Responsibly. New Release, 12 October: 2000.

19. Chuffart A: *Genetics and Life Insurance: A Few Thoughts.* Zurich, Swiss Re; 1997.

20. Le Grys DJ: Actuarial considerations on genetic testing. *Philos Trans R Soc London* 1997; **352**: 1057–1061.

21. Roscam-Abbing HDC: Predictive genetic knowledge, insurance and the legal position of the individual. in Swiss Institute of Comparative Law (ed):*Human Genetic Analysis and the Protection of Personality and Privacy, International Coloquium.* Lausanne: Swiss Institute of

Comparative Law; 1994.

22. Ross T: The likely financial effects on individuals, industry and commerce of the use of genetic information. *Philos Trans R Soc London* 1997; **352**: 1103–1106.

23. Goedvolk VI: The Medical Examinations Act: practical experience from the Netherlands. International Conference on Genetics and Private Life/Health Insurance, Paris, 11–12 February: 1999.

24. Schill AL: Genetic information in the workplace. Implications for occupational health surveillance. *AAOHN J* 2000; **48**: 80–91.

25. Bonn D: Genetic testing and insurance: fears unfounded? *Lancet* 2000;**355**: 1526.

26. Ewald F, Lorenzi JH (eds): *Encyclopédie de l'assurance*. Paris: Economica, 1998, 1780 p.

27. Hall M, Rich S: Laws restricting health insurers' use of genetic information: impact on genetic discrimination. *Am J Hum Genet* 2000; **66**: 293–307. Article PubMed ISI ChemPort |

28. The Human Genetics Advisory Committee: *The Implications of Genetic Testing for Insurance*. London: Office of Science and Technology, 1997.

29. Williams JK, Schutte DL, Evers CA, Forcucci C: Adults seeking presymptomatic gene testing for Huntington disease. *Image J Nurs Sch* 1999; **31**: 109–114.

30. Kaufert PA: Health policy and the new genetics. *Soc Sci Med* 2000; **51**: 821–829.

31. Lapham EV, Kozma C, Weiss JO: Genetic discrimination: perspectives of consumers. *Science* 1996; **274**: 621–624. Article PubMed ChemPort |

32. EUROSCREEN Insurance Sub-Group. Insurance, Newsletter 7, Spring: 1997.

33. British Society for Human Genetics. Statement on Genetics and Life Insurance, May 1998.

34. Pokorski RJ: Insurance underwriting in the genetic era. *Am J Hum Genet* 1997; **60**: 205–216.

35. Zick CD, Smith KR, Mayer RN, Botkin JR: Genetic testing, adverse selection, and the demand for life insurance. *Am J Med Genet* 2000; **93**: 29–39.

36. Kmietowicz Z: Health put at risk by insurers' demands for gene test results. *Br Med J* 1997; **314**: 625.

37. Hauser G, Jenisch A: Laws regarding insurance companies. *J Med*

*Genet*1998; **35**: 526–528.

38. The Human Genetics Commission. The use of genetic information in insurance: interim recommendations, May: 2001.

39. Natowicz MR, Alper JK, Alper JS: Genetic discrimination and the law. *Am J Hum Genet* 1992; **50**: 465–475.

40. Rothenberg K, Fuller B, Rothstein M *et al:*: Genetic information and the workplace: legislative approaches and policy challenges. *Science* 1997;**275**: 1755–1757. Article PubMed ISI ChemPort |

41. Jacobs LA: At-risk for cancer: genetic discrimination in the workplace. *Oncol Nurs Forum* 1998; **25**: 475–480.

42. Nunes R, Pereira de Melo H: Genetic testing in the workplace. Medical, ethical and legal issues. *Law Hum Genome Rev* 2000; **13**: 119–142.

43. Rothstein MA, Knoppers BM: Legal aspects of genetics, work and insurance in North America and Europe. *Eur J Health Law* 1996; **3**: 143–161.

44. Lemmens T: Genetic testing in the workplace. *Polit Life Sci* 1997; **16**: 57–75.

45. Mossialos E, Dixon A: Genetic testing and insurance: opportunities and challenges for society. *Trends Mol Med* 2001; **7**: 323–324. PubMed |

46. Tauer CA: Genetic testing and discrimination. How can we protect job and insurance policy applicants from negative test consequences? *Health Prog*2001; **82**: 48–53, 71.

Chapter 8

PROGRESS AND PROSPECTS: GENETIC ENGINEERING IN XENOTRANSPLANTATION

S Le Bas-Bernardet[1], I Anegon[1] and G Blancho[1,2]

[1]INSERM, U643, Nantes, France; CHU Nantes, Institut de Transplantation et de Recherche en Transplantation, ITERT, Nantes, France; Université de Nantes, Faculté de Médecine, Nantes, France

[2]Service de Néphrologie, Immunologie Clinique et Transplantation, CHU Nantes, Nantes, France

ABSTRACT

In this review, we summarize the work published over the last 2 years using genetic modifications of animals in the field of xenotransplantation. Genetic engineering of the donor has become a powerful tool in xenotransplantation, both for the inactivation of one particular porcine gene and for the addition of human genes with the goal of overcoming xenogeneic barriers. We summarize the work relative to the knockout of the α1,3-galactosyltransferase gene, followed by genetic engineering aimed at reducing the humoral and cellular immune response, complement activation and coagulation. Finally, we report on the genetic modification of pigs to reduce porcine endogenous retrovirus infection risk in the xenogeneic context.

IN BRIEF

Progress

- Genetic engineering approaches leading to a reduction of αGal expression in donor cells/organs and production of α1,3-galactosyltransferase knockout (GalT-KO) pigs.
- Control of complement through the use of transgenic pigs for multiple complement regulatory proteins.
- Genetic engineering for molecules that control coagulation activation.

- Genetic engineering for molecules that modify the cellular xenogeneic immune response
- Control of PERV expression in pig cells/organs by RNA interference.
  Prospects
- Deletion of the *iGb3S* gene in addition to deletion of GalT in donor cells/organs.
- Crossbreeding of GalT-KO pigs with pigs transgenic for EndoGalC.
- Crossbreeding of GalT-KO pigs with pigs transgenic for different human anticoagulant and/or complement regulatory proteins to decrease humoral and cellular immunological barriers.
- Use of pig pancreatic islets genetically modified to express transgenes that control the cellular xenogeneic immune response and coagulation.
- Controlling PERV expression in pig cells/organs used in xenotransplantation to reduce the risk of host infection.

INTRODUCTION

Transplantation is currently facing a problem of organ shortage; waiting lists are increasing with some of these patients dying before any proposition of an organ transplant. One solution could be to increase the donor 'reservoir' by using animal organs: that is, xenotransplantation. In the past, non-human primates (NHPs) were used as organ donors for humans (1960–1992), but are no longer considered as potential donors essentially for ethical reasons and for a high risk of endogenous virus transmission to humans. Now, pigs are considered for the future as animals of choice (porcine valves used for years in humans, may not be considered as xenotransplantation as the tissue is devitalized before implantation). In fact, pigs have been identified as potential donors, primarily because of their physiology and anatomy being similar to humans. Moreover, there are no major ethical considerations regarding their use in medicine. Finally, their potential has been increased by the application of the most recent techniques of genetic manipulation: transgenesis, gene transfer through viral vector and cloning through nuclear transfer. However, some significant problems remain regarding their use: immunological barriers and the risk of viral transmission. Manipulations of the pig genome targeting the human or NHP immune responses, coagulation disorders and viral risk are research approaches that will be addressed in this review (Table 1).

Table 1: Progress in xenotransplantation through genetic engineering approaches

Technique	Molecule	Models		Study results	References
		Donor/target cells	Recipient/responder cells		
Genetic engineering approaches leading to a reduction of αGal expression in donor cells/organs and production of α1,3-galactosyltransferase knockout (GalT-KO) pigs					
Gene transfer	EndoGalC	EndoGalC transgenic mice		Mice transgenic for EndoGalC present reduced αGal expression, but an abnormal phenotype	Watanabe et al.[1]
Cloning, direct mutagenesis	GnT-I	Transfected pig cells with pigGnT-I mutations	Human serum	Transfection with a pigGnT-I(320) mutation reduced the antigenicity to human sera	Matsunami et al.[2]
Gene inactivation, gene transfer	GalT	iGb3 vs GalT-transfected 293 cells	GalT-KO mice	αGal synthesized by iGb3S is as immunogenic as GalT	Milland et al.[3]
Gene inactivation	iGb3S				
Gene inactivation	GalT	GalT-KO pig		Complementary breeding and SCNT in GalT-KO pig production	Nottle et al.[4]
Gene inactivation	GalT	GalT-KO pig heart	Baboon	No improvement in coagulation problems in the GalT-KO context	Ezzelarab et al.[5]
Gene inactivation	GalT	PBMC from WT vs GalT-KO pig	Healthy human vs patients awaiting renal transplantation	Allosensitized compared to unsensitized patients at no greater risk of humoral rejection to pig xenografts	Hara et al.[6]
Control of complement through the use of transgenic pigs for multiple complement regulatory proteins					
DNA sequencing, gene transfer, phagemid vector	hDAF	hDAF transgenic pig heart	Cynomolgus	A restricted family IgV$_H$3 gene encoded for their XNA	Zahorsky-Reeves et al.[7]
Gene transfer	hCD59	hCD59 transgenic Pig		Microinjection of one hCD59 transgene into zygote pronuclei is a suitable technique to generate hCD59 transgenic pigs	Deppenmeier et al.[8]
Gene transfer	hCD55 hCD59	Ex vivo perfused hCD55/59 transgenic porcine lung	Fresh human blood	Xenotransplantation of lungs from hCD55 and hCD59 transgenic pigs induced, respectively, a partial protection and no effect on pulmonary dysfunction compared to nontransgenic lungs	Poling et al.[9]
Gene transfer	hDAF hMCP	Pig heart	Baboon	Deregulated coagulation correlated closely with and probably causes primary failure of pig hearts transgenic for hCRP	Wu et al.[10]
Gene transfer, plasmid	hDAF	Pig fibroblasts	Human and baboon sera	Much higher expression of hDAF might be beneficial in protecting pig organs from one humoral response	Liu et al.[11]
SMGT	hDAF	Ex vivo perfused hDAF transgenic porcine heart	Fresh human blood	Hearts from hDAF transgenic pigs generated by SMGT are protected from HAR after contact with human blood	Smolenski et al.[12]
Gene transfer, plasmid	sCR1	Cultured human sCR1 transgenic porcine ECs	Human AB serum	Transgenic pECs expressing human sCR1 are resistant to in vitro lysis by human sera	Manzi et al.[13]
Cloning, gene transfer, pGL3-basic, gene transfer, pRL-TK	ICAM-2 promoter	Simian Virus 40-transformed pig aortic endothelial cells (SVAP), COS-1		The P8 sites associated with enhancer activity contained in pig ICAM-2 promoters seem to be crucial for achieving strong endothelial-specific transgene expression	Godwin et al.[14]
Genetic engineering for molecules that control coagulation activation					
Gene transfer	hCD39	In vitro islets isolated from hCD39 transgenic mice	Human blood	Expression of transgenic hCD39 on murine islets inhibits the in vitro clotting of human blood	Dwyer et al.[15]
Genetic engineering for molecules that modify the cellular xenogenic immune response					
Gene transfer, lentiviral vector	GalT	(1) Autologous GalT-KO mice BM cells transfected with GalT (2) Rabbit RBC	(1) Immunization of GalT-reconstituted mice (by BMT) with rabbit RBC (2) GalT-BMT mice transplanted with αGal mice hearts	Transduction with lentiviral vectors results in chimerism at sufficient levels to inhibit anti-αGal Ab production and to induce long-term tolerance under nonmyeloablative conditions	Mitsuhashi et al.[16]
Gene transfer, lentiviral vector	GalT	(1) Autologous Rhesus BM cells transfected with GalT (2) Pig fibroblasts	Immunization of αGalT+BM Rhesus macaques with porcine cells	Gene therapy achieves low-level, long-term αGal chimerism sufficient to inhibit production of anti-αGal Ab after immunization with porcine cells in rhesus macaques	Fischer-Lougheed et al.[17]
Gene transfer, lentiviral vector	CTLA4-Ig	WT vs CTLA4-Ig-transduced PIEC	Splenocytes from mice primed with PIEC	Production of CTLA4-Ig inhibits primed indirect T-cell proliferation and cytokine responses in vitro	Mulley et al.[18]
Cloning, gene transfer	PD-L1	Porcine PD-L1-transfected CHO	Human CD4+ T cells	Porcine PD-L1 inhibited human activated CD4+ T-cell proliferation inducing activated T-cell apoptosis	Jeon et al.[19]
Gene transfer	HLA-E SCT	PECs	Human NK cells	Transgenic expression of HLA-E SNCT induced a significant decrease of polyclonal human NK-mediated cytotoxicity against pEC but did not affect NK cell adhesion to pEC	Lilienfeld et al.[20]
Control of PERV expression in pig cells/organs by RNA interference					
Gene transfer, lentiviral vector	PERV	siRNA Pol2 transgenic porcine cells		Inhibition of PERV expression in transduced pig fibroblasts remaining stable over months	Dieckhoff et al.[21]
Gene transfer, lentiviral vector, SCNT	PERV	shRNA Pol2 transgenic pigs		Pol2shRNA was detected in all organs of transgenic piglets, and PERV expression was inhibited by up to 94% in all organs of the transgenic piglets in vivo.	Dieckhoff et al.[22]

Abbreviations: αGal, Galα1,3Galβ1,4GlcNAc-R epitope; Ab, antibodies; BM, bone marrow; BMT, bone marrow transplantation; CDC, complement-dependent cytotoxicity; EndoGalC, endo-β-galactosidase C; GalT, α1,3-galactosyltransferase; GalT-KO, α1,3-galactosyltransferase knockout; GnT-I, N-acetylglucosaminyltransferase I; HAR, hyperacute rejection; hCRP, human complement regulatory proteins; hDAF, decay-accelerating factor or hCD55; HLA-E, human leukocyte antigen-E; hMCP, human membrane cofactor protein or hCD46; HT, α1,2-fucosyltransferase; iGb3S, isoglobotriaosylceramide-3 synthase; PBMC, peripheral blood mononuclear cell; PD-L1, programmed death ligand-1; pECs, porcine endothelial cells; PERV, porcine endogenous retrovirus; PIEC, porcine iliac endothelial cells; RBC, red blood cell; SCNT, somatic cell nuclear transfer; sCR1, soluble complement receptor-1; SCT, single-chain trimer; SMGT, sperm-mediated gene transfer; XNAs, natural xenoantibodies.

IMMUNOLOGICAL AND BIOLOGICAL BARRIERS IN XENOTRANSPLANTATION

The biggest hurdle in xenotransplantation is humoral rejection. This rejection exists in two forms. The first, called hyperacute rejection (HAR), is a very rapid phenomenon occurring within minutes to hours. The second occurs within weeks to months and is referred to as delayed xenograft rejection (DXR). Although both rejections display specific characteristics, both result from the binding of natural xenoantibodies (XNAs) to Galα1,3Galβ1,4GlcNAc-R epitopes (αGal) present on the donor endothelium (reviewed in Yang and Zhong[23]). Such binding results in uncontrolled complement activation, leading to complement-dependent cytotoxicity (CDC), endothelial damage, tissue edema, hemorrhage, fibrin deposition and microvascular thrombosis (Figure 1). αGal is expressed by all mammals except humans and Old World NHP. αGal is synthesized by α1,3-galactosyltransferase (GalT), which is not functional in humans and Old World NHP. The latter have therefore developed anti-αGal XNA.

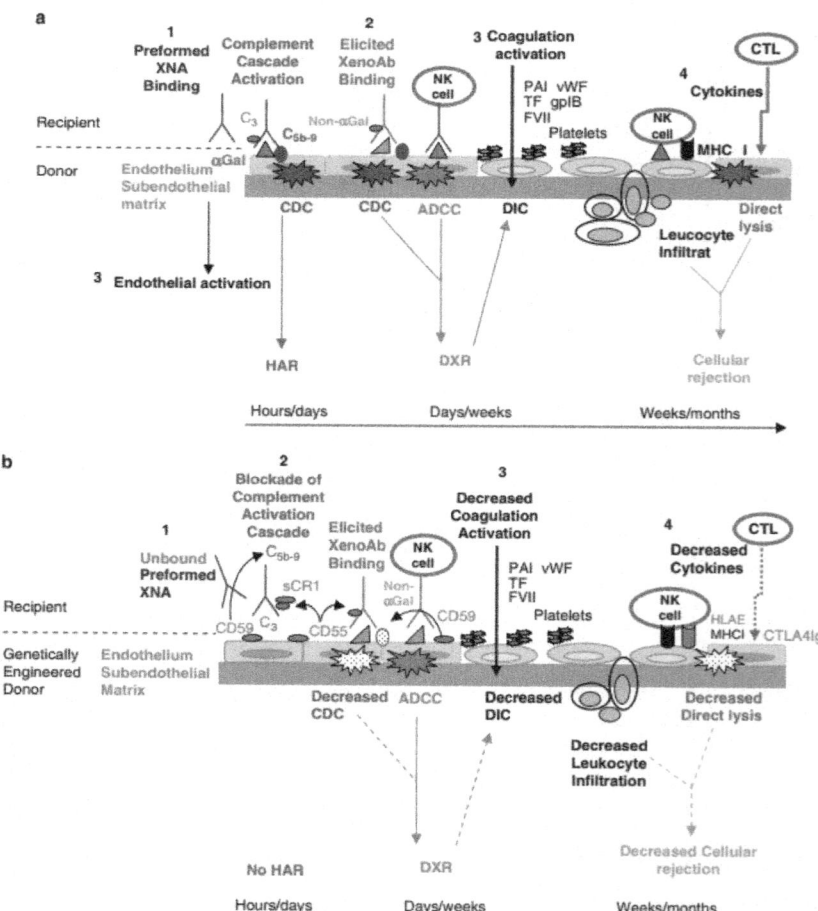

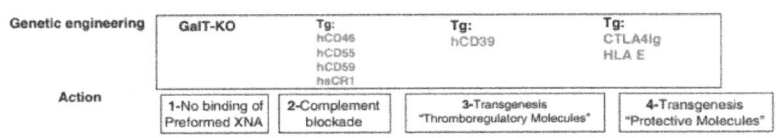

Genetic engineering	GalT-KO	Tg: hCD46 hCD55 hCD59 hsCR1	Tg: hCD39	Tg: CTLA4Ig HLA E
Action	1-No binding of Preformed XNA	2-Complement blockade	3-Transgenesis "Thromboregulatory Molecules"	4-Transgenesis "Protective Molecules"

Figure 1: (a) Mechanisms of vascularized xenograft rejection: (1) preformed natural xenoantibodies (XNAs) bind to αGal leading to the formation of the membrane attack complex (C_{5-9}), complement-dependent cytotoxicity (CDC) and then hyperacute rejection (HAR); (2) elicited XNAs bind to αGal and non-αGal epitopes leading to both CDC and antibody-dependent cell-mediated cytotoxicity (ADCC) and finally to delayed xenograft rejection (DXR); (3) increase in P-selectin, endothelial retraction, subendothelial matrix exposure: plasminogen activator inhibitor (PAI), factor VII (FVII), vonWillebrant factor (vWF), glycoprotein IB (gpIB), tissue factor (TF), platelet coagulation leading to coagulation and finally to DXR, possibly accompanied by disseminated intravascular coagulation (DIC); (4) cellular immune response involving cytotoxic T lymphocytes (CTLs), cytokine release, natural killer (NK) cells and leukocyte infiltration leading to direct cell lysis. **(b)** Genetic engineering strategies to prevent vascularized xenograft rejection: (1) knockout of the GalT gene leading to loss of αGal expression: absence of preformed XNA binding with αGal enabling the prevention of HAR, however, elicited anti-non-αGal antibody (Ab) persist or emerge inducing both CDC and ADCC leading to DXR; (2) complement blockade by human complement regulatory proteins (hCRPs): hCD46, hCD55, hCD59, hsCR1; absence of membrane attack complex formation and decrease of XNA-induced CDC; (3) expression of thromboregulatory molecules such as hCD39 leading to a decrease in thrombosis and DIC; (4) expression of protective molecules such as CTLA4Ig and HLA-E by graft endothelium, thereby reducing the cellular response: CTLA4Ig blocks the CD28:B7 costimulation signal leading to T-cell anergy *in vitro*: no cell proliferation and a decrease in cytokine production; HLA-E binding with CD94/NKG2A receptor-bearing NK cells inhibits direct NK cytotoxicity against the endothelium and subsequently decreases cellular rejection.

Microvascular thrombosis is recognized as a critical element of humoral rejection and remains the major barrier to long-term pig-to-primate xenograft survival (survivals vary from a few weeks to 6 months for the longest one published) (reviewed in Cowan[24]). Furthermore, incompatibility between pig thrombomodulin (TM) and primate thrombin could be involved in microvascular thrombosis during xenograft rejection. Roussel *et al.*[25] recently showed *in vitro* that pig TM bound to human thrombin inhibits its procoagulant activity but is a poor cofactor for human protein C activation (APC) and thrombin-activated fibrinolysis inhibitor, which interacts with various cofactors to shut down coagulation. In addition to this early humoral involvement, cellular responses could also participate in xenogeneic rejection. Among the cells that contribute to the innate immune responses, human natural killer (NK) cells can adhere to and lyse porcine target cells both directly and indirectly with xenoreactive antibody (Ab) by Ab-dependent cell-mediated cytotoxicity (ADCC) (reviewed in Kitchens *et al.*[26]). Moreover, NK cell activity seems to be increased owing to the absence of self-MHC-mediated inhibitory signals and preserved interactions between the activator NKG2D receptor and ligands on porcine cells.[27] Macrophages also seem to play a role in the cellular response in non-vascular xenografts by developing a rapid, local innate response and stimulating the recruitment of T cells that act as direct effectors of rejection. The role of adaptive T-cell responses has been shown by the prolongation of xenograft survival after conventional immunosuppression and gene therapy protocols (reviewed in Pierson[28]).

Another major issue in xenotransplantation is the risk of virus infection, especially by porcine endogenous retrovirus (PERV). PERVs are integrated into the porcine genome and can be released as infectious particles from normal pig cells that could potentially infect human cells after transplantation. PERV-A and -B can infect human cells *in vitro* as well as immunodeficient mice transplanted with pig islets. Until now, no PERV transmission has been reported either in humans (more than 160 patients referred for direct contact with pig tissue in a 1999 retrospective study) or in NHP after xenotransplantation of porcine tissues, but no long-term studies in patients under immunosuppression have been reported (reviewed in Sprangers[29]). Thus, the risk remains and deletion or knockdown of PERV sequences could be a solution.

GENETIC ENGINEERING APPROACHES LEADING TO A REDUCTION OFAGAL EXPRESSION IN DONOR CELLS/ ORGANS AND PRODUCTION OF A1, 3-GALACTOSYLTRANSFERASE KNOCKOUT PIGS

One of the strategies to prevent HAR in xenotransplantation is the production of pigs no longer expressing the major xenoantigen αGal. This major step toward a potential clinical application was achieved in the last few years by generating pigs in which the two *GalT* genes have been invalidated. Thus, in the last few years, transplantations of hearts and kidneys from homozygous GalT knockout (GalT-KO) pigs to NHP have been performed with additional immunosuppression. Although GalT-KO organs are not hyperacutely rejected, powerful immunosuppressive regimens significantly higher than those used in clinical transplantation are needed to reach survivals over a few weeks. In all cases, HAR was prevented in control non-immunosuppressed animals, and variable prolongations of organ survival of up to 179 days for hearts and 83 days for kidneys have been reported (reviewed in [30]Cooper *et al.* and [31]Tai *et al.*). In another study in which a weaker immunosuppressive regimen was administered, kidney graft rejection occurred more rapidly between 8 and 16 days. This was due to an elicited anti-non-αGal humoral response associated with complement activation leading to a strong CDC against GalT-KO cells (reviewed in Yang and Zhong[23]).

These results underline how preventing Gal expression, albeit a major step forward, is insufficient by itself to control the humoral xenogeneic immune response and that continuation of genetic modification of pigs (see below) as well as progress in immunosuppression are necessary further steps.

One recent complementary strategy to control αGal expression was to generate mice transgenic for endo-β-galactosidase C (EndoGalC) (an enzyme able to remove the αGal residue) by DNA pronuclear microinjection. The transgenic mice produced showed a dramatic decrease (from 40% in the pancreas to 90% in the heart, kidney and lung) but not the disappearance of the αGal epitope.[1] Although the generation of *EndoGalC* transgenic animals seems to be a suitable approach in xenotransplantation, systemic EndoGalC expression in mice leads to early death, which may be resolved by using endothelial cell-specific promoters. With a similar strategy of reducing the antigenicity of pig tissues, Matsunami *et al.*[2]isolated the pig *N-acetylglucosaminyltransferase I (GnT-I)* gene coding for the enzyme that initiates the biosynthesis of hybrid- and complex-type *N*-glycans. After having constructed a point-mutated *GnT-I* gene, pigGnT-I(123) and pigGnT-I(320) constructs were introduced into normal porcine endothelial cells (pECs) to identify which *GnT-I* gene mutation was able to significantly alter the antigenicity of the N-linked sugars with respect to human sera. Expression of pigGnT-I(320) in pECs resulted in downregulation of αGal levels and a 30–40% decrease of CDC mediated by human xenoantibodies, suggesting that it is possible to remodel the carbohydrate structures within pig cells to alter their antigenicity.

A second strategy to control αGal expression is to delete or inactivate the *GalT*gene. It has been reported that rats have two functional GalT: the ortholog of GalT and the isoglobotriaosylceramide-3 synthase (iGb3S). Thus, there is a possibility that iGb3S may also be present in other species and may therefore account for the residual αGal expression in GalT-KO animals.[3] Along these lines, low levels of iGb3S mRNA have been detected both in the GalT-KO mouse and the pig thymus and lung, as well as in pECs. The results of immunization of GalT-KO mice with cells transfected with *iGb3S*, in comparison to GalT, suggest thatαGal synthesized by iGb3S is as immunogenic as GalT. The recent description of mice deficient for iGb3S[32] will enable further testing to determine whether this finding could have an important impact on humoral-mediated rejection, such as DXR of GalT-KO organs. To determine the most efficient and safe way to produce $GalT^{-/-}$ pigs, Nottle *et al.*[4] compared somatic cell nuclear transfer (SCNT) and breeding. In the breeding strategy, heterozygous $GalT^{+/-}$-KO females were successively bred twice with wild-type male pigs to obtain homozygous GalT-KO animals. The results showed that *GalT* gene inactivation did not influence survival *in utero* and the GalT-KO crossbred pigs produced normal-size litters, suggesting that αGal may not be essential for normal pig reproduction. In the SCNT strategy, primary cultures of fetal fibroblasts were established from 25-day-old homozygous GalT-KO fetuses obtained by breeding one male and one female $GalT^{+/-}$ pig. After cell fusion of nuclei from homozygous GalT-KO fetuses with enucleated pig oocytes, SCNT $GalT^{-/-}$ embryos were transferred into five synchronized recipient females. Only one became pregnant and had a litter of four viable $GalT^{-/-}$ piglets. These results show that although less efficient than breeding, SNCT is also suitable for $GalT^{-/-}$ pig production. Moreover, the two techniques appear to be complementary in that breeding minimizes any inbreeding problems, whereas SCNT allows for an easier and more rapid introduction of multiple transgenes.

Even though it was shown that HAR was not observed in the context of $GalT^{-/-}$ donors, DXR occurred in baboon recipients of GalT-KO pig kidneys. This rejection was associated with complement activation and coagulation disorders, such as thrombotic microangiopathy and disseminated intravascular coagulation (DIC). To characterize better these two disorders and to assess the magnitude of abnormalities in the clotting system after xenotransplantation, Ezzelarab *et al.*[5] investigated the coagulation profile in baboons receiving GalT-KO pig hearts, compared to healthy baboons. Analyses of the coagulation profile revealed a similarity between healthy baboons and humans. Humoral rejection was observed in heparin-treated transplanted animals between days 8 and 12, with a clinical profile showing evidence of DIC (albeit not detectable in all historical series). As DIC as a consequence of non-Gal Ab-induced activation of endothelial cells may occur even with GalT-KO organs, the control of coagulation remains an important issue.

With the aim of moving xenotransplantation toward a clinical phase, Hara *et al.*[6] investigated the *in vitro* reactivity (IgM/G binding) and cytotoxicity of sera from healthy individuals, as well as unsensitized and allosensitized patients awaiting kidney transplantation, to WT versus GalT-KO pig peripheral blood mononuclear cell (PBMC). The results showed a decrease in serum reactivity as well as in the frequency of reactive and cytotoxic sera, with a lower cytotoxicity against GalT-KO than to WT pig cells in all the three groups of patients. No correlation was found between panel-reactive antibodies directed to human leukocyte antigen (HLA) class I or class II and cytotoxicity to WT or GalT-KO peripheral blood mononuclear cell. Moreover, there were no significant differences between healthy, unsensitized and allosensitized sera cytotoxicity to WT peripheral blood mononuclear cell and GalT-KO peripheral blood mononuclear cell, respectively. Thus, it appears that allosensitized patients would not be at greater risk of rejecting a pig organ than unsensitized patients. Despite dramatic progress in invalidating αGal expression, other issues including antibody responses and complement activation[6] need to be studied further by different genetic manipulations.

CONTROL OF COMPLEMENT THROUGH THE USE OF TRANSGENIC PIGS FOR MULTIPLE COMPLEMENT REGULATORY PROTEINS

One strategy to reduce humoral xenogeneic rejection (HAR and DXR) in heart or kidney pig-to-primate transplantation is the inhibition of complement activation. Over the past 10 years, several pig lines have been generated that are transgenic for human complement regulatory proteins (CRPs), such as CD35 (complement receptor type 1 (CR1)), CD46 (membrane cofactor protein (MCP)), CD55 (decay-accelerating factor (DAF)) and CD59. Although CR1, MCP and CD55 inhibit complement activation at the C3/C5 level, CD59 inhibits formation of the membrane attack complex C_{5b-9}. Despite the existence of GalT-KO animals, some authors have further investigated the humoral response against αGal in a CRP transgenic context.

Thus, Zahorsky-Reeves et al.[7] identified IgV_H genes encoding XNA in cynomolgus monkey recipients of pig hearts transgenic for hDAF under a treatment of GAS914 (soluble αGal-conjugate that depletes anti-αGal Ab) and various immunosuppressive agents. The authors showed that hDAF transgenic organs transplanted into monkeys induced the same increase in anti-αGal and -non-Gal IgM/G Ab. The genetic basis for this humoral response in the immunosuppressed monkeys that rejected their graft indicated that the XNAs were encoded by a restricted family of V_H3 genes, with the potential of being specifically targeted using directed anti-B-cell therapeutics.

Deppenmeier et al.[8] analyzed pathomorphological lesions in organs and tissues from hCD59 transgenic pigs compared to non-transgenic pigs. Transgenic pigs were generated by microinjection of foreign DNA into zygote pronuclei. Variable transgene expression levels among both animals and tissues/organs were observed despite the use of the ubiquitously active cytomegalovirus promoter. The authors did not detect any specific pathological phenotype in pigs that could be associated with expression of the hCD59 transgene construct, leading them to conclude that this technology is suitable for biomedical applications. In pulmonary xenotransplantation, complement inhibition by CRPs seems to be insufficient. Pöling *et al.*[9] used an *ex vivo* perfusion model with fresh human blood to investigate the potential of *hCD55* or *hCD59* transgenic lungs to prevent pulmonary dysfunction and HAR. The results showed a partial protection from pulmonary dysfunction with *hCD55* transgenic pig lungs, even though the typical histopathological features of HAR were still detectable. In the same model, no significant improvement was observed in lungs expressing hCD59 compared to non-transgenic lungs.

Wu *et al.*[10] studied the role of coagulation cascade activation in the pathogenesis of HAR and early graft failure in baboons transplanted with pig hearts transgenic for hDAF or hMCP. They concluded that dysregulated coagulation correlated closely with and probably caused primary failure of pig hearts transgenic for hCRP. Liu *et al.*[11] examined *in vitro* the relationship between the level of hDAF expression and its inhibitory effect on human serum cytotoxicity. A pCAGGS (plasmid) expression vector containing the cytomegalovirus enhancer, the chicken β-actin promoter and the *hDAF* gene was transfected into pig fibroblasts. The results suggested that a much higher expression of hDAF might be beneficial in protecting pig organs from the humoral xenogeneic response.

Smolenski *et al.*[12] evaluated the ability of hearts from *hDAF* transgenic pigs generated by sperm-mediated gene transfer (SMTG=minigene introduction into sperm cells followed by artificial insemination) to protect from structural damage, metabolic changes and mechanical dysfunction during *ex vivo* perfusion with fresh human blood. In this context, the hearts were relatively stable metabolically and protected from HAR. Therefore, SMGT also seems to be suitable for the generation of genetically engineered donors for preclinical xenotransplantation.

Manzi *et al.*[13] showed *in vitro* that pECs transfected with a truncated form of human soluble CR1 (sCR1) were protected from CDC induced by human AB serum, suggesting that generating pigs expressing sCR1 could constitute an additional strategy to prevent HAR. The use of transgenic donors, specifically pigs, in xenotransplantation requires targeting of transgenes to produce transgenic donors with a strong and specific endothelial expression. With this aim in mind, Godwin *et al.*[14] characterized the pig *ICAM-2* promoter by sequencing. Although their results showed many similarities between human and pig *ICAM-2* promoters, the P_8 intronic site detected, which is associated with enhancer activity in the pig gene, may be crucial to achieve strong endothelial-specific transgene expression.

GENETIC ENGINEERING FOR MOLECULES THAT CONTROL COAGULATION ACTIVATION

In the same way as human islets exposed to human blood, exposure of porcine islets to primate blood triggers an instant blood-mediated inflammation reaction (IBMIR) characterized by platelet consumption and activation of the coagulation and complement systems. The thromboregulatory molecule CD39 is an ectonucleotidase that degrades the platelet agonist ATP. Dwyer *et al.*[15] isolated pancreatic islets from hCD39 transgenic mice, incubated them with human blood and analyzed their impact on IBMIR in comparison to WT islets. The results showed that expression of hCD39 on murine islets significantly delayed clotting in human blood.

GENETIC ENGINEERING FOR MOLECULES THAT MODIFY THE CELLULAR XENOGENEIC IMMUNE RESPONSE

Several studies have been conducted to analyze whether hematopoietic chimerism, achieved using recipient bone marrow (BM) expressing αGal through transduction with lentiviral vectors, was able to inhibit anti-αGal XNA responses and induce tolerance to xenografts in mice[16] or to inhibit anti-αGal XNA responses in rhesus monkeys.[17] Mitsuhashi *et al.*[16] used lentiviral transfer of the *GalT* gene to autologous BM from GalT-KO mice and then transplanted the BM cells into submyeloablative irradiated GalT-KO mice. Following BM transplantation, the mice were immunized with rabbit red blood cells and subsequently grafted with αGal+ mice hearts. Anti-αGal Abs were not produced in GalT-reconstituted mice, in contrast to anti-non-Gal xenoAb. Fischer-Lougheed *et al.*[17] reproduced these results in rhesus macaques: several months post-BM transplantation, monkeys were immunized with porcine cells, and the anti-αGal IgM induction was considerably attenuated in αGal+BM recipients compared to immunized controls (no transplantation was performed at this stage). In these studies, the authors reported that this type of gene therapy approach was able to inhibit the production of anti-αGal Ab both in mice and in rhesus macaques, suggesting that B-cell tolerance achieved in mice could also be achieved in NHP.

To test the efficacy of local production of CTLA4-Ig in blocking the CD28:B7 costimulation pathway and more generally inhibiting xenogeneic T-cell responses, pECs were transduced with a lentiviral vector encoding CTLA4-Ig.[18] Murine splenocytes primed with pECs were used as responder cells in *in vitro* proliferation and cytokine assays against cells expressing or lacking CTLA4-Ig compared to WT pECs. The results showed a significant inhibition of splenocyte proliferation (92%) and a marked reduction in interferon-γ (82%), interleukin (IL)-4 (89%) and IL-10 (61%), suggesting the efficacy of such a strategy, even in the context of xenotransplantation.

Programmed death ligand-1 (PD-L1) is a B7 family molecule expressed by T cells, B cells, dendritic cells and macrophages. Programmed death-1 (PD-1), the physiological receptor of PD-L1, is expressed by activated T cells. Targeting of PD-1 signaling has been shown to promote allograft survival, and when combined with costimulatory blockade, it induced long-term allogenic islet graft survival. To evaluate the possible role of PD-1/PD-L1 interaction in pig-to-primate xenotransplantation, Jeon et al.[19] cloned porcine *PD-L1* and characterized its function toward human CD4 T-cell proliferation stimulated by porcine *PD-L1*-transfected cells. *In vitro* functional analysis showed that porcine PD-L1 inhibited human xenogeneic CD4 T-cell proliferation, inducing T-cell apoptosis. The PD-1/PD-L1 pathway may therefore play a regulatory role in the host cellular immune response and improve graft survival in xenotransplantation.

It is hypothesized that pEC susceptibility to human NK-mediated lysis could be due to the inability of porcine MHC class I to signal through human NK cell inhibitory receptors. In this context, the non-classical HLA-E molecule is a potent activator of the inhibitory NK cell receptor CD94/NKG2A. To investigate whether pig organs expressing HLA-E could be protected in a xenogeneic context, Lilienfeld et al.[20] used the HLA-E single-chain trimer originating from linked sequences encoding for an HLA-E-binding peptide, the HLA-E α-chain and the human β2 microglobulin, to transfect porcine cells. Their results showed a significant decrease in polyclonal human NK-mediated cytotoxicity against pEC by the transgenic expression of HLA-E single-chain trimer. Furthermore, this HLA-E single-chain trimer-mediated protection was specifically reversed by blocking CD94/NKG2A. However, HLA-E expression affected neither the adhesion of human NK cells to pEC nor the heteroconjugate formation between human NK and porcine cells. Lilienfeld et al.[20] suggested that although insufficient to totally inhibit xenogeneic NK cell reactivity, transgenic HLA-E expression on pig organs might contribute to the success of clinical xenotransplantation in combination with other protective strategies.

CONTROL OF PERV EXPRESSION IN PIG CELLS/OR-GANS BY RNA INTERFERENCE

To attenuate the risk of PERV transmission in xenotransplantation, Dieckhoff et al.[21, 22] assessed the efficiency of RNA interference approach to inhibit expression of PERV in pigs. In an initial step, fetal pig fibroblasts were transduced with two different lentiviral vectors, RRL-PGK-GFP-*pol2* and pLVTHM-*pol2*, which expressed short hairpin (sh)RNA corresponding to the viral *pol2*sequence. They showed that transduction by both vectors induced an inhibition of PERV expression in fetal pig fibroblasts (~80%) that remained stable over months.[21] Subsequently, pigs transgenic for Pol2-shRNA were generated by SCNT, with the live-born piglets showing no malformations and a normal weight.[22] The transgene was present in all six piglets, and expression of pol2-shRNA was detected in all organs. PERV expression was significantly inhibited by up to 94% in all organs of two tested transgenic piglets, demonstrating the efficiency of the RNA interference approach *in vivo*. This strategy could clearly lead to safer porcine xenotransplants with regard to the PERV risk.

PROSPECTS

The application of gene transfer and transgenic technologies to organ transplantation has already led to important breakthroughs in transplantation immunology and may have clinical applications in the future, both in the allotransplantation and in the xenotransplantation settings. Concerning αGal expression, even though *GalT*$^{-/-}$ pig organs were shown to be not susceptible to HAR in NHP models, these organs are still rejected at later time points by antibodies and complement activation. The targets of these antibodies are potentially other forms of Gal epitopes and a myriad of xenoantigens that generate induced antibodies. The deletion of the *iGb3S* gene[3, 33] or the expression of EndoGalC[1] in the *GalT*$^{-/-}$ background could result in further reduction in Gal epitope expression.

As far as elicited anti-non-Gal Abs directed against non-Gal xenoantigens are concerned, the next major step is to generate genetically engineered pigs combining coexpression of different human genes (anticoagulant, CRP and/or molecules able to provide local immunosuppression) in the GalT-KO pig background.[4, 6, 12, 30, 33, 34] For example, during the last International Xenotransplantation Association Meeting, there were reports on the production of GalT-KO pigs coexpressing hCD39[35] and triple transgenic pigs expressing hCD55, hCD59 and hTM,[36] which should become very useful organ/cell donors for preclinical xenotransplantation studies. For xenotransplantation of pancreatic islets and neurons, it will be of interest to generate pigs genetically modified to express transgenes that control cellular xenogeneic immune responses, such as those expressing CTLA4-Ig by neurons. The final goal will be to control PERV expression in genetically engineered pigs to reduce the risk of host infection in the case of pig cell/organ xenotransplantations.

Regarding clinical studies, the international xenotransplantation community has so far put a non-official moratorium until the safety issue has been better clarified. However, future clinical trials using pig islets are seriously being considered and could probably be proposed soon.

Ethical considerations are of major importance in this field for many reasons. One of them is with regard to the rights of humans to use animals for medical purposes. Because of the genetic proximity of NHP with humans, the general consensus has been to ban their use in xenotransplantation (beyond the issue of viral risk). Besides, other issues such as informing the public, informed consents and so on, another ethical issue is related to the purpose of this article: Is it right for humans to modify the genetic background of another species?

CONCLUSIONS

As this review shows, most of the improvements achieved in the xenotransplantation field have been generated by genetic modifications of the donor, either by invalidating and/or adding genes related to antigen expression, humoral and cellular responses, complement activation, the coagulation cascade or even trans-species infectious risk. The most consensual and promising next step is to continue in this way by coexpressing in GalT-KO pigs multiple genes with different immunological impacts, which could improve graft survival in a safe way.

REFERENCES

1.   Watanabe S, Misawa M, Matsuzaki T, Sakurai T, Muramatsu T, Yokomine TA *et al*. Production and characterization of transgenic mice systemically expressing endo-beta-galactosidase C. *Glycobiology* 2008; **18**: 9–19.
2.   Matsunami K, Miyagawa S, Nakagawa K, Otsuka H, Shirakura R. Molecular cloning of pigGnT-I and I.2: an application to xenotransplantation. *Biochem Biophys Res Commun* 2006; **343**: 677–683.
3.   Milland J, Christiansen D, Lazarus BD, Taylor SG, Xing PX, Sandrin MS. The molecular basis for galalpha(1,3)gal expression in animals with a deletion of the alpha1,3galactosyltransferase gene. *J Immunol* 2006; **176**: 2448–2454.
4.   Nottle MB, Beebe LF, Harrison SJ, McIlfatrick SM, Ashman RJ, O'Connell PJ *et al*. Production of homozygous alpha-1,3-galactosyltransferase knockout pigs by breeding and somatic cell nuclear transfer. *Xenotransplantation*2007; **14**: 339–344.
5.   Ezzelarab M, Cortese-Hassett A, Cooper DK, Yazer MH. Extended coagulation profiles of healthy baboons and of baboons rejecting GT-KO

pig heart grafts. *Xenotransplantation* 2006; **13**: 522–528.

6. Hara H, Ezzelarab M, Rood PP, Lin YJ, Busch J, Ibrahim Z *et al.* Allosensitized humans are at no greater risk of humoral rejection of GT-KO pig organs than other humans. *Xenotransplantation* 2006; **13**: 357–365.

7. Zahorsky-Reeves JL, Kearns-Jonker MK, Lam TT, Jackson JR, Morris RE, Starnes VA *et al.* The xenoantibody response and immunoglobulin gene expression profile of cynomolgus monkeys transplanted with hDAF-transgenic porcine hearts. *Xenotransplantation* 2007; **14**: 135–144.

8. Deppenmeier S, Bock O, Mengel M, Niemann H, Kues W, Lemme E *et al.* Health status of transgenic pigs expressing the human complement regulatory protein CD59. *Xenotransplantation* 2006; **13**: 345–356.

9. Poling J, Oezkur M, Kogge K, Mengel M, Niemann H, Winkler M *et al.* Hyperacute rejection in *ex vivo*-perfused porcine lungs transgenic for human complement regulatory proteins. *Transpl Int* 2006; **19**: 225–232.

10. Wu G, Pfeiffer S, Schroder C, Zhang T, Nguyen BN, Kelishadi S *et al.* Coagulation cascade activation triggers early failure of pig hearts expressing human complement regulatory genes. *Xenotransplantation* 2007; **14**: 34–47.

11. Liu D, Kobayashi T, Onishi A, Furusawa T, Iwamoto M, Suzuki S *et al.* Relation between human decay-accelerating factor (hDAF) expression in pig cells and inhibition of human serum anti-pig cytotoxicity: value of highly expressed hDAF for xenotransplantation. *Xenotransplantation* 2007; **14**: 67–73.

12. Smolenski RT, Forni M, Maccherini M, Bacci ML, Slominska EM, Wang H *et al.* Reduction of hyperacute rejection and protection of metabolism and function in hearts of human decay accelerating factor (hDAF)-expressing pigs. *Cardiovasc Res* 2007; **73**: 143–152.

13. Manzi L, Montano R, Abad MJ, Arsenak M, Romano E, Taylor P. Expression of human soluble complement receptor 1 by a pig endothelial cell line inhibits lysis by human serum. *Xenotransplantation* 2006; **13**: 75–79.

14. Godwin JW, Fisicaro N, d'Apice AJ, Cowan PJ. Towards endothelial cell-specific transgene expression in pigs: characterization of the pig ICAM-2 promoter. *Xenotransplantation* 2006; **13**: 514–521.

15. Dwyer KM, Mysore TB, Crikis S, Robson SC, Nandurkar H, Cowan PJ *et al.* The transgenic expression of human CD39 on murine islets inhibits clotting of human blood. *Transplantation* 2006; **82**: 428–432.

16. Mitsuhashi N, Fischer-Lougheed J, Shulkin I, Kleihauer A, Kohn DB, Weinberg KI *et al*. Tolerance induction by lentiviral gene therapy with a nonmyeloablative regimen. *Blood* 2006; **107**: 2286–2293.

17. Fischer-Lougheed JY, Tarantal AF, Shulkin I, Mitsuhashi N, Kohn DB, Lee CC*et al*. Gene therapy to inhibit xenoantibody production using lentiviral vectors in non-human primates. *Gene Therapy* 2007; **14**: 49–57.

18. Mulley WR, Wee JL, Christiansen D, Milland J, Ierino FL, Sandrin MS. Lentiviral expression of CTLA4Ig inhibits primed xenogeneic lymphocyte proliferation and cytokine responses. *Xenotransplantation* 2006; **13**: 248–252.

19. Jeon DH, Oh K, Oh BC, Nam DH, Kim CH, Park HB *et al*. Porcine PD-L1: cloning, characterization, and implications during xenotransplantation. *Xenotransplantation* 2007; **14**: 236–242.

20. Lilienfeld BG, Crew MD, Forte P, Baumann BC, Seebach JD. Transgenic expression of HLA-E single chain trimer protects porcine endothelial cells against human natural killer cell-mediated cytotoxicity. *Xenotransplantation*2007; **14**: 126–134.

21. Dieckhoff B, Karlas A, Hofmann A, Kues WA, Petersen B, Pfeifer A *et al*. Inhibition of porcine endogenous retroviruses (PERVs) in primary porcine cells by RNA interference using lentiviral vectors. *Arch Virol* 2007; **152**: 629–634.

22. Dieckhoff B, Petersen B, Kues WA, Kurth R, Niemann H, Denner J. Knockdown of porcine endogenous retrovirus (PERV) expression by PERV-specific shRNA in transgenic pigs. *Xenotransplantation* 2008; **15**: 36–45.

23. Yang H, Zhong R. Satellite Symposium on Xenotransplantation-D. Session 'Renal Transplantation'-(1) The role of anti-non-Gal antibodies in xenograft rejection in a pig-to-baboon kidney transplantation model. *Xenotransplantation* 2007; **14**: 184–186. | Article |

24. Cowan PJ. Coagulation and the xenograft endothelium. *Xenotransplantation*2007; **14**: 7–12.

25. Roussel JC, Moran CJ, Salvaris EJ, Nandurkar HH, d'Apice AJ, Cowan PJ. Pig thrombomodulin binds human thrombin but is a poor cofactor for activation of human protein C and TAFI. *Am J Transplant* 2008; **8**: 1101–1112.

26. Kitchens WH, Uehara S, Chase CM, Colvin RB, Russell PS, Madsen JC. The changing role of natural killer cells in solid organ rejection and tolerance.*Transplantation* 2006; **81**: 811–817.

27. Lilienfeld BG, Garcia-Borges C, Crew MD, Seebach JD. Porcine UL16-binding protein 1 expressed on the surface of endothelial cells triggers human NK cytotoxicity through NKG2D. *J Immunol* 2006; **177**: 2146–2152.

28. Pierson RN. Primate T-cell responses to porcine antigens: implications for clinical xenotransplantation. *Xenotransplantation* 2006; **13**: 14–18.

29. Sprangers B, Waer M, Billiau AD. Xenotransplantation: where are we in 2008? *Kidney Int* 2008; **74**: 14–21.

30. Cooper DK, Dorling A, Pierson III RN, Rees M, Seebach J, Yazer M *et al*. Alpha1,3-galactosyltransferase gene-knockout pigs for xenotransplantation: where do we go from here? *Transplantation* 2007; **84**: 1–7.

31. Tai HC, Ezzelarab M, Hara H, Ayares D, Cooper DK. Progress in xenotransplantation following the introduction of gene-knockout technology.*Transpl Int* 2007; **20**: 107–117.

32. Porubsky S, Speak AO, Luckow B, Cerundolo V, Platt FM, Grone HJ. Normal development and function of invariant natural killer T cells in mice with isoglobotrihexosylceramide (iGb3) deficiency. *Proc Natl Acad Sci USA* 2007;**104**: 5977–5982.

33. Sandrin MS. Gal knockout pigs: any more carbohydrates? *Transplantation*2007; **84**: 8–9.

34. d'Apice AJ, Cowan PJ. Building on the GalKO platform. *Transplantation*2007; **84**: 10–11.

35. Salvaris EJ, Harrison S, Beebe LF, Dwyer KM, Crikis S, Hawthorne WJ *et al*. Generation of a Gal knockout pig expressing human CD39. *Xenotransplantation* 2007; **14**: 417–418.

36. Petersen B, Lucas-Hahn A, Lemme E, Hermann D, Barg-Kues B, Carnwath JW *et al*. Production and characterization of pigs transgenic for human thrombomodulin. *Xenotransplantation* 2007; **14**: 371.

Chapter 9

AGROBACTERIUM: NATURE'S GENETIC ENGINEER

Eugene W. Nester

Department of Microbiology, University of Washington, Seattle, WA, USA

ABSTRACT

Agrobacterium was identified as the agent causing the plant tumor, crown gall over 100 years ago. Since then, studies have resulted in many surprising observations. Armin Braun demonstrated that *Agrobacterium* infected cells had unusual nutritional properties, and that the bacterium was necessary to start the infection but not for continued tumor development. He developed the concept of a tumor inducing principle (TIP), the factor that actually caused the disease. Thirty years later the TIP was shown to be a piece of a tumor inducing (Ti) plasmid excised by an endonuclease. In the next 20 years, most of the key features of the disease were described. The single-strand DNA (T-DNA) with the endonuclease attached is transferred through a type IV secretion system into the host cell where it is likely coated and protected from nucleases by a bacterial secreted protein to form the T-complex. A nuclear localization signal in the endonuclease guides the transferred strand (T-strand), into the nucleus where it is integrated randomly into the host chromosome. Other secreted proteins likely aid in uncoating the T-complex. The T-DNA encodes enzymes of auxin, cytokinin, and opine synthesis, the latter a food source for *Agrobacterium*. The genes associated with T-strand formation and transfer (*vir*) map to the Ti plasmid and are only expressed when the bacteria are in close association with a plant. Plant signals are recognized by a two-component regulatory system which activates *vir* genes. Chromosomal genes with pleiotropic functions also play important roles in plant transformation. The data now explain Braun's old observations and also explain why *Agrobacterium* is nature's genetic engineer. Any DNA inserted between the border sequences which define the T-DNA will be transferred and integrated into host cells. Thus, *Agrobacterium* has become the major vector in plant genetic engineering.

INTRODUCTION

Agrobacterium is a truly remarkable organism. Its study over the last 100 years has revolutionized plant molecular genetics, and has given birth to a whole new industry dedicated to the genetic modification of plants. Initially, studies were aimed solely at identifying the cause of destructive galls on ornamental plants and fruit trees. In the United States, two plant pathologists, Smith and Townsend (1907) reported that the causative agent of the disease, crown gall, was a bacterium that they named *Bacterium tumefaciens*. More than 30 years later, Armin Braun, a scientist at the Rockefeller Institute in Princeton, New Jersey demonstrated that this was a very unusual plant disease with properties never seen before. His observations raised several intriguing questions. How could a bacterium cause a disease that changed the nutritional properties of the infected cells (Braun, 1958)? And most surprisingly, how could these changes occur in the absence of the bacterium (White and Braun, 1941)? To answer these questions required technologies not available to Braun. In the 1960s, a number of laboratories skilled in the techniques of bacterial genetics and nucleic acid chemistry began to study the system. In a relatively short time, several key discoveries were made. An unusually large plasmid was discovered and its association with gall formation demonstrated (Zaenen et al., 1974). This was followed by the discovery that a piece of the plasmid was transferred and randomly integrated into the chromosome of the plant cell (Chilton et al., 1977; Lemmers et al., 1980; Thomashow et al., 1980; Zambryski et al., 1980). Over the next 10 years, studies from laboratories around the world answered these major questions. What do the genes transferred to the plant cell encode? What signals are exchanged between plants and bacteria? And why has *Agrobacterium* developed the complex machinery required to form tumors on plants? The answers to many of these questions have resulted in several paradigms of general biological importance which relate not only to bacterial–plant interactions but also to bacterial-animal interactions. An understanding of the basic biology of this unique system made possible the development of *Agrobacterium* as the key player in the genetic modification of plants. However, this bacterium has capabilities that extend beyond plant cell transformation. In the laboratory, *Agrobacterium* can transfer its T-DNA into representative algae (Kumar et al., 2004), fungi (Bundock et al., 1995), and even human cells (Kunik et al., 2001). Thus, what *Agrobacterium* has made possible in plant cell studies should now be possible in the study of other eukaryotic cells.

The story of crown gall must acknowledge the features of *Agrobacterium* which have contributed to the rapid progress made in understanding this system. The organism grows rapidly on a simple medium, is amenable to genetic manipulations developed in *Escherichia coli* and the assays for gene transfer are inexpensive and rapid. Further, it has a relatively small genome that lent itself to sequencing and genome analysis long before sequencing became routine (Goodner et al., 2001; Wood et al., 2001). Those in the field of agrobiology know it's an organism which is a pleasure to study!

The Early Years

In 1907, two American plant pathologists, Erwin Smith and Charles Townsend reported that the agent causing the common, destructive disease of a variety of ornamental plants called crown gall was a bacterium (Smith and Townsend, 1907). The person most responsible for bridging the gap between the description of the pathogen and the modern era of crown gall research was Armin Braun. His seminal contributions over 35 years, beginning in the 1940s, set the stage for the molecular analysis beginning in the 1960s (Binns, 2005). Braun observed how unique this disease was. He demonstrated that although living bacteria were necessary to start the infection, once initiated, the tumor developed in their absence (White and Braun, 1941). Further, Braun (1958) discovered that tumor cells could be cultured in media lacking the plant hormones, auxin and cytokinin, which are necessary for growth of normal cells. He recognized that some product of *Agrobacterium*, not the bacteria themselves, was altering the properties of plant cells. He developed the concept that this product was the actual tumor inducing principle or TIP (Braun, 1947; Braun and Mandel, 1948). These observations were very important because they provided valuable clues as to the mechanism by which *Agrobacterium* transforms plant cells. It's interesting that Braun, 1947 suggested that the TIP might be DNA. However, the proof of his prescient suggestion did not come until 30 years later and Braun himself was reluctant to accept this conclusion until he saw convincing data (Binns, 2005).

Important observations continued to be made in a number of laboratories in many countries over the next 20 years. However, it wasn't until bacterial and molecular genetics became part of the routine thinking of scientists that the modern era of crown gall research became possible. Further, new techniques associated with molecular genetics and nucleic acids had to be developed in order to carry out the experiments which identified and characterized the TIP.

A recent historical, well researched and informative review on *Agrobacterium* covers some of the same ground as the present review

(Kado, 2014). Dr. Kado has been a pioneer in the study of *Agrobacterium* and has made numerous important contributions over his many years of research.

The Molecular Era Begins

Three papers published in the years 1969–1971 strongly suggested that the TIP was most likely DNA that was transferred from *Agrobacterium* into plant cells. In 1969, the Australian Allen Kerr reported that virulence could be transferred between bacteria when a virulent strain of *Agrobacterium* was inoculated onto tomato plants and several weeks later an avirulent strain was inoculated onto the developing tumor. Using appropriate genetic markers, he showed that virulence was transferred from the virulent to the avirulent strain (Kerr, 1969). However, the mechanism of transfer was not indicated although Kerr commented that DNA transformation of *Agrobacterium* and *Rhizobium* had been reported. In the following year, the laboratory of George Morel in France demonstrated that tumors elicited by different strains of *Agrobacterium* contained low molecular weight compounds not found in normal plant tissues (Petit et al., 1970). The two compounds this laboratory identified were the secondary amines, octopine and nopaline given the general name, opine. Octopine is a condensation product of arginine and pyruvic acid and nopaline a condensation product of arginine and alpha-ketoglutaric acid. Years earlier, another opine, lysopine, a condensation product of lysine and pyruvic acid, had been identified in tumors (Lioret, 1957). The authors suggested that tumors acquired new genetic information from the bacterium. Interestingly, in all three cases, the strain of *Agrobacterium* that induced the synthesis of a particular opine had the ability to degrade the same opine. Unfortunately, several other independent studies, which are likely incorrect, reported that these opines were present in normal plant tissue (Seitz and Hochster, 1964; Johnson et al., 1974). Nevertheless, they clouded the significance of Morel's observations. The last paper in this set suggesting that DNA was the TIP came from the laboratory of Robert Hamilton at Pennsylvania State University in the USA (Hamilton and Fall, 1971). He observed that cells of strain C58 of *Agrobacterium* when incubated at the elevated temperature of 36°C gave rise to a high proportion of cells that were stably avirulent. Hamilton suggested that heating resulted in the loss of a virulence factor. Since it was known that heating at an elevated temperature could cure strains of their plasmids and plasmids could be transferred, both Kerr and Hamilton looked for evidence of plasmids to explain their respective data but their results were equivocal (Hamilton, 2005; Kerr, 2005).

Being aware of these papers, Bruce Watson, a graduate student of Milton Gordon in Seattle examined strain C58 for plasmids. Despite his concerted efforts, he was unable to demonstrate that this strain contained any (Watson and Nester, 2005).

The Breakthrough

The study which initiated the molecular analysis of crown gall tumor formation came in a report of Ivo Zaenen, a graduate student of Jeff Schell in Belgium. Using alkaline sucrose gradients of bacterial cell lysates, he observed megaplasmids ranging in size from 96 to 156 MDa in 11 virulent strains and none in eight avirulent strains of *Agrobacterium* (Zaenen et al., 1974). Plasmids of such large size had not been reported before. Very quickly, the Belgian group led by Schell and Marc Van Montagu and the Seattle group of Mary-Dell Chilton, Milton Gordon, and Eugene Nester demonstrated that the virulent strain contained a megaplasmid that was not present in the avirulent, heat treated C58 strain (Van Larabeke et al., 1975; Watson et al., 1975). Further, transferring this megaplasmid to the avirulent strain converted it to virulence. The Belgian group termed this plasmid the tumor inducing or Ti plasmid.

Genome Organization

Before continuing the crown gall saga, it is necessary to point out another important feature of many strains of *Agrobacterium*, its unusual genome organization. The genome of *Agrobacterium tumefaciens* C58 consists of four replicons (Allardet-Servent et al., 1993). These are a circular chromosome, a linear chromosome and two megaplasmids of similar size, pTiC58 and pAtC58. Linear chromosomes are very rare in prokaryotes and even in *Agrobacterium* only biovar I strains (classified on the basis of phenotypic characteristics) contain a linear chromosome.

The C58 genome was sequenced independently by two groups in the United States by a collaborative effort of two academic institutions collaborating with two companies interested in the genetic engineering of plants. Investigators at Hiram College in Ohio collaborated with a sequencing group at the Monsanto Company (Goodner et al., 2001) and the Crown Gall group and the genome sequencing team at the University of Washington in Seattle collaborated with scientists at E.I. DuPont de Nemours Company (Wood et al., 2001). Both sequences, which agreed with each other surprisingly well, were published simultaneously. These academic-industry collaborations were necessary because the genome was sequenced at a time when sequencing and annotating of even one relatively small bacterial genome required a major commitment of time, as well as significant intellectual and financial resources. The sequence revealed an extensive similarity of the circular chromosomes of *A. tumefaciens* and the plant symbiont, *Sinorhizobium meliloti*. This suggests that both organisms evolved from a common ancestor from which they recently diverged.

Characterization of DNA in Tumors

How does the Ti plasmid cause tumors? Because of the observations which strongly suggested that the TIP could be transferred between bacteria, the Seattle group searched for plasmid genes in tumor tissue. The technique they used is relatively simple in theory but very tedious in practice. Heat denatured ^{32}P labeled DNA of the plasmid (probe) was mixed at a low concentration with a high concentration of unlabeled single-strand DNA isolated from either tumor or normal tissue (driver DNA). If plasmid genes were present in the tumor tissue, the concentration of these sequences would be elevated in the mixture. Consequently, the rate of reassociation of the single-strand probe DNA would be accelerated when mixed with tumor DNA as compared to the reassociation of probe DNA mixed with DNA from normal tissue. What was observed was a slight but reproducible increase in the rate of reassociation of the probe when a mixture of probe with tumor DNA was compared to probe mixed with normal plant DNA. These data could be explained if only a small fraction of the Ti plasmid was present in the tumor. This possibility was pursued through experiments designed to identify the putative piece of the Ti plasmid in tumor tissue (Chilton et al., 1977).

The plasmid was labeled with ^{32}P then cleaved with a restriction enzyme. Sixteen fragments were separated by electrophoresis, then eluted from the gel and denatured. Each fragment was then mixed with normal and tumor DNA. This time the results were unequivocal. The reassociation of one doublet band mixed with tumor DNA showed a significant increase in the rate of reassociation. When this band was separated into two, one band showed no increase in reassociation whereas the other band increased the rate of reassociation even more than the doublet (Chilton et al., 1977). These data proved that the TIP of Braun was indeed DNA as others had previously suspected. This DNA was termed T-DNA for transferred DNA.

The identification of the TIP raised many interesting questions. Many laboratories studied the Ti plasmid. Some focused on the Ti plasmid from strains that induced tumors that synthesized octopine (octopine strains like A6 and B6). Others studied strains that induced nopaline synthesizing tumors, like C58 (nopaline strains). A few laboratories focused on *Agrobacterium rhizogenes*, a species of *Agrobacterium* that produces tumors, called hairy root because of a preponderance of roots at the site of inoculation (Moore et al., 1979; White and Nester, 1980). The virulence plasmid from this strain was termed the Ri plasmid for root inducing. The more common strains of *A. tumefaciens* like A6 and C58 result in unorganized tumor growth at the site of inoculation. It was quickly determined that the T-DNA in tumors is transcribed, the first evidence that bacterial DNA can be transcribed in a eukaryotic cell (Drummond et al.,

1977). Studying the patterns of transcription of the entire Ti plasmid revealed that the T-DNA was transcribed at a low rate in*Agrobacterium* whereas other regions of the Ti plasmid were transcribed to high levels (Gelvin et al., 1981). At least some of the T-DNA transcripts detected in the tumor originated and terminated within the T-DNA and not from promoters and terminators in the plant. The transcript levels of various regions of the T-DNA varied. All transcripts were polyadenylated (Gelvin et al., 1982; Willmitzer et al., 1982). Since the biological functions encoded by these genes were not yet known, the full significance of these data could not be appreciated. However, the use of promoters encoded in the T-DNA proved invaluable in the genetic engineering of plants.

Once a restriction map of an octopine Ti plasmid was generated (Chilton et al., 1978), studies on the Ti plasmid were aimed at identifying the various functions encoded on the plasmid. An early study was carried out by Holsters et al. (1980) on a nopaline plasmid and Ooms et al. (1981) on an octopine plasmid. They isolated insertion and deletion mutants using different transposons and mapped regions required for tumor formation including the T-DNA as well as another region distinct from the T-DNA. Some insertions in the T-DNA region resulted in tumors which did not synthesize an opine. They also demonstrated that regions required for tumor formation in the nopaline strain are homologous to regions required for tumor formation in an octopine strain.

Simultaneously and independently of these studies, the entire genome of an octopine strain was being mutagenized using a different transposon (Garfinkel and Nester, 1980). Mutants which mapped to the T-DNA gave rise to tumors with altered morphologies, either extensive root or shoot proliferation. These tumors still synthesized octopine but some insertions in the *T-DNA* did not. Other mutations distinct from the T-DNA region resulted in cells unable to catabolize octopine, proof that proteins required to synthesize octopine are not involved in its degradation. Interestingly, none of the single mutants in the T-DNA resulted in an avirulent strain. However, mutations in a region on the Ti plasmid distinct from the T-DNA did give rise to avirulent mutants. Further, about one-third of the mutations that affected tumorigenesis mapped to the chromosome. Thus, the genes associated with virulence are located at three different sites in the genome. Two sets map to the Ti plasmid. One set includes the T-DNA. The other includes genes we now know are required for the processing and transfer of the T-DNA, the *vir* genes. In addition, many genes in the circular chromosome, the *chv* genes, strongly affect virulence. The genes on the Ti plasmid function in various stages of plant cell transformation. Their major role is associated with plant cell transformation. However, most *chv* genes are pleiotropic. Their protein products are required for optimal plant cell transformation but they also play various roles in the physiology of the bacteria in the absence of the host plant.

Functional Analysis of the T-DNA

To understand the mechanism by which *Agrobacterium* confers transformed properties on the plant requires an understanding of what the T-DNA encodes. To this end, Garfinkel et al. (1981)generated a fine-structure map of the T-DNA through site-directed mutagenesis in an octopine strain. Insertions in one region resulted in tumors that no longer synthesized octopine. Insertions in three other regions affected tumorigenesis.

Insertions in one region resulted in tumors forming a massive amount of roots emanating from the tumor callus (*tmr* mutations). Insertions in another region resulted in tumors with shoots growing from the tumor callus (*tms* mutations). Insertions in another region resulted in unusually large tumors on certain plants (*tml* mutations). Insertions between each of these regions had no effect on tumorigenesis. Ooms et al. (1981) independently generated mutations in a similar Ti plasmid and made similar observations. In addition, they observed that supplying an auxin (naphthalene acetic acid) to developing tumors incited by a *tms*mutant or a cytokinin (kinetin) to a *tmr* mutant stimulated unorganized tumor formation on tomato plants. This suggested that the proliferation of shoots in the *tms* mutant and roots in the*tmr* mutant resulted from an imbalance of the two phytohormones.

The analysis of phytohormone levels in uninfected tobacco stem tissues, wild type tumors and tumors induced by *tmr* and *tms* mutants on tobacco stems supported this idea (Akiyoshi et al., 1983). Whereas the ratio of cytokinin (*trans*-ribosylzeatin) to auxin (indoleacetic acid) levels in wild-type tumors was 0.2, the same ratio was much lower in *tmr* tumors and much higher in *tms*mutants. A simple explanation was that the T-DNA encodes the enzymes of auxin and cytokinin synthesis, an interpretation shown to be correct when Akiyoshi et al. (1984) showed that the *tmr*gene encodes an enzyme of cytokinin synthesis, dimethylallyltransferase. Sequencing the region of the *tms* gene revealed that this region encodes two transcripts, both of which are involved in auxin synthesis (Klee et al., 1984). One locus, *tms1*, encodes a tryptophan mono-oxygenase and *tms2*encodes indole-3-acetamide hydrolase (Thomashow et al., 1984). Together, both enzymes convert tryptophan to the auxin indole-3-acetic acid (Thomashow et al., 1986).

Organization of T-DNA in Tumors

Does the T-DNA replicate as a plasmid or is it integrated into the chromosome?

If the latter, are there specific sites at which the T-DNA is integrated? Are the putative inserts stably integrated or can they jump to different locations? The studies of Thomashow et al. (1980) answered these questions. They defined the T-DNA in four tumor lines and reached the following conclusions: (1) the T-DNA is integrated into numerous sites in the plant DNA; (2) each line contains a "core" DNA which is co-linear with the Ti plasmid; (3) a given Ti plasmid does not always give rise to the same insertions; and (4) the number of insertions of "core" DNA varies in different tumor lines. We now know that the "core" DNA includes those genes responsible for the tumor phenotype.

Virulence (Vir) Region-Overview

The one region on the Ti plasmid in which mutations result in a complete loss of virulence is the *vir*region. It comprises ~30 genes of which about 20 are essential for tumor formation. They are organized into operons which together comprise a regulon under a common control mechanism (Stachel and Nester, 1986; Gelvin, 2003). To study the genetic and transcriptional organization of this region, Stachel et al. (1985a) developed a Tn3-*lacZ* transposon and generated a random series of mutations. After analyzing 124 insertions for tumor formation and beta-galactosidase in bacteria grown in the presence of plant cells, they divided the *vir* region into six complementation groups:*virA*, *virB*, *virC*, *virD*, *virE*, and *virG* (Stachel and Nester, 1986). Mutations in *virA*, *B*, *D*, and *G*resulted in a complete loss of virulence. Mutations in *virC* and *virE* led to attenuation. Another *vir*gene, *virF*, is required for robust tumor formation on some, but not on other plants (Melchers et al., 1990). Another locus, *virH* (*pin*), is not required for virulence on any plant (Stachel and Nester, 1986). Additional *vir* genes that are not required for tumor formation on several plants are *virK*,*virL*, and *virM* (Kalogeraki and Winans, 1998) and a gene involved in cytokinin synthesis (*tzs*) found only in nopaline strains (Holsters et al., 1980). To observe any beta-galactosidase activity from the *lacZ* gene, bacteria had to be co-cultivated with plant cells because the expression of this region is under the tight control of plant metabolites (Stachel et al., 1986a).

Three signal molecules, all associated with the wound site on a plant are important in *vir* gene induction. These include a number of different phenolic compounds, a variety of monosaccharides which are components of plant cell walls and act through a binding protein encoded on the bacterial chromosome (ChvE) and acidic conditions which are required at several steps in the induction process. A two-component regulatory system is critical for recognizing all three signal molecules.

Mechanism of Activation of *vir* Genes-Two-Component System

Once the *vir* regulon was defined genetically, studies were aimed at understanding how these critical genes are regulated by the three types of signal molecules. The first clue came when it was found that mutations in either *virA* or *virG* eliminated induction of all *vir* genes (Stachel and Zambryski, 1986; Winans et al., 1988). Insight into how these two genes functioned came through gene sequencing and comparing the sequences with other regulatory gene pairs. Simultaneously with these studies on *Agrobacterium*, Fred Ausubel at Harvard was sequencing the nitrogen assimilation regulatory genes, *ntrB* and *ntrC* of *Bradyrhizobium*. After comparing the sequences of these two systems, both groups concluded independently that many regulatory systems that respond to environmental stimuli share strongly conserved domains (Nixon et al., 1986; Winans et al., 1986; Ronson et al., 1987).

How *VirA* and *VirG* function in regulating the *vir* genes was helped enormously by data generated in two similar systems in other bacteria, NtrB (nitrogen metabolism) and CheA (chemotaxis; Nixon et al., 1986). Taking cues from these other systems, VirA can be designated as the sensor protein which recognizes the plant signal molecules and VirG, the response regulator, which activates all genes in the regulon. As in these other systems, it was shown that VirA is an autophosphorylase which phosphorylates a specific histidine moiety (Jin et al., 1990a) and then transfers the phosphate to a specific aspartic acid in VirG, thereby activating the molecule (Jin et al., 1990b). The activated VirG then binds to a conserved 12 base pair sequence upstream of each of the *vir* genes (Das et al., 1986; Jin et al., 1990c).

VirA

Insight into how the sensor protein VirA functions was provided by its sequence (Leroux et al., 1987). The deduced protein has two hydrophobic regions which suggests it is imbedded in a membrane. Using antibodies, the protein was shown to be anchored in the inner membrane with ~275 amino acids near the amino terminus localized in the periplasmic space and the rest of the protein located in the cytoplasm. Other members of the family of homologous proteins have a similar hydropathy profile (Charles et al., 1992). VirA exists as a preformed dimer in the cell (Pan et al., 1993).

Later genetic studies divided the VirA protein into several additional domains which function independently of one another and to which functions were assigned (Chang and Winans, 1992). The periplasmic domain is required for the sensing of monosaccharides (Cangelosi et al., 1990a). A linker domain joins the transmembrane region to the cytoplasmic region which includes the kinase and receiver domains. The linker domain is required for the sensing of phenolic compounds and acidity whereas the kinase domain contains the phosphorylatable histidine moiety. The receiver domain serves as an enhancing region of VirA and is required for *vir* gene expression (Wise et al., 2010).

VirG

The *virG* locus is transcriptionally activated by plant signal molecules acting on one of two promoters, P1 and P2 located downstream of P1 (Mantis and Winans, 1992). The P1 promoter functions with phenolic inducers and phosphate starvation and requires the VirA/G system. In contrast, the P2 promoter is activated solely by acidic conditions which serves to raise the level of VirG to the level required to achieve maximum induction of the *vir* regulon by phenolic and monosaccharide inducers. Acid induction is independent of the VirA/G system (Mantis and Winans, 1992).

Plant Signal Molecules

That it was necessary to co-cultivate *Agrobacterium* with plant cells in order to observe expression of the *vir* genes, suggested that plant cells were secreting molecules that induced the *vir* regulon (Stachel et al., 1986a). The identification and functional characterization of the three inducing molecules represents one of the first examples of our understanding, albeit incomplete, of how a bacterial cell responds to its complex, natural environment.

Phenolic Inducers

Two low molecular weight phenolic compounds, dimethoxyphenol [acetosyringone (AS) and hydroxyacetosyringone (OH-AS)], secreted by tobacco cells and at biologically relevant amounts were shown to induce the *vir* genes (Stachel et al., 1985b). This was a key discovery in crown gall research and made many other investigations possible. A whole host of naturally occurring phenolic plant metabolites were later shown to be inducers (Brencic et al., 2004). These included vanillin, coniferyl alcohol, sinapyl alcohol, syringaldehyde, and eugenol.

Although the model of phenolic signaling strongly suggests that the phenolic molecule must bind, directly or indirectly, to VirA, such binding has never been demonstrated biochemically. However, genetic evidence is consistent with this model (Lee et al., 1995). By transferring different Ti plasmids having different specificities for *vir* gene inducing phenolic compounds into isogenic chromosomal backgrounds, it was shown that the specificity of *vir* gene activation by these different phenolic compounds tracks with the *virA* locus.

Monosaccharides-Neutral and Acidic

A surprisingly wide range of monosaccharide components of plant cell walls act in concert with phenolic inducers to increase the level of induction achieved by phenolic compounds alone. Many different neutral and acidic sugars are recognized by a chromosomally encoded protein, ChvE (Ankenbauer and Nester, 1990; Cangelosi et al., 1990a; Shimoda et al., 1990; He et al., 2009; Hu et al., 2013). Once bound to a sugar, ChvE can bind to the periplasmic region of VirA (Cangelosi et al., 1990a; Shimoda et al., 1993), relieve repression and transduce this information through the cytoplasmic membrane domain to activate the kinase domain which then transfers its phosphate to VirG (Nair et al., 2011). Mutations in *chvE* resulted in the *vir* genes being poorly inducible both in the maximum level of induction achieved and also in their level of induction at low concentrations of the phenolic inducer. Also, such mutants were less virulent (Cangelosi et al., 1990a; Shimoda et al., 1990; Nair et al., 2011). ChvE mutants were also defective in chemotaxis and grew poorly on a variety of sugars which are involved in *vir* gene induction (Cangelosi et al., 1990a; He et al., 2009). Sequencing the gene which encodes the ChvE protein revealed that it is homologous to the glucose/galactose binding protein of *E. coli*, which plays a role in the uptake of sugars as well as chemotaxis. Mutants defective in ChvE have an altered host range. They remain virulent on some plants but avirulent or weakly virulent on others. These differences are likely a consequence of the level of *vir* gene induction required for plant infection (Banta et al., 1994; Nair et al., 2011). Some plants are more susceptible to infection than others and the degree of susceptibility is reflected in the level of *vir* gene products required for a successful infection (Cangelosi et al., 1990a). The *chvE*locus represents an excellent example of the pleiotropic nature of a chromosomal gene which is important in virulence but also plays a role in the physiology of the organism growing in the absence of the plant. This pleiotropic nature of ChvE will be covered in more detail shortly.

Acidic Conditions

Acidic conditions (pH 5.5) play a critical role in *vir* gene induction through a number of different mechanisms some of which are not well understood or even recognized. One important function is to raise the level of the response regulator, VirG, to the level necessary for maximum *vir* gene induction (Mantis and Winans, 1992). This occurs in the absence of the phenolic and sugar inducers. Apparently, the elevated level of VirG then becomes sufficient to induce all genes of the*vir* regulon, including *virA* and *virG*, through signal molecules at the wounded plant site. Acidic conditions are also required for the binding of sugars to ChvE (Hu et al., 2013).

To expand the repertoire of genes affected by an acidic environment, a transcriptomic analysis of cells grown at pH 5.5 and pH 7.0 was carried out (Yuan et al., 2008). Seventy eight genes were significantly induced and 74 repressed at pH 5.5. Of the genes induced, 17 were involved in the synthesis of the cell envelope. This may reflect the need of an altered cell surface to associate with the plant cell. Of special interest was another two-component regulatory system that was strongly induced, ChvG/I. This system was identified previously by screening avirulent mutants that mapped to the chromosome (Charles and Nester, 1993). It was isolated independently by another screen (Mantis and Winans, 1993). This complex regulatory system will be discussed later when chromosomal mutants associated with virulence are considered.

Another set of genes that was strongly induced were the genes involved with a Type VI secretion system (Yuan et al., 2008; Wu et al., 2012). This system mediates interaction between a wide variety of bacteria by acting as an export channel for the transfer of various kinds of toxins into neighboring cells following cell–cell contact (Russell et al., 2014). The significance of this system to the association of *Agrobacterium* and plants is not clear. However, work in *Agrobacterium* and other systems suggests that this secretion system may provide a mechanism for *Agrobacterium* to achieve a competitive advantage with other organisms in the acidic environment of the rhizosphere (Ma et al., 2014; Russell et al., 2014).

Vir Proteins

As discussed already, two of these proteins VirA and VirG are concerned with the activation of all*vir* genes. The other Vir proteins are required for the processing and transfer of the T-DNA.

VirB

Perhaps the most intriguing operon of the *vir* regulon is *virB*, in part because of its large size. The entire operon was sequenced and 11 open reading frames identified (virB1-11; Ward et al., 1988). Many encode presumed gene products with secretion signals and membrane spanning domains. The sequence analysis suggested that this operon encodes a transmembrane structure that likely mediates the passage of the T-DNA and certain Vir proteins into the plant cell. VirB is now the paradigm for the intensely studied Type IV secretion system found in a wide variety of prokaryotic cells. More is known about the structure-function relationships of the Type IV secretion apparatus in *Agrobacterium* than in any other organism (Cascales and Christie, 2004; Alvarez-Martinez and Christie, 2009).

Recently, the structure of another well-studied Type IV secretion system in *E. coli* has been elucidated in great detail (Low et al., 2014). The secretion system from the conjugative plasmid R388 was over-expressed in *E. coli*. The structure consists of a core complex which is joined to the inner membrane complex by a stalk. Many but not all of the constituent proteins were localized in the structure. VirB1 which was not included encodes a transglycosylase which cleaves beta-1,4 glycosidic bonds (Mushegian et al., 1996). VirB2 encodes the synthesis of pilin, the subunit of the T-pilus (Lai and Kado, 1998). The synthesis of T-pilin is temperature sensitive (Fullner et al., 1996;Lai and Kado, 1998). Interestingly, the intact T-pilus is not required for transfer of T-DNA but its subunit is (Kerr and Christie, 2010).

VirD

The *virD* operon consists of five open reading frames. Interestingly, the encoded proteins play quite different roles in the infection process. VirD2 is an endonuclease that nicks one of the two strands of the Ti plasmid at two sites which flank and delineate the T-DNA (Yadav et al., 1982; Yanofsky et al., 1986). These 25 base pairs occur as direct repeats and their cleavage results in the formation of a single-strand T-DNA molecule, the T-DNA (Stachel et al., 1986b; Albright et al., 1987; Veluthambi et al., 1988). The VirD2 protein remains covalently attached at the 5′ end through a phosphotyrosine bond (Ward and Barnes, 1988; Young and Nester, 1988; Pansegrau et al., 1993). This protects the 5′ end from exonucleolytic degradation (Durrenberger et al., 1989). In addition to the indispensable border sequences, efficient T-DNA transmission requires an additional sequence, termed overdrive which is located to the right of the right border (Peralta et al., 1986) and serves to enhance the production of T-strands (Veluthambi et al., 1988).

Following nicking, the VirD2 protein which also contains nuclear localization signals (NLSs) directs the transport of the T-DNA into the nucleus of the plant cell (Herrera-Estrella et al., 1990; Howard et al., 1992). In addition, this protein is in part responsible for the efficiency of transformation and the preservation of the ends of the integrated DNA (Tinland et al., 1995; Pelczar et al., 2004).

The VirD2 protein also carries the translocation signal for T-strand docking with the VirD4 coupling protein. This latter protein has ATPase activity and delivers the T-strand to the mating pair channel, the *trans*-envelope secretion system encoded by the *virB* operon. The binding of the T-DNA to VirD4 and VirB11, which also has ATPase activity, stimulates their ATPase activities. This energy in turn activates VirB10 through a structural transition which opens up the channel to the cell surface. The ATPase activities of VirD4 and VirB11 as well as the binding of the T-DNA to VirD4 are required to activate VirB10 through a structural modification (Cascales et al., 2013).

VirC

The virC operon consists of two open reading frames, *virC1* and *virC2* (Stachel and Nester, 1986). Mutations in either result in attenuated virulence on some but not all plants (Yanofsky et al., 1985). Presumably the plants on which the mutations do not effect virulence are those most susceptible to infection. This operon appears to function at several early stages in the transformation process. VirC1 binds to the overdrive sequence and enhances the site-specific nicking by the VirD endonuclease thereby resulting in increased T-strand production (Toro et al., 1988, 1989). Although the *virC* operon is not essential for endonuclease nicking, VirC1 does enhance nicking, most likely because of the interaction between VirC1 and overdrive. More recent studies demonstrated that VirC2 increases the number of copies of T-strands per cell as a result of the pair-wise interactions with VirD2, VirC1, and VirD1 which most likely exist as multimers (Atmakuri et al., 2007). In addition to its role in T-strand formation, VirC1 recruits the cytosolic T-strands to the type IV secretion channel.

VirE

The *virE* operon consists of three open reading frames, of which the most intensely studied is*virE2*. *virE2* encodes a non-specific single-strand DNA binding protein which likely covers the length of the T-DNA (Gietl et al., 1987; Christie et al., 1988; Citovsky et al., 1988; Das, 1988;Yusibov et al., 1994). It is delivered via the Type IV secretion system into the plant cell independent of T-DNA transfer where it presumably protects the T-DNA against nuclease degradation and maintains the integrity of the 3' end of the T-DNA prior to integration (Rossi et al., 1996). Unlike*virE2* and *virE1*, *virE3* encodes a host range locus and will be considered in a later section.

Although VirE2 also contains NLS, data conflict as to whether these signals play a significant role in the nuclear import of T-DNA. Early studies strongly suggested that the NLS of VirE2 were very important in nuclear import. These studies indicated that they were important both in localizing the T-DNA into the nucleus (Zupan et al., 1996) and also in virulence (Gelvin, 1998). In the latter study, a VirE2 mutant lacking NLS was avirulent but regained virulence on tobacco plants that expressed VirE2. Another report distinguished between nuclear targeting and nuclear import (Ziemienowicz et al., 2001). These authors concluded that VirD2 was necessary to target the DNA to the nucleus but VirE2 was necessary for its import. A recent study however did not show that VirE2 localized to the nuclei of yeast and tobacco cells (Sakalis et al., 2014). Using several different visualization techniques, these investigators demonstrated that VirE2 traveled from*Agrobacterium* into plant cells where it associated with microtubuli. However, the interaction with the microtubules might be merely one step on the way to the nucleus. This is only the latest study of many which did not show localization of VirE2 in the nucleus (Reviewed in Gelvin, 2012; Lacroix and Citovsky, 2013). Some of the conflicting localization data reported might be resolved by results reported by Bhattacharjee et al. (2008). They observed that VirE2 localized to the cytoplasm of*Arabidopsis* cells but that over-expression of a specific isoform of importin, IMPa-4, a nuclear transport factor which interacts with VirE2, resulted in the VirE2 now localizing in the nucleus. This suggests that the level of this specific isoform could determine whether VirE2 localizes to the cytoplasm or the nucleus. Since different plants and different parts of plants would likely have different levels of this importin, this could explain conflicting results from different studies. Further, reducing the level of IMPa-4 reduced the level of plant cell transformation, also suggesting that this specific importin, and by implication VirE2, is important in nuclear transport.

VirE1

Another gene in this operon is *virE1*. This gene encodes a chaperone which keeps the VirE2 protein from aggregating with itself inside *Agrobacterium* (Sundberg et al., 1996; Deng et al., 1999). It cements two independent domains of VirE2 into a locked form, thereby creating a soluble heterodimer (Dym et al., 2008). Further, VirE1 competes with the T-strand for binding to VirE2 inside the bacterial cell but unlike VirE2, VirE1 apparently does not enter the plant cell (Dym et al., 2008).

Non-Essential and Host Range Vir Proteins

The mutational and sequence analysis of the *vir* regulon revealed many open reading frames whose products seemed to be non-essential for tumor formation on the plants tested (Kalogeraki and Winans, 1998; Kalogeraki et al., 2000). In some cases, however, assays on additional plants showed that they were important for tumor formation. However, some loci still seem to play no role in tumor formation on any plants at least in laboratory assays.

VirH

The *virH* region consists of two genes which encode proteins which resemble P-450-type monooxygenases (Kanemoto et al., 1989). Such genes are usually associated with detoxification of a variety of compounds. A strain mutant for both genes was still tumorigenic on *Kalanchoe daigremontiana* and carrot disks (Kalogeraki and Winans, 1998). Each gene encodes a protein that can convert a strong phenolic *vir* gene inducer, ferulic acid, to a non-inducer, caffeate by demethylating a methoxyl group. Most *vir* gene inducers are toxic to *Agrobacterium* and in general their conversion to non-inducers relieves this toxicity. However, not all *vir* gene inducers are demethylated at an appreciable rate so it is unclear what role these proteins play in the plant environment (Brencic et al., 2004).

VirF

Mutants lacking this gene in an octopine strain formed normal appearing tumors on *Nicotiana tabacum* and *Kalanchoe*. However, the same mutants were highly attenuated on *Nicotiana glauca*and tomato (Regensburg-Tuink and Hooykaas, 1993). Importantly, transgenic plants expressing the *virF* gene became transformable by these same mutants, thus indicating that VirF functions in the plant cell. Vir F is an F-box protein, representatives of which are part of the SCF complex which mediates ubiquitination of proteins targeted for degradation by the proteasome. This suggests that VirF may be involved in proteolysis of proteins such as VirE2. Any protein associated with the T-DNA presumably must be stripped prior to DNA integration. In support of this idea, VirF can mediate targeted proteolysis of VirE2 both in yeast and *in planta* (Tzfira et al., 2004). Interestingly, VirF has a functional ortholog (VBF) in some plants whose synthesis is induced by *Agrobacterium*infection. This ortholog can supply VirF function in *Agrobacterium* strains lacking VirF (Zaltsman et al., 2010). Reducing the level of this ortholog in plants increases their resistance to*Agrobacterium* infection.

VirD5

Another exported protein which originally was underappreciated is VirD5. The VirF protein is unstable in plants apparently because of degradation by the plant's ubiquitin proteasome system (Magori and Citovsky, 2011). To overcome this instability, *Agrobacterium* transfers VirD5 into the plant where it binds to VirF and prevents its rapid turnover.

VirE3

VirE3 is presumably another host range virulence factor. Like VirF, VirE3 has a functional ortholog in plants, VIP1. This transcription factor with NLS was reported to be important in plant cell transformation because it interacts with VirE2 and functions in the nuclear import of T-DNA (Tzfira et al.,

2001). However, the importance of VIP1 in plant cell transformation has been brought into question by a recent study which presents convincing evidence that VIP1 is not required for transformation of at least several plants (Shi et al., 2014). Therefore, VIP1 and presumably its ortholog VirE3 must function in some capacity other than helping VirE2 enter the nucleus. Perhaps relevant to this conundrum is a study which reported that VirE2 does not interact with importin IMPa and therefore must require another factor to interact with other than VirE2 for nuclear import (Ballas and Citovsky, 1997). Presumably this is VIP1. However, a more recent study showed that VirE2 does indeed associate with this specific importin. Therefore there is no need to postulate that another plant factor must be involved in nuclear import of VirE2 (Bhattacharjee et al., 2008). The significance of studies which reported that VirE3 can complement a VIP1 mutation both in nuclear import of T-DNA as well as susceptibility to transformation is unclear (Lacroix et al., 2005). Meanwhile, the function(s) of VIP1 and VirE3 remains unresolved.

VirJ

The VirJ protein is encoded by the *virJ* gene which maps between *virA* and *virB* and is the only copy present in nopaline strains (Pan et al., 1995). Mutants lacking this gene are avirulent (Wirawan et al., 1993). Octopine strains contain a functional chromosomal copy of *virJ*, labeled *acvB* which is constitutively expressed as well as the copy on the Ti plasmid (Pan et al., 1995). Agroinfection studies indicated that these mutants cannot transfer T-DNA into host cells (Pan et al., 1995). However, the exact role of this protein in T-DNA transfer is unclear. The protein is located in the periplasm and by immunoprecipitation assays bound to all VirB proteins (Pantoja et al., 2002). However, neither VirJ nor AcvB interact with the T-strand and do not appear to be subunits of the Type IV secretion system (Cascales and Christie, 2004). Its role in T-DNA transfer remains an intriguing mystery.

Small Heat Shock Protein (HspL)

A proteomics analysis of AS induced genes in *A. tumefaciens* C58 (a nopaline strain) revealed two new Vir proteins (Lai et al., 2006). One, Y4mC, which maps in the *vir* region, is not found in octopine strains. Whether it plays any role in virulence is unknown. The other is a small heat shock protein which is chromosomally encoded. Although it requires the VirA/G two-component system for its synthesis, it does not have a canonical vir box in its promoter region. This suggests that its synthesis depends on the expression of other genes under the control of VirA/G. This specific heat shock protein is located in the membrane where it functions as a chaperone for the VirB8 protein, a component of the structure of the Type IV secretion system (Tsai et al., 2012).

Chromosomal Vir Proteins (Chv)

Mutations in the chromosome as well as in the Ti plasmid can lead to alterations in virulence. The *chv* genes are not induced by AS, although many are induced by acidic conditions (Yuan et al., 2008). Analyzing many of these mutations has led to the conclusion that most of these genes encode proteins that play important roles in the physiology of the organism growing independently of the plant cell environment. However, these proteins directly or indirectly also play important roles in the interaction of *Agrobacterium* with plants. The pleiotropic nature of these functions make their exact role in tumor formation generally difficult to unravel. For purposes of discussion, *chv* mutations can be divided into several different categories, recognizing that these are subject to change as additional information is gained about them. These categories are: (1) mutations affecting expression of *vir* genes, (2) mutations involved in membrane structure, and (3) mutations involved in plant response.

MUTATIONS AFFECTING EXPRESSION OF *VIR* GENES

ChvE

The *chv* gene encodes a periplasmic binding protein and probably represents the most intensely studied of all the Chv proteins. Recent studies show just how important this protein is in plant cell transformation. Its role in signaling in the VirA/G system has already been discussed. However, it also binds to sugar transport as well as chemotaxis proteins (Cangelosi et al., 1990a; Kemner et al., 1997; Zhao and Binns, 2011). In these multiple functions, the N-terminal and C-terminal domains of ChvE that interact with VirA partially overlap the surface required for binding to the sugar transport protein (He et al., 2009) and thus the sugar uptake system competes with VirA as a receptor for ChvE. This competition has important consequences in determining the strength of phenolic compounds to serve as inducers of the *vir* genes (Hu et al., 2013). Thus, a specific inducing sugar can increase the inducing capability of a weak phenolic inducer if the level of ChvE is increased. Consequently, the ability of various phenolic compounds to serve as inducers depends qualitatively and quantitatively on the level of ChvE interacting with VirA (Peng et al., 1998).

Mutating the *chvE* locus at several different sites resulted in mutant ChvE proteins which responded differently to different neutral and acidic sugars in *vir* gene expression (Hu et al., 2013). Testing these mutants for their virulence on several different plants demonstrated that different sugars limit tumor formation depending on the host plant. Also, in at least one common and well studied strain of *Agrobacterium*, C58, ChvE is essential for *vir* gene induction even in the presence of high levels of AS. (Doty et al., 1996) and ChvE is required for successful agroinfection of maize (Raineri et al., 1993). Thus, ChvE can be considered a host range determinant whose biological importance in the physiology and tumor-inducing capabilities of *Agrobacterium* still remains incomplete.

ChvG/I

An intriguing *chv* region encodes the two-component system, ChvG/I. Mutants were isolated and characterized independently in two laboratories and were shown to be avirulent (Charles and Nester, 1993; Mantis and Winans, 1993). Mutants in either the sensor protein (ChvI) or response regulator (ChvG) are pleiotropic resulting in cells which are weakly inducible, are unable to grow on a rich medium, are sensitive to detergents, wound sap and acidic conditions. Further, the *virG* locus is no longer induced by acid (Charles and Nester, 1993; Mantis and Winans, 1993) although acid induces the expression of *chvG/I* (Yuan et al., 2008). Elevating the level of VirG by placing this locus under the control of an inducible *lac* promoter in a *chvI* mutant did not rescue *vir* gene expression (Mantis and Winans, 1993) suggesting that one effect of the *chvI* mutation must be downstream of the expression of *virG*.

The ChvG/I system is a global regulator of many acid-inducible genes many of which are required for virulence (Li et al., 2002). These include genes on the Ti plasmid and the two chromosomes and include the acid induction of *virG*. This is consistent with data that a chromosomal gene is responsible for the acid induction of *virG* (Mantis and Winans, 1992). This system in turn is under the control of ExoR, a periplasmic, acid sensitive protein which apparently binds to ChvG and prevents phosphate transfer to ChvI (Wu et al., 2012; Heckel et al., 2014). However, under acidic conditions, ExoR is subject to specific proteolysis which allows ChvI to be activated which promotes binding to upstream regions of acid-inducible genes. This model, in part, is based on data from a homologous system in *Sinorhizobium* (Chen et al., 2008, 2009; Lu et al., 2012).

ChvH

A mutation in the *chvH* locus results in pleiotropic mutants that are highly attenuated in virulence, synthesize much lower levels of VirG and VirE as well as VirB8, VirB9, VirB10, and VirB11 (Peng et al., 2001). Further, the mutants are highly sensitive to detergents and carbenicillin, and grow more slowly than the parental strain. In contrast to the decrease in the levels of Vir proteins which were synthesized, a 32 kDa protein accumulated which was not identified. Thus, the ChvH protein appears to selectively alter the level of certain proteins in the cell and specifically those that contribute to certain stress responses, such as acidic conditions, sensitivity to detergents and penicillin. DNA sequencing revealed that the *chvH* locus encodes an elongation factor P (EF-P), a translation factor that stimulates the formation of the first peptide bond in protein synthesis (Blaha et al., 2009). The mutation is highly conserved since the mutation can be complemented by the homologous *ef-p* locus from distantly related *E. coli* (Peng et al., 2001).

Recent studies on the EF-P protein in *E. coli* and *Salmonella* revealed that this post-transcriptional regulatory pathway is also essential for virulence of *Salmonella* and the synthesis of functional membranes, a story similar to what was observed in *Agrobacterium* (Zou et al., 2011, 2012). Thus, the *chvH* locus encodes a novel control system that regulates a limited number of proteins some of which are important in virulence and the stress response of *Agrobacterium*. What is the mechanism for this regulation? The protein EF-P interacts with the ribosome and specifically facilitates the synthesis of proteins with the specific amino acid motifs PPP and PPG (Hersch et al., 2013). The loss of EF-P would result in a limited synthesis of such proteins, apparently a feature of many of the Vir proteins. The factors that regulate ChvH synthesis are not known but do not appear to include acidic conditions (Yuan et al., 2008).

ChvD

This mutant protein was identified in a screen for mutants with altered expression of *virG*. The chromosomal mutant demonstrated reduced acid induction of *virG* gene expression and was also attenuated in virulence (Winans et al., 1988). Gene sequencing revealed that the predicted protein sequences were homologous to a family of proteins involved in active transport (Winans et al., 1988). Introduction of a constitutive *virG* locus into the *chvD* mutant restored virulence, indicating that the effect of the mutation on virulence is associated with its effect on *virG* expression (Liu et al., 2001). However, how a mutation in this putative transport system affects *virG* expression is not at all clear.

Citrate Synthase

This mutant, blocked in the synthesis of citrate in the tricarboxylic acid (TCA) cycle was isolated as a Tn*PhoA* insertion mutant that is attenuated in virulence (Suksomtip et al., 2005). The mutation resulted in a 10-fold reduction in *vir* gene expression. Introduction of a constitutive *virG* locus into the mutant restored both *vir* gene expression and virulence. Thus the mutant defect must be upstream of the VirA/G regulatory system but the basis of this effect is not known. The addition of several intermediates of the TCA cycle had no effect on *vir* gene expression. Surprisingly, the mutant cells grew almost as well as the wild type cells on minimal medium and restoring *vir* gene induction had no effect on the growth rate. Thus the reduced growth rate cannot account for the attenuation of virulence of the mutant.

Phosphoenolpyruvate Carboxykinase (PckA)

The *pckA* gene encodes a protein which catalyzes the reversible decarboxylation and phosphorylation of oxaloacetate to form phosphoenolpyruvate. The locus maps to the circular chromosome and interestingly is located immediately downstream of the *chvG/I* genes but is transcribed in the opposite direction. This gene is acid inducible, and like other acid inducible genes is under the control of ChvG/I. *pckA* mutants grew more slowly under acidic conditions than did the parent cells (Liu et al., 2005). They were highly attenuated in virulence and also in *vir* gene expression. However, introduction of a constitutive *virG* locus into the mutant restored *vir* gene induction to wild-type levels but only partially restored virulence. Neither were the growth rates restored at pH 5.5 or pH 7.0. The continued sensitivity of the mutant to growth inhibition under acidic conditions may account for the lack of complete restoration of virulence in the mutant with a constitutive *virG* gene. Why a mutation in the *pckA* gene should reduce *vir* gene expression and why its expression is acid inducible remain a mystery.

MUTATIONS INVOLVED IN MEMBRANE STRUCTURE

ChvA and ChvB

Other *chv* mutants with pleiotropic effects are those unable to synthesize or transport beta-1,2-glucan molecules into the periplasm. They are designated *chvB* and *chvA* respectively. They were isolated as mutants unable to attach to plant cells and therefore are avirulent (Douglas et al., 1982,1985). On low osmotic strength media, the mutants grew more slowly than wild-type cells and exhibited an altered periplasmic and cytoplasmic protein content as well as reduced motility. When returned to a high osmotic strength medium, their growth rate and protein content were restored to wild-type levels. However, the mutants were still avirulent and had reduced motility regardless of the media (Cangelosi et al., 1990b). These data suggest that the periplasmic glucan plays a role in osmoadaptation although its role in virulence and motility may be only indirectly related to its role in osmoadaptation. Addition of beta-1,2-glucan to both mutants did not affect their ability to bind to plant cells. It seems highly unlikely that the glucan is directly involved in binding. That *Agrobacterium* can attach to so many different hosts suggests that binding may not be very specific.

Agrobacterium Outer Membrane Protein (AopB)

AopB is an outer membrane protein in which mutations lead to a highly attenuated virulence phenotype (Jia et al., 2002). The encoding locus maps to the circular chromosome, is acid inducible and shares high homology with a gene from *Rhizobium leguminosarum*. Like other acid inducible loci, this gene is also under the control of ChvG/I (Yuan et al., 2008). The *vir* genes are expressed normally. There is some sequence similarity to genes encoding porin proteins suggesting that this protein may be involved in transport functions, like ChvD, but its role in tumorigenesis is unknown.

MUTATIONS INVOLVED IN PLANT RESPONSE

tRNA: Isopentenyltransferase (MiaA)

This Tn5 induced mutation involves the locus that is responsible for the specific modification of a specific residue in a codon of a tRNA species (*miaA* gene). The mutation reduced *vir* gene expression 2- to 10-fold when induced by AS (Gray et al., 1992). Virulence was reduced on some but not on other plants. Sequence analysis revealed an open reading frame with a strong homology to the *miaA* gene from *E. coli*.

The understanding of how a mutation in the *miaA* gene reduces tumorigenesis in some plants took more than 20 years to solve. In addition to the *miaA* locus, the story involves the *tzs* gene of the *vir* regulon found in nopaline but not octopine strains and a plant product, MTF, a transcription factor. The gene products of both *tzs* and *miaA* involve cytokinin synthesis, a phytohormone which has been implicated in promoting tumorigenesis (Zhan et al., 1990; Chateau et al., 2000; Hwang et al., 2010). Mutations in the *miaA* gene or loss of the *tzs* locus result in reduced levels of secreted cytokinin (Regier and Morris, 1982; Gray et al., 1996). Recently, the molecular basis of cytokinin action on the plant has been elucidated. Cytokinin decreases the expression of a plant transcription factor, MTF, which normally inhibits plant cell transformation (Sardesai et al., 2013). Thus, any mutation that reduces the level of secreted cytokinin decreases the susceptibility of the plant to transformation by *Agrobacterium*. Apparently, the MTF transcription factor may control the synthesis of plant receptors to which *Agrobacterium* may attach. As predicted, mutating MTF increases plant cell transformation (Sardesai et al., 2013). However, these impressive studies do not explain why *vir* gene expression is reduced in a MiaA mutant. Does cytokinin modulate *vir* gene expression?

Catalase (KatA)

A transposon induced mutation screen for acid inducible genes identified a gene (*katA*) encoding a catalase (Xu and Pan, 2000). This enzyme catalyzes the dismutation of hydrogen peroxide to water and oxygen and its synthesis is acid inducible (Li et al., 2002). Hydrogen peroxide must provide an important defense to the *Kalanchoe* plant because mutants lacking catalase are weakly virulent. This is the only *chv* gene thus far identified whose gene product apparently serves a single function.

Host Range

Agrobacterium has a remarkably broad host range especially considering other plant pathogens. Initially, plant infection was assayed by tumor formation, a measure of both T-DNA transfer and integration into the plant chromosome. Using this assay, a very large number of monocots and dicots were tested with the conclusion that although most dicots were susceptible to varying degrees, no monocots were (DeCleene and DeLey, 1976). This conclusion was shown to be an overgeneralization when it was demonstrated that *Agrobacterium* could form tumors on the monocot, *Asparagus* (Bytebier et al., 1987). Also, it was clear that the host range of *Agrobacterium*could be expanded to include certain poorly transformable plants by increasing the level of expression of the *vir* genes (Jin et al., 1987; Gelvin, 2003). "Super virulent" strains were constructed that proved useful in infecting many recalcitrant dicots (Hood et al., 1986; Jin et al., 1987; Reviewed in Banta and Montenegro, 2008). It was also clear that the defense a plant mounts against the invading *Agrobacterium* plays an important role is determining whether an infection is successful (Reviewed in Pitzschke, 2013). However, for whatever reason, all attempts to show tumor formation on cereal plants proved negative. Some plant scientists in the mid 1980s believed that the biology of monocots and dicots might be so different that monocots were inherently non-transformable.

A number of laboratories continued their attempts to understand why monocots were resistant to infection by *Agrobacterium* using several different assays. It was clear that tumor formation was a poor assay since monocots do not respond to increased levels of phytohormones as do dicots and therefore gene transfer might occur without gall formation. Therefore genes encoding color markers such as beta-glucuronidase and antibiotic resistance were substituted as markers for T-DNA transfer, but not necessarily integration. These assays were fast, and could be carried out in Petri dishes testing a variety of bacterial strains, plant tissues, and environmental conditions.

An especially clever and elegant assay allowed Barbara Hohn and her colleagues to demonstrate that *Agrobacterium* can transfer T-DNA into maize plants (Grimsley et al., 1987). In this assay, a viral genome, maize steak virus, was inserted into the T-DNA and maize plants were inoculated with *Agrobacterium* containing the viral genome (Grimsley et al., 1986). The appearance of viral symptoms at the site of inoculation indicates that the T-DNA has been transferred, the viral genome excised from the T-DNA and the virus has replicated. This assay, termed agroinfection, is a very sensitive measure of gene transfer because virus replication magnifies a single T-DNA transfer event. Thus, it was possible to readily study the features of *Agrobacterium* important for successful gene transfer independent of integration. Using this assay, it was shown that only certain strains of *Agrobacterium* could agroinfect. Thus, octopine strains could not, but nopaline strains could (Boulton et al., 1989). Introducing the *virA* gene from the nopaline into the octopine strain converted the latter strain into an agroinfected (Raineri et al., 1993). Dissecting the VirA protein into distinct regions revealed that the linker region of VirA is especially important for successful agroinfection. Further, the ChvE protein is also essential, suggesting that maximum *vir* gene induction is important (Heath et al., 1997).

The final chapter in the *Agrobacterium* transformation and regeneration of a whole host of cereals came from a series of papers from Japan Tobacco Inc. These highly significant papers make it abundantly clear that successful transformation and subsequent regeneration requires the collaboration of numerous investigators skilled in plant biology, tissue culture, and agrobiology (Hiei et al., 1994; Ishida et al., 1996; Gelvin, 2008).

Remarkably, the host range of *Agrobacterium* extends well beyond plants. In the laboratory, representative algae (Kumar et al., 2004), fungi (Bundock et al., 1995), and even human cells (Kunik et al., 2001) have all been infected. Additional non-plant organisms transformed by *Agrobacterium* are reviewed in Soltani et al. (2008). These expanded abilities now allow the revolution in plant molecular biology and genetics that *Agrobacterium* made possible in plants to be extended to a whole host of other eukaryotic systems.

Relevant Topics Not Covered

Restrictions on the length of this review preclude discussions of many aspects of crown gall tumorigenesis which deserve attention. One such area is the story of the journey of the T-strand, once inside the host cell, to its site of integration and the plant factors involved. Fortunately, several recent reviews cover this aspect (Gelvin, 2010a,b, 2012; Lacroix and Citovsky, 2013). Another area of great interest and importance in a discussion of plant tumorigenesis is that of plant defense against *Agrobacterium*. This subject has also been covered recently (Pitzschke, 2013).

A third aspect of great importance in the interaction of *Agrobacterium* with its hosts relates to the two-way signal exchange between *Agrobacterium* and its host plants. A recently published review covers many aspects of this subject (Subramoni et al., 2014).

Another inadequately covered area in this review is that of *Agrobacterium* and its role in plant biotechnology. This subject was covered in an excellent review (Banta and Montenegro, 2008).

The subject of horizontal gene transfer from *Agrobacterium* to plants in nature is an intriguing subject. This subject was recently reviewed (Matveeva and Lutova, 2014).

My apologies to the many crown gall investigators whose contributions I have not been able to cover or their studies cited because of space limitations.

Summary

It is unusual that the study of a single organism can reveal a unique biological system, contribute to an understanding of fundamental biological principles and lead to the development of an entirely new industry. In addition, *Agrobacterium* has revolutionized plant molecular genetics and has the capabilities to do the same for many other eukaryotic organisms. Certainly, Smith and Townsend (1907) studying the cause of a destructive plant disease and perhaps even Armin Braun who recognized the unique character of this disease could not have imagined how exciting and important the study of *Agrobacterium* and crown gall would turn out to be.

However, as much as we have learned about the mechanism by which *Agrobacterium* transforms plant cells, many features of this bacterium-host relationship deserve much more study. Clearly, we need to learn more about the world of *Agrobacterium* in its natural environment. What is the microbiome inside and outside the rhizosphere of a susceptible plant? What mechanisms does *Agrobacterium* use to successfully maintain its niche in the rhizosphere and what environmental cues and plant signals does it co-opt to promote its success? How does it recognize these signals? Does *Agrobacterium* send signals to the plant which promote the secretion of signals beneficial to *Agrobacterium*? What plant genes are affected by molecules secreted by the bacterium and do some contribute to host defense? What is the mechanism and consequences of attachment of *Agrobacterium* to plant cells? Does this affect the physiology of the bacterium? Answers to some of these questions may reveal the functions of some of the *chv* mutations which lead to attenuated virulence.

The natural environment of a tumor should also be studied with time-course studies of tumor development. Do the products of transformation, phytohormones and opines, modify the physiology of *Agrobacterium*? Is the *vir* regulon turned off in bacteria within a tumor? If so, what turns it off?

Another exciting area that should receive increasing attention is the detailed analysis of the travels of the T-DNA from the time *Agrobacterium* attaches to the host cell until the DNA becomes integrated into the host chromosome. Conflicting reports in some important aspects of this journey need to be resolved. The identification and availability of plant mutants will make such studies increasingly amenable to detailed analysis and understanding.

The study of *Agrobacterium*–plant interactions continues to point out just how clever this pathogen is. Future studies will no doubt reveal additional examples of how *Agrobacterium* takes advantage of its natural environment to further its own goals.

CONFLICT OF INTEREST STATEMENT

The author declares that the research was conducted in the absence of any commercial or financial relationships that could be construed as a potential conflict of interest.

ACKNOWLEDGMENTS

Several individuals, no longer with us, contributed enormously to the body of knowledge summarized in this review. These include Jeff Schell, Milt Gordon, Rob Schilperoort, and Scott Stachel. They were personal colleagues and friends. Their contributions should be remembered by all who continue to study *Agrobacterium* and its unique ability to transform plants.

REFERENCES

1. Akiyoshi, D. E., Klee, H., Amasino, R. M., Nester, E. W., and Gordon, M. P. (1984). T-DNA of *Agrobacterium* encodes an enzyme of cytokinin biosynthesis. *Proc. Natl. Acad. Sci. U.S.A.* 81, 5994–5998. doi: 10.1073/pnas.81.19.5994

2. Akiyoshi, D. E., Morris, R. O., Hinz, R., Mische, B. S., Kosuge, T., Garfinkel, D. J., et al. (1983). Cytokinin/auxin balance in crown gall tumors is regulated by specific loci in the T-DNA. *Proc. Natl. Acad. Sci. U.S.A.* 80, 407–411. doi: 10.1073/pnas.80.2.407

3.   Albright, L., Yanofsky, M. F., Leroux, B., Ma, D., and Nester, E. W. (1987). Processing of the T-DNA of *Agrobacterium tumefaciens* generates border knicks and linear, single stranded T-DNA. *J. Bacteriol.* 169, 1046–1055.

4.   Allardet-Servent, A., Michaux-Charachon, S., Jumas-Bilak, M., Karayan, L., and Ramuz, M. (1993). Presence of one linear and one circular chromosome in the *Agrobacterium tumefaciens* C58 genome. *J. Bacteriol.* 175, 7869–7874.

5.   Alvarez-Martinez, C. E., and Christie, P. J. (2009). Biological diversity of prokaryotic type IV secretion systems. *Microbiol. Mol. Biol. Rev.* 73, 775–808. doi: 10.1128/MMBR.00023-09

6.   Ankenbauer, R. G., and Nester, E. W. (1990). Sugar mediated induction of *Agrobacterium* virulence genes: structural specificity and activities of monosaccharides. *J. Bacteriol.* 172, 6442–6446.

7.   Atmakuri, K., Cascales, E., Burton, O. T., Banta, L. M., and Christie, P. J. (2007). *Agrobacterium* ParA/MinD-like VirC1 spatially coordinate early conjugative DNA transfer reactions. *EMBO J.* 26, 2540–2551. doi: 10.1038/sj.emboj.7601696

8.   Ballas, N., and Citovsky, V. (1997). Nuclear localization signal binding protein from *Arabidopsis* mediates nuclear import of *Agrobacterium* VirD2 protein. *Proc. Natl. Acad. Sci. U.S.A.* 94, 10723–10728. doi: 10.1073/pnas.94.20.10723

9.   Banta, L. M., Joeger, R. D., Horvitz, V. R., Campbell, A. M., and Binns, A. N. (1994). Glu-255 outside the predicted ChvE binding site in VirA is crucial for sugar enhancement of acetosyringone perception by *Agrobacterium*. *J. Bacteriol.* 176, 3242–3249.

10.   Banta, L. M., and Montenegro, M. (2008). "*Agrobacterium* and plant biotechnology," in *Agrobacterium: From Biology to Biotechnology*, eds T. Tzfira and V. Citovsky (New York, NY: Springer Science+Business Media), 73–147.

11.   Bhattacharjee, S., Lee, L., Oltmanns, H., Cao, H., Veena, J. C., and Gelvin, S. B. (2008). IMPa-4, an *Arabidopsis* importin isoform, is preferentially involved in *Agrobacterium*-mediated plant transformation. *Plant Cell* 20, 2661–2680. doi: 10.1105/tpc.108.060467

12.   Binns, A. N. (2005). "Armin C. Braun and the discovery of *Agrobacterium*-mediated transformation of plant cells," in *Agrobacterium tumefaciens: From Plant Pathology To Biotechnology*, eds E. Nester, M. P. Gordon, and A. Kerr (St. Paul, MN: APS Press), 7–10.

13.   Blaha, G., Stanley, R. E., and Steitz, T. A. (2009). Formation of the first peptide bond: the structure of EF-P bound to the 70S ribosome. *Science* 325,

966–970. doi: 10.1126/science.1175800

14. Boulton, M. I., Buchholz, W. G., Marks, M. S., Markam, P. G., and Davies, J. (1989). Specificity of *Agrobacterium*-mediated delivery of maize streak virus DNA to members of the Gramineae. *Plant Mol. Biol.* 12, 31–40. doi: 10.1007/BF00017445

15. Braun, A. C. (1947). Thermal studies on tumor inception in the crown gall disease. *Am. J. Bot.* 30, 674–677.

16. Braun, A. C. (1958). A physiological basis for autonomous growth of the crown-gall tumor cell. *Proc. Natl. Acad. Sci. U.S.A.* 44, 344–349. doi: 10.1073/pnas.44.4.344

17. Braun, A. C., and Mandel, R. J. (1948). Studies on the inactivation of the tumor inducing principle in crown gall. *Growth* 12, 14–28.

18. Brencic, A., Eberhard, A., and Winans, S. C. (2004). Signal quenching, detoxification and mineralization of vir gene-inducing phenolics by the VirH2 protein of *Agrobacterium tumefaciens*. *Mol. Microbiol.* 51, 1103–1115. doi: 10.1046/j.1365-2958.2003.03887.x

19. Bundock, P., denDulk-Ras, A., Beijersbergen, A., and Hooykaas, P. J. J. (1995). Trans-kingdom T-DNA transfer from*Agrobacterium tumefaciens* to *Saccharomyces cerevisiae*. *EMBO J.* 14, 3206–3214.

20. Bytebier, B., Deboeck, F., DeGreve, H., Van Montagu, M., and Hernalsteens, J. P. (1987). T-DNA organization in tissue culture and transgenic plants of the monocotyledon *Asparagus officinalis*. *Proc. Natl. Acad. Sci. U.S.A.* 83, 3282–3286.

21. Cangelosi, G. A., Ankenbauer, R. G., and Nester, E. W. (1990a). Sugars induce the *Agrobacterium* virulence genes through a periplasmic binding protein and a transmembrane signal protein. *Proc. Natl. Acad. Sci. U.S.A.* 87, 6708–6712. doi: 10.1073/pnas.87.17.6708

22. Cangelosi, G, A., Martinetti, G., and Nester, E. W. (1990b). Osmosensitivity phenotypes of mutants of *Agrobacterium tumefaciens* that lack periplasmic beta-1,2 glucan. *J. Bacteriol.* 172, 2172–2174.

23. Cascales, E., Atmakuri, K., Sakar, M. K., and Christie, P. J. (2013). DNA substrate – induced activation of the *Agrobacterium*VirB/VirD4 type IV secretion system. *J. Bacteriol.* 195, 2691–2704. doi: 10.1128/JB.00114-13

24. Cascales, E., and Christie, P. J. (2004). Definition of a bacterial type IV secretion pathway for a DNA substrate. *Science* 304, 1170–1173. doi: 10.1126/science.1095211

25. Chang, C.-H., and Winans, S. C. (1992). Functional roles assigned to

the periplasmic, linker and receiver domains of the*Agrobacterium tumefaciens* VirA protein. *J. Bacteriol.* 174, 7033–7039.

26. Charles, T., Jin, S., and Nester, E. (1992). Two-component transduction systems in phytobacteria. *Ann. Rev. Phytopathol.* 30, 463–484. doi: 10.1146/annurev.py.30.090192.002335

27. Charles, T. C., and Nester, E. W. (1993). A chromosomally-encoded two-component sensory transduction system is required for virulence of *Agrobacterium tumefaciens. J. Bacteriol.* 175, 6614–6625.

28. Chateau, S., Sangwan, R. S., and Sangwan-Norreel, B. S. (2000). Competence of *Arabidopsis thaliana* genotypes and mutants for *Agrobacterium tumefaciens*-mediated gene transfer: role of phytohormones. *J. Exp. Bot.* 51, 1961–1968. doi: 10.1093/jexbot/51.353.1961

29. Chen, E. J., Fisher, R. F., Perovich, V. M., Sabio, E. A., and Long, S. R. (2009). Identification of direct transcriptional target genes of ExoS/ChvI two-component signaling in *Sinorhizobium meliloti. J. Bacteriol.* 191 6833–6842. doi: 10.1128/JB.00734-09

30. Chen, E. J., Sabio, E. A., and Long, S. R. (2008). The periplasmic regulator ExoR inhibits ExoS/ChvI two-component signaling in*Sinorhizobium meliloti. Mol. Microbiol.* 69, 1290–1303. doi: 10.1111/j.1365-2958.2008.06362.x

31. Chilton, M.-D., Drummond, M. H., Merlo, D. J., Sciaky, D., Montoya, A. L., Gordon, M. P., et al. (1977). Stable incorporation of plasmid DNA into higher plant cells: the molecular basis of crown gall tumorigenesis. *Cell* 11, 263–271. doi: 10.1016/0092-8674(77)90043-5

32. Chilton, M.-D., Montoya, A. L., Merlo, D. J., Drummond, M. H., Nutter, R., Gordon, M. P., et al. (1978). Restriction endonuclease mapping of a plasmid that confers oncogenicity upon *Agrobacterium tumefaciens* strain B6-806. *Plasmid* 1, 254–269. doi: 10.1016/0147-619X(78)90043-4

33. Christie, P. J., Ward, J. E., Winans, S. C., and Nester, E. W. (1988). The *Agrobacterium tumefaciens* gene product VirE2 is a single-stranded DNA-binding protein that associates with T-DNA. *J. Bacteriol.* 170, 2659–2667.

34. Citovsky, V., DeVos, G., and Zambryski, P. (1988). Single-stranded DNA binding protein encoded by the virE2 locus of*Agrobacterium. Science* 240, 501–504. doi: 10.1126/science.240.4851.501

35. Das, A. (1988). *Agrobacterium tumefaciens* virE operon encodes a single-stranded DNA binding protein. *Proc. Natl. Acad. Sci. U.S.A.* 85, 2909–2913. doi: 10.1073/pnas.85.9.2909

36. Das, A., Stachel, S., Ebert, P., Allenza, P., Montoya, A., and Nester, E. (1986). Promoters of *Agrobacterium tumefaciens* Ti-plasmid virulence genes. *Nucleic Acids Res.* 14, 1355–1364. doi: 10.1093/nar/14.3.1355

37. DeCleene, M., and DeLey, J. (1976). The host range of crown gall. *Bot. Rev.* 42, 389–466.

38. Deng, W., Chen, L., Peng, W.-T., Metcalf, T, T., Liang, X., Gordon, M. P., et al. (1999). Vir E1 is a specific molecular chaperone for the exported single-stranded-DNA-binding protein VirE2 in *Agrobacterium*. *Mol. Microbiol.* 31, 1795–1807. doi: 10.1046/j.1365-2958.1999.01316.x

39. Doty, S. L., Yu, M. C., Lundin, J. I., Heath, J. D., and Nester, E. W. (1996). Mutational analysis of the input domain of the VirA protein of *Agrobacterium tumefaciens*. *J. Bacteriol.* 178, 961–970.

40. Douglas, C. J., Halperin, W., and Nester, E. W. (1982). *Agrobacterium tumefaciens* mutants affected in attachment to plant cells. *J. Bacteriol.* 152, 1265–1275.

41. Douglas, C. J., Staneloni, R. J., Rubin, R. A., and Nester, E. W. (1985). Identification and genetic analysis of an *Agrobacterium tumefaciens* chromosomal virulence region. *J. Bacteriol.* 161, 850–860.

42. Drummond, M. H., Gordon, M. P., Nester, E. W., and Chilton, M.-D. (1977). Foreign DNA of bacterial origin is transcribed in crown gall tumors. *Nature* 269, 535–536. doi: 10.1038/269535a0

43. Durrenberger, F., Crameri, A., Hohn, B., and Koukolikova-Nicola, Z. (1989). Covalently bound VirD2 protein of *Agrobacterium tumefaciens* protects the T-DNA from exonucleolytic degradation. *Proc. Natl. Acad. Sci. U.S.A.* 86, 9154–9158. doi: 10.1073/pnas.86.23.9154

44. Dym, O., Albeck, S., Unger, T., Jacobovitch, J., Branz-burg, A., Michael, Y., et al. (2008). Crystal structure of the *Agrobacterium* virulence complex VirE1-VirE2 reveals a flexible protein that can accomodate different partners. *Proc. Natl. Acad. Sci. U.S.A.* 105, 11170–11175. doi: 10.1073/pnas.0801525105

45. Fullner, K. J., Lara, J. C., and Nester, E. W. (1996). Pilus assembly by *Agrobacterium* T-DNA transfer genes. *Science* 273, 1107–1109. doi: 10.1126/science.273.5278.1107

46. Garfinkel, D., and Nester, E. W. (1980). *Agrobacterium tumefaciens* mutants affected in crown gall tumorigenesis and octopine catabolism. *J. Bacteriol.* 144, 732–743.

47. Garfinkel, D. J., Simpson, R. P., Ream, L. W., White, F. F., Gordon, M. P., and Nester, E. W. (1981). Genetic analysis of crown gall: fine structure

map of the T-DNA by site-directed mutagenesis. *Cell* 27, 143–15323. doi: 10.1016/0092-8674(81)90368-8

48. Gelvin, S. B. (1998). *Agrobacterium* VirE2 protein can form a complex with T-strands in the plant cytoplasm. *J. Bacteriol.* 180, 4300–4302.

49. Gelvin, S. B. (2003). *Agrobacterium*-mediated plant transformation: the biology behind the "gene-jockeying tool." *Microbiol. Mol. Bio. Rev.* 67, 16–37. doi: 10.1128/MMBR.67.1.16-37.2003

50. Gelvin, S. B. (2008). "Function of host proteins in the *Agrobacterium*-mediated plant transformation process," in*Agrobacterium: From Biology to Biotechnology*, eds T. Tzfira and V. Citovsky (New York, NY: Springer Science+Business Media), 489–522.

51. Gelvin, S. B. (2010a). Finding a way to the nucleus. *Curr. Opin. Microbiol.* 13, 53–58. doi: 10.1016/j.mib.2009.11.003

52. Gelvin, S. B. (2010b). Plant proteins involved in *Agrobacterium*-mediated genetic transformation. *Annu. Rev. Phytopathol.* 48, 45–68. doi: 10.1146/annurev-phyto-080508-081852

53. Gelvin, S. B. (2012). Traversing the cell: *Agrobacterium* T-DNA's journey to the host genome. *Front. Plant Sci.* 3:52. doi: 10.3389/fpls.2012.00052

54. Gelvin, S. B., Gordon, M. P., Nester, E. W., and Aronson, A. I. (1981). Transcription of *Agrobacterium* Ti-plasmid in the bacterium and crown gall tumors. *Plasmid* 6, 17–29. doi: 10.1016/0147-619X(81)90051-2

55. Gelvin, S. B., Thomashow, M. F., McPherson, J. C., Gordon, M. P., and Nester, E. W. (1982). Sizes and map positions of T-DNA encoded transcripts in octopine-type crown gall tumors. *Proc. Natl. Acad. Sci. U.S.A.* 79, 76–80. doi: 10.1073/pnas.79.1.76

56. Gietl, C., Koukolikova-Nicola, Z., and Hohn, B. (1987). Mobilization of T-DNA from *Agrobacterium* to plant cells involves a protein that binds single-stranded DNA. *Proc. Natl. Acad. Sci. U.S.A.* 84, 9006–9010. doi: 10.1073/pnas.84.24.9006

57. Goodner, B., Hinkle, G., Gattung, S., Miller, N., Blanchard, M., Qurollo, B., et al. (2001). Genome sequence of the plant pathogen and biotechnology agent *Agrobacterium tumefaciens* C58. *Science* 294, 2323–2328. doi: 10.1126/science.1066803

58. Gray, J., Gelvin, S. B., Meilan, R., and Morris, R. O. (1996). Transfer RNA is the source of extracellular isopentenyladenine in a Ti-plasmidless strain of *Agrobacterium tumefaciens*. *Plant Physiol.* 110, 431–438.

59. Gray, J., Wang, J., and Gelvin, S. B. (1992). Mutation of the *miaA* gene of *Agrobacterium tumefaciens* results in reduced vir gene expression. *J.*

Bacteriol. 174, 1086–1098.

60. Grimsley, N., Hohn, B., Hohn, T., and Walden, R. (1986). Agroinfection, a novel route for plant viral infection using Ti plasmid.*Proc. Natl. Acad. Sci. U.S.A.* 83, 3282–3286. doi: 10.1073/pnas.83.10.3282

61. Grimsley, N., Hohn, T., Davies, J. W., and Hohn, B. (1987).*Agrobacterium*-mediated delivery of infectious maize streak virus into maize plants. *Nature* 325, 177–179. doi: 10.1038/325177a0

62. Hamilton, R. H. (2005). "Loss of tumor-inducing ability," in *Agrobacterium tumefaciens: From Plant Pathology to Biotechnology*, eds E. W. Nester, M. P. Gordon, and A. Kerr (St. Paul, MN: APS Press), 45–46.

63. Hamilton, R. H., and Fall, M. Z. (1971). The loss of tumor-initiating ability in *Agrobacterium tumefaciens* by incubation at high temperature. *Experentia* 27, 229–230. doi: 10.1007/BF02145913

64. He, F., Nair, G. R., Soto, C. S., Chang, Y., Hsu, L., Ronzone, E., et al. (2009). Molecular basis of ChvE function in sugar binding, sugar utilization and virulence in *Agrobacterium tumefaciens*. *J. Bacteriol.* 191, 5802–5813. doi: 10.1128/JB.00451-09

65. Heath, J. D., Boulton, M. I., Raineri, D. M., Doty, S. L. Mushegian, A. R., Charles, T. C., et al. (1997). Discrete regions of the sensor protein VirA determine the strain specific ability of *Agrobacterium* to infect maize. *Mol. Plant Microbe Interact.* 10, 221–227. doi: 10.1094/MPMI.1997.10.2.221

66. Heckel, B. C., Tomlinson, A. D., Morton, E. R., Choi, J.-H., and Fuqua, C. (2014). *Agrobacterium tumefaciens* ExoR controls acid response genes and impacts exopolysaccharide synthesis, horizontal gene transfer and virulence gene expression. *J. Bacteriol.* 196, 3221–3233. doi: 10.1128/JB.01751-14

67. Herrera-Estrella, A., Van Montagu, M., and Wang, K. (1990). A bacterial peptide acting as a nuclear targeting signal: the amino terminal portion of *Agrobacterium* VirD2 protein directs a -galactosidase fusion protein into tobacco nuclei. *Proc. Natl. Acad. Sci. U.S.A.* 87, 9534–9537. doi: 10.1073/pnas.87.24.9534

68. Hersch, S. J., Wang, M., Zou, S. B., Moon, K.-M., Foster, L. J., Ibba, M., et al. (2013). Divergent protein motifs direct elongation factor P-mediated translational regulation in *Salmonella enterica* and *Escherichia coli*. *MBio* 4, 1–10. doi: 10.1128/mBio.00180-13

69. Hiei, Y., Ohta, S., Komari, T., and Kumashiro, T. (1994). Efficient transformation of rice (*Oryza sativa* L.) mediated by*Agrobacterium* and sequence analysis of the boundaries of the T-DNA. *Plant J.* 6, 271–282.

doi: 10.1046/j.1365-313X.1994.6020271.x

70. Holsters, M., Silva, B., Van Vleit, F., Genetello, C., De Block, M., Dhaese, P., et al. (1980). The functional organization of the nopaline *A. tumefaciens* plasmid pTiC58. *Plasmid* 3, 212–230. doi: 10.1016/0147-619X(80)90110-9

71. Hood, E. E., Helmer, G. L., Fraley, R. T., and Chilton, M.-D. (1986). The hypervirulence of *Agrobacterium tumefaciens* A281 is encoded in a region of pTiBo542 outside of T-DNA. *J. Bacteriol.* 168, 1291–1301.

72. Howard, E. A., Zupan, J. R., Citovsky, V., and Zambryski, P. C. (1992). The VirD2 protein of *A. tumefaciens* contains a C-terminal bipartite nuclear localization signal: implications for nuclear uptake of DNA in plant cells. *Cell* 68, 109–118. doi: 10.1016/0092-8674(92)90210-4

73. Hu, X., Zhao, J., DeGrado, W. F., and Binns, A. N. (2013). *Agrobacterium tumefaciens* recognizes its host plant environment using ChvE to bind diverse plant sugars as virulence signals. *Proc. Natl. Acad. Sci. U.S.A.* 110, 678–683. doi: 10.1073/pnas.1215033110

74. Hwang, H.-H., Wang, M.-H., Lee, Y.-L., Tsai, Y. L., Li, Y.-H., Yang, F. H., et al. (2010). *Agrobacterium*-produced and exogenous cytokinin-modulated *Agrobacterium*-mediated plant transformation. *Mol. Plant Pathol.* 11, 677–690. doi: 10.1111/j.1364-3703.2010.00637.x

75. Ishida, Y., Saito, H., Ohta, S., Hiei, Y., Komari, T., and Kumashiro, T. (1996). High efficiency transformation of maize (*Zea mays* L.) mediated by *Agrobacterium tumefaciens*. *Nat. Biotechnol.* 14, 745–750. doi: 10.1038/nbt0696-745

76. Jia, Y. H., Li, L. P., Hou, Q. M., and Pan, S. Q. (2002). An *Agrobacterium* gene involved in tumorigenesis encodes an outer membrane protein exposed on the bacterial cell surface. *Gene* 284, 113–124. doi: 10.1016/S0378-1119(02)00385-2

77. Jin, S., Komari, T., Gordon, M. P., and Nester, E. W. (1987). Genes responsible for the supervirulent phenotype of *Agrobacterium tumefaciens* strain A281. *J. Bacteriol.* 169, 4417–4425.

78. Jin, S., Roitsch, T., Ankenbauer, R. G., Gordon, M. P., and Nester, E. W. (1990a). The VirA protein of *Agrobacterium tumefaciens* is autophosphorylated and is essential for vir gene regulation. *J. Bacteriol.* 172, 525–530.

79. Jin, S., Prusti, R., Roitsch, T., Ankenbauer, R. G., and Nester, E. W. (1990b). Phosphorylation of the VirG protein of *Agrobacterium tumefaciens* by the autophosphorylated VirA protein: essential role in biological activity of VirG. *J. Bacteriol.* 172, 4945–495042.

80. Jin, S., Roitsch, T., Christie, P. J., and Nester, E. W. (1990c). The regulatory VirG protein specifically binds to *cis*-acting regulatory sequence involved in transcriptional activation of *Agrobacterium tumefaciens* virulence genes. *J. Bacteriol.* 172, 531–537.

81. Johnson, R., Guiderian, R. H., Eden, F., Chilton, M.-D., Gordon, M. P., and Nester, E. W. (1974). Detection and quantification of octopine in normal plant tissue and crown gall tumors. *Proc. Natl. Acad. Sci. U.S.A.* 71, 536–539.

82. Kado, C. (2014). Historical account on gaining insights on the mechanism of crown gall tumorigenesis induced by *Agrobacterium tumefaciens*. *Front. Microbiol.* 5:340. doi: 10.3389/fmicb.2014.00340

83. Kalogeraki, V. S., and Winans, S. C. (1998). Wound released chemical signals may elicit multiple responses from an*Agrobacterium tumefaciens* strain containing an octopine-type Ti plasmid. *J. Bacteriol.* 180, 5660–5667.

84. Kalogeraki, V. S., Zhu, J., Stryker, J. L., and Winans, S. C. (2000). The right end of the vir region of an octopine-type Ti plasmid contains four new members of the vir regulon that are not required for pathogenesis. *J. Bacteriol.* 182, 1774–1778. doi: 10.1128/JB.182.6.1774-1778.2000

85. Kanemoto, R. H., Powell, A. T., Akiyoshi, D. E., Regier, D. A., Kerstetter, R. A., Nester, E. W., et al. (1989). Nucleotide sequence and analysis of the plant-inducible locus, pinF, from *Agrobacterium tumefaciens*. *J. Bacteriol.* 171, 2506–2512.

86. Kemner, J. M., Liang, X., and Nester, E. W. (1997). The *Agrobacterium tumefaciens* virulence gene chvE is part of a putative ABC-type sugar transport operon. *J. Bacteriol.* 179, 2452–2458.

87. Kerr, A. (1969). Transfer of virulence between isolates of *Agrobacterium*. *Nature* 223, 1175–1176. doi: 10.1038/2231175a0

88. Kerr, A. (2005). "Treasure the unexpected," in *Agrobacterium tumefaciens: From Plant Pathology to Biotechnology*, eds E. W. Nester, M. P. Gordon, and A. Kerr (St. Paul, MN: APS Press), 29–30.

89. Kerr, J. E., and Christie, P. J. (2010). Evidence for VirB4-mediated dislocation of membrane-integrated VirB2 pilin during biogenesis of the *Agrobacterium* VirB/VirD4 type IV secretion system. *J. Bacteriol.* 192, 4923–4934. doi: 10.1128/JB.00557-10

90. Klee, H., Montoya, A., Horodyski, F., Lichtenstein, C., Garfinkel, D., Fuller, S., et al. (1984). Nucleotide sequences of the *tms*genes of the pTiA6NC octopine Ti-plasmid: two gene products involved in plant tumorigenesis. *Proc. Natl. Acad. Sci. U.S.A.* 81, 1728–1732. doi: 10.1073/

pnas.81.6.1728

91. Kumar, S. V., Misquitta, R. W., Reddy, V. S., Rao, B. J., and Rajam, M. V. (2004). Genetic transformation of the green alga*Chlamydomonas reinhardtii* by *Agrobacterium tumefaciens*. *Plant Sci.* 166, 731–738. doi: 10.1016/j.plantsci.2003.11.012

92. Kunik, T., Tzfira, T., Kapulnik, Y., Gafni, Y., Dingwall, C., and Citovsky, V. (2001). Genetic transformation of HeLa cells by*Agrobacterium*. *Proc. Natl. Acad. Sci, U.S.A.* 98, 1871–1876. doi: 10.1073/pnas.98.4.1871

93. Lacroix, B., and Citovsky, V. (2013). The roles of bacterial and host plant factors in *Agrobacterium*-mediated genetic transformation. *Int. J. Dev. Biol.* 57, 467–481. doi: 10.1387/ijdb.130199bl

94. Lacroix, B., Vaidya, M., Tzfira, T., and Citovsky, V. (2005). The VirE3 protein of *Agrobacterium* mimics a host cell function required for plant genetic transformation. *EMBO J.* 24, 428–437. doi: 10.1038/sj.emboj.7600524

95. Lai, E.-M., and Kado, C. I. (1998). Processed VirB2 is the major subunit of the promiscuous pilus of *Agrobacterium tumefaciens*.*J. Bacteriol.* 180, 2711–2717.

96. Lai, E.-M., Shih, H.-W., Wen, S.-W., Cheng, M.-W., Hwang, H.-H., and Chiu, S.-H. (2006). Proteomic analysis of *Agrobacterium tumefaciens* response to the vir gene inducer acetosyringone.*Proteomics* 6, 4130–4136. doi: 10.1002/pmic.200600254

97. Lee, Y.-W., Jin, S., Sim, W.-S., and Nester, E. W. (1995). Genetic evidence for direct sensing of phenolic compounds by the VirA protein of *Agrobacterium tumefaciens*. *Proc. Natl. Acad. Sci., U.S.A.* 92, 12245–12249.

98. Lemmers, M., Holsters, M., Zambryski, P., DePicker, A., Hernalsteens, J. P., Van Montagu, M., et al. (1980). Internal organization, boundaries and integration of Ti-plasmid DNA in nopaline crown gall tumors. *J. Mol. Biol.* 144, 353–376. doi: 10.1016/0022-2836(80)90095-9

99. Leroux, B., Yanofsky, M. F., Winans, S. C., Ward, J. E., Ziegler, S. F., and Nester, E. W. (1987). Characterization of the virA locus of *Agrobacterium tumefaciens*: a transcriptional regulator and host range determinant. *EMBO J.* 6, 849–856.

100. Li, L., Jia, Y., Hori, Q., Charles, T. C., Nester, E. W., and Pan, S. (2002). A global pH sensor: *Agrobacterium* sensor protein ChvG regulates acid-inducible genes on its two chromosomes and Ti plasmid. *Proc. Natl. Acad. Sci. U.S.A.* 99, 12369–12374.

101. Lioret, C. (1957). Les acides amines' libres de tissues de crown-gall. Mise en évidence d'un acide aminé particulier à ces tissues. *C. R. Hebd. Seances Acad. Sci.* 244, 2171–2174.

102. Liu, P., Wood, D. W., and Nester, E. W. (2005). Phosphoenolpyruvate carboxykinase is an acid induced, chromosomally encoded, virulence factor in *Agrobacterium tumefaciens. J. Bacteriol.* 187, 6039–6045. doi: 10.1128/JB.187.17.6039-6045.2005

103. Liu, Z., Jacobs, M., Schaff, D. A., McCullen, C. A., and Binns, A. N. (2001). ChvD, a chromosomally encoded ATP binding cassette transporter-homologous protein involved in regulation of virulence gene expression in *Agrobacterium tumefaciens. J. Bacteriol.* 183, 3310–3317. doi: 10.1128/JB.183.11.3310-3317.2001

104. Low, H. H., Gubellini, F., Rivera-Calzada, A., Braun, N., Connery, S., Dujeancourt, A., et al. (2014). Structure of a type IV secretion system. *Nature* 508, 550–553. doi: 10.1038/nature13081

105. Lu, H.-X., Luo, L., Yang, M.-H., and Cheng, H.-P. (2012). *Sinorhizobium meliloti* ExoR is the target of periplasmic proteolysis. *J. Bacteriol.* 194, 4029–4040. doi: 10.1128/JB.00313-12

106. Ma, L. S., Hachani, A., Lin, J. S., Filloux, A., and Lai, E. M. (2014). *Agrobacterium tumefaciens* deploys a superfamily of type VI secretion DNase effectors as weapons for interbacterial competition in plants. *Cell Host Microbe* 16, 94–104. doi: 10.1016/j.chom.2014.06.002

107. Magori, S., and Citovsky, V. (2011). *Agrobacterium* counteracts host-induced degradation of its effector F-box protein. *Sci. Signal.* 4, ra69. doi: 10.1126/scisignal.2002124

108. Mantis, N. J., and Winans, S. C. (1992). The *Agrobacterium* vir gene transcriptional activator VirG is transcriptionally induced by acid pH and other stress stimuli. *J. Bacteriol.* 174, 1189–1196.

109. Mantis, N. J., and Winans, S. C. (1993). The chromosomal response regulatory gene chvI of *Agrobacterium tumefaciens*complements an *Escherichia coli* phoB mutation and is required for virulence. *J. Bacteriol.* 175, 6626–6636.

110. Matveeva, T. V., and Lutova, L. A. (2014). Horizontal gene transfer from *Agrobacterium* to plants. *Front. Plant Sci.* 5:326. doi: 10.3389/fpls.2014.00326

111. Melchers, L. S., Maroney, M. J., den Dulk-Ras, A., Thompson, D. V., van Vuuren, H. A. J., Schilperoort, R. A., et al. (1990). Octopine and nopaline strains of *Agrobacterium tumefaciens* differ in virulence; molecular characterization of the virF locus.*Plant Mol. Biol.* 14, 249–259. doi:

10.1007/BF00018565

112. Moore, L., Warren, G., and Strobel, G. (1979). Involvement of a plasmid in the hairy root disease of plants. *Plasmid* 2, 617–626. doi: 10.1016/0147-619X(79)90059-3

113. Mushegian, A. R., Fullner, K. J., Koonin, E. V., and Nester, E. W. (1996). A family of lysozyme-like virulence factors in bacterial pathogens of plants and animals. *Proc. Natl. Acad. Sci. U.S.A.* 93, 7321–7326. doi: 10.1073/pnas.93.14.7321

114. Nair, G. R., Lai, X., Wise, A., Wonjae, B. W., Jacobs, M., and Binns, A. N. (2011). The integrity of the periplasmic domain of the VirA sensor kinase is critical for optimal coordination of the virulence signal response in *Agrobacterium tumefaciens*. *J. Bacteriol.* 193, 1436–1448. doi: 10.1128/JB.01227-10

115. Nixon, B. T., Ronson, C. W., and Ausubel, F. M. (1986). Two-component regulatory systems responsive to environmental stimuli share strongly conserved domains with the nitrogen assimilation regulatory genes ntrB and ntrC. *Proc. Natl. Acad. Sci. U.S.A.* 83, 8278–8282. doi: 10.1073/pnas.83.20.7850

116. Ooms, G., Hooykaas, P. J. J., Noleman, G., and Schilperoort, R. A. (1981). Crown gall tumors of abnormal morphology induced by *Agrobacterium tumefaciens* carrying mutated Ti plasmids: analysis of T-DNA functions. *Gene* 14, 33–50. doi: 10.1016/0378-1119(81)90146-3

117. Pan, S. Q., Charles, T., Jin, S., Wu, Z.-L., and Nester, E. W. (1993). Preformed dimeric state of the sensor protein VirA is involved in plant-*Agrobacterium* signal transduction. *Proc. Natl. Acad. Sci. U.S.A.* 90, 9939–9943. doi: 10.1073/pnas.90.21.9939

118. Pan, S. Q., Jin, S., Boulton, M. I., Hawes, M., Gordon, M. P., and Nester, E. W. (1995). An *Agrobacterium* virulence factor encoded by a Ti plasmid gene or a chromosomal gene is required for T-DNA transfer into plants. *Mol. Microbiol.* 17, 259–269. doi: 10.1111/j.1365-2958.1995.mmi_17020259.x

119. Pansegrau, W., Schoumacher, F., Hohn, B., and Lanka, E. (1993). Site-specific cleavage and joining of single-stranded DNA by ViirD2 protein of *Agrobacterium tumefaciens* Ti plasmids: analogy to bacterial conjugation. *Proc. Natl. Acad. Sci. U.S.A.* 90, 11538–11542. doi: 10.1073/pnas.90.24.11538

120. Pantoja, M., Chen, L., Chen, Y., and Nester, E. W. (2002). *Agrobacterium* type IV secretion is a two-step process in which export substrates associate with the virulence protein VirJ in the periplasm. *Mol. Microbiol.* 45,

1325–1335. doi: 10.1046/j.1365-2958.2002.03098.x

121. Pelczar, P., Kalck, V., Gomez, D., and Hohn, B. (2004). *Agrobacterium* proteins VirD2 and VirE2 mediate precise integration of synthetic T-DNA complexes in mammalian cells. *EMBO Rep.* 5, 632–637. doi: 10.1038/sj.embor.7400165

122. Peng, W.-T., Banta, L. M., Charles, T. C., and Nester, E. W. (2001). The chvH locus of *Agrobacterium* encodes a homolog of an elongation factor involved in protein synthesis. *J. Bacteriol.* 183, 36–45. doi: 10.1128/JB.183.1.36-45.2001

123. Peng, W. T., Lee, Y. W., and Nester, E. W. (1998). The phenolic recognition profiles of the *Agrobacterium tumefaciens* VirA protein are broadened by a high level of the sugar binding protein, ChvE. *J. Bacteriol.* 180, 5632–5638.

124. Peralta, E. G., Hellmis, R., and Ream, W. (1986). Overdrive, a T-DNA transmission enhancer on the *A. tumefaciens* tumor-inducing plasmid. *EMBO J.* 5, 1137–1142.

125. Petit, A., Delhaye, S., Tempé, J., and Morel, G. (1970). Reserches sur les guanidines des tissues de crown gall. Mise en évidence d'une relation biochemique spécifique entre les souches d'*Agrobacterium tumefaciens* et les tumeurs qu'eles induisent. *Physiol. Vég.* 8, 205–2174.

126. Pitzschke, A. A. (2013). *Agrobacterium* and plant defense-transformation success hangs by a thread. *Front. Plant Sci.* 4:519 doi: 10.3389/fpls.2013.00519

127. Raineri, D. M., Boulton, M. I., Davies, J. W., and Nester, E. W. (1993). VirA, the plant signal receptor, is responsible for the strain-specific transfer of DNA to maize by *Agrobacterium. Proc. Natl. Acad. Sci. U.S.A.* 90, 3549–3553. doi: 10.1073/pnas.90.8.3549

128. Regensburg-Tuink, A. J., and Hooykaas, P. J. J. (1993). Transgenic *N. glauca* plants expressing bacterial virulence gene virF are converted into hosts for nopaline strains of *A. tumefaciens. Nature* 363, 69–71. doi: 10.1038/363069a0

129. Regier, D. A., and Morris, R. O. (1982). Secretion of *trans*-zeatin by *Agrobacterium tumefaciens*: a function determined by the nopaline Ti-plasmid. *Biochem. Biophys. Res. Commun.* 104, 1560–1566. doi: 10.1016/0006-291X(82)91429-2

130. Ronson, C, V., Nixon, B. T., and Ausubel, F. M. (1987). Conserved domains in bacterial regulatory proteins that respond to environmental stimuli. *Cell* 49, 579–581. doi: 10.1016/0092-8674(87)90530-7

131. Rossi, L., Hohn, B., and Tinland, B. (1996). Integration of complete transferred T-DNA units is dependent on the activity of virulence E2 proteins of *Agrobacterium tumefaciens*. *Proc. Natl. Acad. Sci. U.S.A.* 93, 126–130. doi: 10.1073/pnas.93.1.126

132. Russell, A. B., Peterson, S. B., and Mougous, J. D. (2014). Type VI secretion system effectors: poisons with a purpose. *Nat. Rev. Microbiol.* 12, 137–148. doi: 10.1038/nrmicro3185

133. Sakalis, P. A., van Heusden, G. P. H., and Hooykaas, P. J. J. (2014). Visualization of VirE2 protein translocation by the *Agrobacterium* type IV secretion system into host cells. *Microbiol. Open* 3, 104–117. doi: 10.1002/mbo3.152

134. Sardesai, N., Lee, L.-Y., Chen, H., Yi, H., Olbricht, R., Stinberg, A., et al. (2013). Cytokinins secreted by *Agrobacterium* promote transformation by repressing a plant Myb transcription factor. *Sci. Signal.* 6, 1–11. doi: 10.1126/scisignal.2004518

135. Seitz, E. W., and Hochster, R. M. (1964). Lysopine in normal and in crown-gall tumor tissue of tomato and tobacco. *Can. J. Bot.* 42, 999–1004. doi: 10.1139/b64-091

136. Shi, Y., Lee, L.-Y., and Gelvin, S. B. (2014). Is VIP1 important for *Agrobacterium*-mediated transformation? *Plant J.* 79, 848–860. doi: 10.1111/tpj.12596

137. Shimoda, N., Toyoda-Yamamoto, A., Aoki, S., and Machida, Y. (1993). Genetic evidence for an interaction between the VirA sensor protein and the ChvE sugar-binding protein of *Agrobacterium*. *J. Biol. Chem.* 268, 26552–26558.

138. Shimoda, N., Toyoda-Yamamoto, A., Nagamine, J., Usami, S., Katayama, M., Sakagami, Y., et al. (1990). Control of expression of *Agrobacterium vir* genes by synergistic actions of phenolic signal molecules and monosaccharides. *Proc. Natl. Acad. Sci. U.S.A.* 87, 6684–6688. doi: 10.1073/pnas.87.17.6684

139. Smith, E. F., and Townsend, C. O. (1907). A plant tumor of bacterial origin. *Science* 25, 671–673. doi: 10.1126/science.25.643.671

140. Soltani, J., van Heusden, G. P. H., and Hooykaas, P. J. J. (2008). "*Agrobacterium*-mediated transformation of non-plant organisms," in *Agrobacterium: From Biology to Biotechnology*, eds T. Tzfira and V. Citovsky (New York, NY: Springer Science+Business Media), 649–675.

141. Stachel, S. E., An, G., Flores, C., and Nester, E. W. (1985a). A Tn3 lacZ transposon for the random generation of -galactosidase gene fusions: application to the analysis of gene expression in *Agrobacterium*. *EMBO*

J. 4, 277–284.

142. Stachel, S. E., Messens, E., Van Montagu, M., and Zambryski, P. (1985b). Identification of the signal molecules produced by wounded plant cells that activate T-DNA transfer in *Agrobacterium tumefaciens*. *Nature* 318, 624–628. doi: 10.1038/318624a0

143. Stachel, S. E., and Nester, E. W. (1986). The genetic and transcriptional organization of the vir region of the A6 Ti-plasmid of*Agrobacterium tumefaciens*. *EMBO J.* 5, 1445–1454.

144. Stachel, S. E., Nester, E. W., and Zambryski, P. (1986a). A plant cell factor induces *Agrobacterium tumefaciens vir* gene expression. *Proc. Natl. Acad. Sci. U.S.A.* 83, 379–383. doi: 10.1073/pnas.83.2.379

145. Stachel, S. E., Timmerman, B., and Zambryski, P. (1986b). Generation of single-stranded T-DNA molecules during the initial stages of T-DNA transfer from *Agrobacterium tumefaciens* to plant cells. *Nature* 322, 706–712. doi: 10.1038/322706a0

146. Stachel, S. E., and Zambryski, P. C. (1986). VirA and VirG control the plant induced activation of the T-DNA transfer process of*A. tumefaciens*. *Cell* 46, 325–333. doi: 10.1016/0092-8674(86)90653-7

147. Subramoni, S., Nathoo, N., Klimov, E., and Yuan, Z.-C. (2014). *Agrobacterium tumefaciens* responses to plant-derived signaling molecules. *Front. Plant Sci.* 5:322. doi: 10.3389/fpls.2014.00322

148. Sundberg, C. D., Meek, L., Carroll, K., Das, A., and Ream, W. (1996). VirE1 protein mediates export of the single-stranded DNA binding protein VirE2 from *Agrobacterium tumefaciens* into plant cells. *J. Bacteriol.* 178, 1207–1212.

149. Suksomtip, M., Liu, P., Wood, D., and Nester, E. W. (2005). Citrate synthase mutants of *Agrobacterium* are attenuated in virulence and display reduced *vir* gene induction. *J. Bacteriol.* 187, 4844–4852.

150. Thomashow, L. S., Reeves, S., and Thomashow, M. F. (1984). Crown gall oncogenesis: evidence that a T-DNA gene from the*Agrobacterium* Ti plasmid pTiA6 encodes an enzyme that catalyzes synthesis of indole-3-acetic acid. *Proc. Natl. Acad. Sci. U.S.A.* 81, 5071–5075. doi: 10.1073/pnas.81.16.5071

151. Thomashow, M. F., Hughly, S., Buchlolz, W. G., and Thomashow, L. S. (1986). Molecular basis for the auxin-independent phenotype of crown gall tumor tissue. *Science* 231, 616–618. doi: 10.1126/science.3511528

152. Thomashow, M. F., Nutter, R., Montoya, A. L., Gordon, M. P., and Nester, E. W. (1980). Integration and organization of Ti plasmid

sequences in crown gall tumors. *Cell* 19, 729–739. doi: 10.1016/S0092-8674(80)80049-3

153. Tinland, B., Schoumacher, F., Gloeckler, V., Bravo-Angel, A. M., and Hohn, B. (1995). The *Agrobacterium tumefaciens* virulence D2 protein is responsible for precise integration of T-DNA into the plant genome. *EMBO J.* 14, 3585–3595.

154. Toro, N., Datta, A., Carmi, O. A., Young, C., Prusti, R. K., and Nester, E. W. (1989). The *Agrobacterium tumefaciens* virC1 gene product binds to overdrive, a T-DNA transfer enhancer. *J. Bacteriol.* 171, 6845–6849.

155. Toro, N., Datta, A., Yanofsky, M., and Nester, E. (1988). Role of overdrive sequence in T-DNA border cleavage. *Proc. Natl. Acad. Sci. U.S.A.* 85, 8558–8562. doi: 10.1073/pnas.85.22.8558

156. Tsai, Y. L., Chaing, Y. R., Wu, C. F., Narberhaus, F., and Lai, E. M. (2012). One out of four: HspL but no other heat shock protein of *Agrobacterium tumefaciens* acts as efficient virulence-promoting VirB8 chaperone. *PLoS ONE* 7:e49685. doi: 10.1371/journal.pone.0049685

157. Tzfira, T., Vaidya, M., and Citovsky, V. (2001). VIP1, an *Arabidopsis* protein that interacts with *Agrobacterium* VirE2, is involved in VirE2 nuclear import and *Agrobacterium* infectivity. *EMBO J.* 20, 3596–3607. doi: 10.1093/emboj/20.13.3596

158. Tzfira, T., Vaidya, M., and Citovsky, V. (2004). Involvement of targeted proteolysis in plant genetic transformation by *Agrobacterium*. *Nature* 431, 87–92.

159. Van Larabeke, N., Schell, J., Schilperoort, R. A., Hermans, A. K., Hernalsteens, J.-P., and Montagu, M. (1975). Acquisition of tumor-inducing ability by non-oncogenic agrobacteria as a result of plasmid transfer. *Nature* 255, 742–743. doi: 10.1038/255742a0

160. Veluthambi, K., Ream, W., and Gelvin, S. B. (1988). Virulence genes, borders, and overdrive generate single-stranded T-DNA molecules from the A6 Ti plasmid of *Agrobacterium tumefaciens*. *J. Bacteriol.* 170, 1523–1532.

161. Ward, E. R., and Barnes, W. M. (1988). VirD2 protein of *Agrobacterium tumefaciens* very tightly linked to the 5′ end of T-strand DNA. *Science* 242, 927–930. doi: 10.1126/science.242.4880.927

162. Ward, J. E., Akiyoshi, D. E., Regier, D., Datta, A., Gordon, M. P., and Nester, E. W. (1988). Characterization of the virB operon from an *Agrobacterium tumefaciens* Ti plasmid. *J. Biol. Chem.* 263, 5804–5814.

163. Watson, B., Currier, T. C., Gordon, M. P., Chilton, M.-D., and Nester, E. W. (1975). Plasmid required for virulence of *Agrobacterium tumefaciens*. *J. Bacteriol.* 123, 255–264.

164. Watson, B., and Nester, E. W. (2005). "A plasmid was there after all," in *Agrobacterium tumefaciens: From Plant Pathology to Biotechnology*, eds E. W. Nester, M. P. Gordon, and A. Kerr (St. Paul, MN: APS Press), 53–55.

165. White, F. F., and Nester, E. W. (1980). Relationships of plasmids responsible for hairy root and crown gall tumorigenicity. *J. Bacteriol.* 144, 710–720.

166. White, P. R., and Braun, A. C. (1941). Crown gall production by bacteria-free tumor tissues. *Science* 94, 239–241. doi: 10.1126/science.94.2436.239

167. Willmitzer, L., Simons, G., and Schell, J. (1982). The TL-DNA in octopine crown gall tumors codes for seven well-defined polyadenylated transcripts. *EMBO J.* 1, 139–146.

168. Winans, S. C., Ebert, P. R., Stachel, S. E., Gordon, M. P., and Nester, E. W. (1986). A gene essential for *Agrobacterium* virulence is homologous to a family of positive regulatory loci. *Proc. Natl. Acad. Sci. U.S.A.* 83, 8278–8282. doi: 10.1073/pnas.83.21.8278

169. Winans, S. C., Kerstetter, R. A., and Nester, E. W. (1988). Transcriptional regulation of the *virA* and *virG* genes of *Agrobacterium tumefaciens*. *J. Bacteriol.* 170, 4047–4054.

170. Wirawan, I. G. P., Kang, H. W., and Kojima, M. (1993). Isolation and characterization of a chromosomal virulence gene of *Agrobacterium tumefaciens*. *J. Bacteriol.* 175, 3208–3212.

171. Wise, A. A., Fang, F., Lin, Y.-H., He, F., Lynn, D., and Binns, A. N. (2010). The receiver domain of hybrid histidine kinase VirA: an enhancing for gene expression in *Agrobacterium tumefaciens*. *J. Bacteriol.* 192, 1534–1542. doi: 10.1128/JB.01007-09

172. Wood, D. W., Setubal, J. C., Kaul, R., Monks, D. E., Kitajima, J. P., and Okura, V. K., et al. (2001). The genome of the natural genetic engineer *Agrobacterium tumefaciens* C58. *Science* 294, 2317–2323. doi: 10.1126/science.1066804

173. Wu, C.-F., Lin, J.-S., Shaw, G.-C., and Lai, E.-M. (2012). Acid-induced type VI secretion system is regulated by ExoR-ChvG/ChvI signaling cascade in *Agrobacterium tumefaciens*. *PLOS Pathog.* 8:e1002938. doi: 10.1371/journal.ppat.1002938

174. Xu, X. Q., and Pan, S. Q. (2000). An *Agrobacterium* catalase is a

virulence factor involved in tumorigenesis. *Mol. Microbiol.* 35, 407–414. doi: 10.1046/j.1365-2958.2000.01709.x

175. Yadav, N. S., Vanderleyden, J., Bennett, D. R., Barnes, W. M., and Chilton, M.-D. (1982). Short direct repeats flank the T-DNA on a nopaline Ti-plasmid. *Proc. Natl. Acad. Sci. U.S.A.* 79, 6322–6326. doi: 10.1073/pnas.79.20.6322

176. Yanofsky, M., Lowe, B., Montoya, A., Rubin, R., Krul, W., Gordon, M., et al. (1985). Molecular and genetic analysis of factors controlling host-range in *Agrobacterium tumefaciens*. *Mol. Gen. Genet.* 201, 237–246. doi: 10.1007/BF00425665

177. Yanofsky, M., Porter, S. C., Young, C., Albright, L. M., Gordon, M. P., and Nester, E. W. (1986). The virD operon of *Agrobacterium tumefaciens* encodes a site-specific endonuclease. *Cell* 47, 471–477. doi: 10.1016/0092-8674(86)90604-5

178. Young, C., and Nester, E. W. (1988). Association of the VirD2 protein with the 5′ end of T-strands of *Agrobacterium tumefaciens*. *J. Bacteriol.* 170, 3367–3374.

179. Yuan, Z. C., Liu, P., Saenkham, P., Kerr, K., and Nester, E. W. (2008). Transcriptome profiling and functional analysis of *Agrobacterium tumefaciens* reveals a general conserved response to acidic conditions (pH 5.5) and a complex acid-mediated signaling involved in *Agrobacterium*–plant interactions. *J. Bacteriol.* 190, 494–507. doi: 10.1128/JB.01387-07

180. Yusibov, V. M., Steck, T. R., Gupta, V., and Gelvin, S. B. (1994). Association of single – stranded transferred DNA from *Agrobacterium tumefaciens* with tobacco cells. *Proc. Natl. Acad. Sci. U.S.A.* 91, 2994–2998. doi: 10.1073/pnas.91.8.2994

181. Zaenen, I., Van Larabeke, N., Teuchy, H., Van Montagu, M., and Schell, J. (1974). Supercoiled circular DNA in crown gall inducing *Agrobacterium* strains. *J. Mol. Biol.* 86, 109–127. doi: 10.1016/S0022-2836(74)80011-2

182. Zaltsman, A., Krichevsky, A., Loyter, A., and Citovsky, V. (2010). *Agrobacterium* induces expression of a host F-box protein required for tumorigenicity. *Cell Host Microbe* 7, 197–209. doi: 10.1016/j.chom.2010.02.009

183. Zambryski, P., Holsters, M., Kruger, K., Depicker, A., Schell, J., Van Montagu, M., et al. (1980). Tumor DNA structure in plant cells transformed by *A. tumefaciens*. *Science* 209, 1385–1391. doi: 10.1126/science.6251546

184. Zhan, X., Jones, D. A., and Kerr, A. (1990). The pTiC58 *tzs* gene promotes

high-efficiency root induction by agropine strain 1855 of *Agrobacterium rhizogenes*. *Plant Mol. Biol.* 14, 785–792. doi: 10.1007/BF00016511

185. Zhao, J., and Binns, A. (2011). Characterization of the mmsAB-araD1 (gguABC) genes of *Agrobacterium tumefaciens*. *J. Bacteriol.* 193, 6586–6596. doi: 10.1128/JB.05790-11

186. Ziemienowicz, A., Merkle, T., Schoumacher, F., Hohn, B., and Rossi, L. (2001). Import of *Agrobacterium* T-DNA into plant nuclei: two distinct functions of VirD2 and VirE2 proteins. *Plant Cell* 13, 369–383. doi: 10.1105/tpc.13.2.369

187. Zou, S. B., Hersch, S. J., Roy, H., Wiggers, J. B., Leung, A. S., Buranyi, S., et al. (2012). Loss of elongation factor P disrupts bacterial outer membrane integrity. *J. Bacteriol.* 194, 413–425. doi: 10.1128/JB.05864-11

188. Zou, S. B., Roy, H., Ibba, M., and Navarre, W. W. (2011). Elongation factor P mediates a novel post-transcriptional regulatory pathway critical for bacterial virulence. *Virulence* 2, 147–151. doi: 10.4161/viru.2.2.15039

189. Zupan, J. R., Citovsky, V., and Zambryski, P. (1996). *Agrobacterium* VirE2 protein mediates nuclear uptake of single-stranded DNA in plant cells. *Proc. Natl. Acad. Sci. U.S.A.* 93, 2392–2397. doi: 10.1073/pnas.93.6.2392

Chapter 10

HISTORY OF ONCOLYTIC VIRUSES: GENESIS TO GENETIC ENGINEERING

Elizabeth Kelly[1] and Stephen J Russell[1]

[1]Molecular Medicine Program, Mayo Clinic College of Medicine, Rochester, Minnesota, USA

ABSTRACT

Since the turn of the nineteenth century, when their existence was first recognized, viruses have attracted considerable interest as possible agents of tumor destruction. Early case reports emphasized regression of cancers during naturally acquired virus infections, providing the basis for clinical trials where body fluids containing human or animal viruses were used to transmit infections to cancer patients. Most often the viruses were arrested by the host immune system and failed to impact tumor growth, but sometimes, in immunosuppressed patients, infection persisted and tumors regressed, although morbidity as a result of the infection of normal tissues was unacceptable. With the advent of rodent models and new methods for virus propagation, there were numerous attempts through the 1950s and 1960s to force the evolution of viruses with greater tumor specificity, but success was limited and many researchers abandoned the field. Technology employing reverse genetics later brought about a renewal of interest in virotherapy that allowed the generation of more potent, tumor-specific oncolytics. Here, examination of early oncolytic virotherapy before genetic engineering serves to highlight tremendous advances, yet also hints at ways to penetrate host immune defenses, a significant remaining challenge in modern virotherapy research.

INTRODUCTION

The years leading up to the twentieth century are traditionally considered the beginning of modern medicine. This was the time of Virchow, Lister, Koch, Pasteur, and of dramatic advances in science and technology. In the treatment of malignant disease, however, there were no appreciable developments. Cancer therapy as a field in effect simply referred to surgery, and even that was primitive. Anesthesia was just beginning to be standard practice and chemotherapy meant arsenic or castor oil. It appears that the use of viruses in the treatment of cancer was not the result of some perspicacious theory of an alternative therapy but rather just stemmed from the observation that, occasionally, cancer patients who contracted an infectious disease went into brief periods of clinical remission. In the case of leukemia, it was well recognized that contraction of influenza sometimes produced beneficial effects.[1,2] Although no cases were reported where an accompanying infectious disease led to complete cure of leukemia, it was anticipated that a treatment based upon the causal agent of infection would provide an alternative to the «hopelessness of the ordinary treatment of leukemia.»[1]

For more than a hundred years, viruses have been pursued as experimental agents of cancer destruction. Interest in the field has fluctuated during this time, reaching fever pitch in the 1950s and 1960s, followed by near-abandonment in the 1970s and 1980s, and a resurgence of interest in the past two decades, culminating in the first marketing approval of an oncolytic virus, granted by Chinese regulators for the genetically modified oncolytic adenovirus H101 in November 2005.[3]

Historical perspective here is provided on the numerous approaches that were explored before the modern era of virus engineering through reverse genetics to develop non-pathogenic viruses selectively destructive to human tumor tissue (Figure 1).

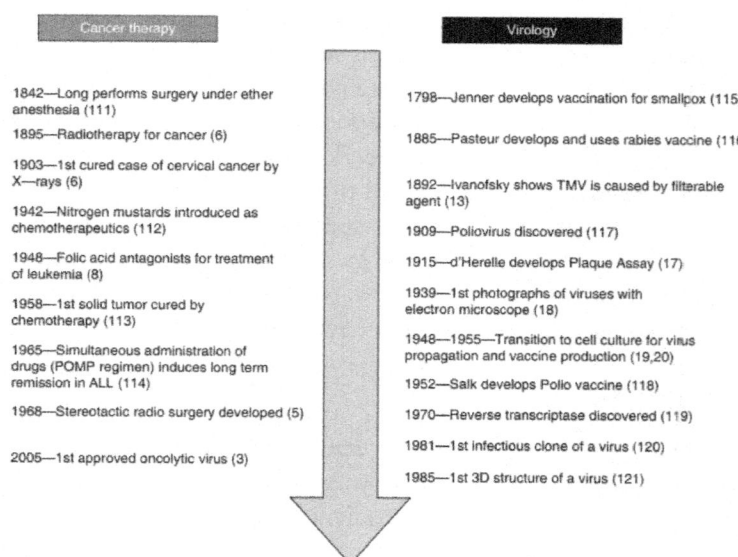

Figure 1: Milestones in the fields of virology and cancer therapy.

CANCER THERAPY: THE PAST 150 YEARS

Until the early twentieth century, cancer therapy referred to excision of the tumor by surgery. However, new treatments comprising what we would now classify as radiotherapy, chemotherapy, and immunotherapy were shortly to be introduced. Progress in the field of virotherapy came, to a large extent, from the appreciation of deficiencies in these other forms of cancer treatment, and, perhaps because of the absence of viable alternatives, promising new therapies often superseded their antecedents as the preferred method of treatment shortly after introduction.

Surgery was the predominant form of treatment until the 1900s. The evolution of surgery, however, was a slow one, stalemated until the introduction of anesthesia in 1846 and progressing only slowly thereafter.[4] Overwhelming were the advances in pathology, histology, cytology, and physiology, and although surgery had certainly become more aseptic, it remained in technique chiefly static. Though a surgical cure to cancers was often reported, the likelihood of success depended greatly upon a tumor being quickly diagnosed and somewhat readily accessible. For those patients for whom this was not the case, the outlook was bleak. Perhaps for this reason, upon the discovery of X-rays by Roentgen in 1895 and radioactivity by Marie Curie in 1898 and their subsequent application in treating cancer, radiation therapy became all the rage.[5]

Only three months after Roentgen published his report on X-rays, there was an account of X-rays in the treatment of breast cancer.[6] So fervent was the support for radiotherapy that it eclipsed surgery as the preferred method of treatment in many cancers, including head and neck.[7] Radiotherapy evolved from a horribly inexact science in which the large doses of radiation given were quantified by the extent of tissue damage to a targeted approach where the tumor was irradiated more specifically.[6] However, by the 1940s much of the initial enthusiasm for radiotherapy had diminished. Despite significant technical advances and some promising results, treatment of malignant disease with radiation had done little to improve long-term survival. Surgery, after several years on the backburner, was once again being advocated-as was chemotherapy.

Chemotherapy blossomed seemingly overnight with the introduction by Farber *et al.* in 1948 of the folic acid antagonist aminopterin for the treatment of leukemia.[8] Previously, heavy metals and nitrogen mustards had been used, but with no striking effects. As Farber›s therapy was the first to bring about regular remissions, chemical compounds for the treatment of cancer became much sought after. Steroids such as cortisone[9] and prednisone[10] were produced shortly thereafter, and antagonists of nucleic acids followed.[11] Chemotherapy was indeed an effective modality for leukemia treatment. It did not follow that this was the case for other cancers, however. Although hematological malignancies appeared susceptible to many chemotherapeutics, solid tumors were negligibly affected. As with its predecessor technologies, by the middle of the 1950s, there was uncertainty that chemotherapy alone would prove a panacea as a cancer treatment.

The early 1900s produced significant advances in cancer therapy-but no cures that were applicable to all tumors at all stages. This was a time of many scientific advances, but a growing skepticism about the future of cancer therapy. Tumors could not be removed all the time; they could not always be killed with high-energy beams of radiation or with poisons delivered intravenously; even the combination of these modalities proved insufficient. It was time for more investigation and a return to the pathogens that produced some of the first reports of remission.

VIROLOGY: THE PAST 150 YEARS

Viruses began to be employed for cancer therapy at the end of the nineteenth century, but despite the high number of infectious diseases known to be of viral etiology at that time, there was no real concept of the nature of a virus. The "discovery" of viruses is difficult to pinpoint. Martinus Beijerinck reported in 1898 that after Chamberland candle filtration (through which bacteria could not pass), the agent causing tobacco mosaic disease could amplify itself in living, growing plant tissue.[12] His work built on the prior observations of Adolf Mayer and Dimitri Ivanofsky[13,14] and provided at least an operational definition of a virus, which he labeled a *contagium vivium fluidum*. Foot and mouth virus, in 1898, was the first "filterable agent" to be implicated in animal disease,[15] and yellow fever virus, in 1901, was the first to be implicated in human disease.[16] Biochemical analysis of viruses proceeded apace, but it was not until the advent of the plaque assay in 1917 that their particulate nature could be proven,[17] and electron microscopic images of viral particles were not obtained until 1939.[18] The first half of the twentieth century was something of a dark age for virology. Although agents that could pass through filters had been implicated in many diseases (*e.g.*, polio, rabies, influenza), their precise identity was still unclear.

Understanding of viruses accelerated rapidly in the 1950s and 1960s, in large measure because of the advent of cell and tissue culture systems that allowed *ex vivo* virus propagation.[19,20] It is no coincidence that this was also a time of intensive virotherapy research when the oncolytic properties of numerous viruses were evaluated, first in human tumor cell lines, often implanted in immunosuppressed rodents, and subsequently in humans. During the past fifty years, viruses have been studied with such unparalleled intensity that their biology is now understood more thoroughly than that of any other organism in nature. Their genomes and proteins have been sequenced; their physical structures are known, as are many of the mechanisms whereby their genomes are regulated; their diversity is recognized; their replication cycles and pathogenetic strategies have been elucidated; and methods have been developed to manipulate their genome sequences to permit their further refinement as anticancer agents.

EXPERIMENTS OF NATURE: TUMOR REGRESSIONS DURING NATURALLY ACQUIRED VIRAL INFECTIONS

Since the mid-1800s, there has been a steady trickle of case reports where tumor regressions have coincided with natural virus infections.[1,2,5,21] Most often the patients in question were suffering from hematological malignancies such as leukemia or lymphoma, known to be associated with significant suppression of immune function. In addition, the remissions were short-lived, typically lasting only one or two months. In one very widely cited example, Dock[1] described a 42-year-old woman with «myelogenous leukemia» that went into remission after a presumed influenza infection. The report was made in 1896, 37 years before it was determined that influenza was a virus infection. The woman had a greatly enlarged liver and spleen, which shrank to nearly normal size, and a grossly elevated leukocyte count, which dropped more than 70-fold after the infection. In another case, chickenpox led to the regression of lymphatic leukemia in a 4-year-old boy.[5] Before the onset of chickenpox he was not receiving anti-leukemic therapy, his liver and spleen were enormous, he had enlarged cervical lymph nodes, and his leukocyte count was grossly elevated (200 lymphoid cells/μl). Within a few days of his developing the classical varicella rash, his spleen and liver returned to near-normal size, his white count fell to normal (4.1 cells/μl), the differential count normalized, his platelet count and hemoglobin increased significantly, and examination of his bone marrow confirmed that he was in remission. Unfortunately, the remission lasted only one month, and his leukemia progressed rapidly until death.

More recent clinical reports have described the regression of leukemia,[22,23] Hodgkin's disease,[24,25] and Burkitt›s lymphoma[26] concomitantly with measles infection. One can ascertain from these and similar case reports where remissions coincided with other naturally acquired viral infections, such as hepatitis, glandular fever, and measles,[22,23,24,25,26,27,28] the following: (i) under the right circumstances, certain viruses can destroy tumors without causing undue harm to the patient; (ii) virus-mediated tumor regression has most often been seen when the patient is young and has a compromised immune system, *i.e.,* is suffering from leukemia or lymphoma; (iii) virus-induced remissions have generally been short-lived and incomplete.

HUMAN PATHOGENS

Exploiting viruses for therapeutic gain was first suggested at the beginning of the twentieth century, yet attempts at implementation remained sporadic[29,30,31]until nearly fifty years later, when clinical testing began in earnest. In the context of current ethical standards, some of the clinical studies performed at that time seem quite alarming, as the therapeutic material administered to cancer patients often consisted of infectious body fluids or infected tissue harvested from patients with ongoing virus infections, some of which were quite serious. However, these were desperate times for those afflicted with cancer.[32,33,34]

Table 1: Clinical virotherapy: four historically significant clinical trials

Year(s)	Virus	Disease	No. of patients	Administration	Outcome	Side effect
1949	Hepatitis B virus[32]	Hodgkin's disease	22	Parenteral injection of unpurified human serum, tissue extract	14/22 developed hepatitis; 7/22 improved in clinical aspect of disease; 4/22 reduction in tumor size	Fever, malaise, death (1 confirmed)
1952	Egypt 101 virus (early passage West Nile)[36]	Advanced, unresponsive neoplastic disease	34	IV, intramuscular injection of bacteriologically sterile mouse brain, chick embryo, human tissue	27/34 infected; 14/34 oncotropism; 4/34 (transient) tumor regression	Fever, malaise; mild encephalitis (2 confirmed)
1956	Adenovirus adenoidal-pharyngeal-conjuctival virus (APC)[50]	Cervical carcinoma	30	IT, IA, IV injection of TC supernatant	26/40 inoculations resulted in localized necrosis	Vaginal hemorrhage; infrequent (3/30) fever, malaise
1974	Mumps virus (wild-type, non-attenuated)[56]	Terminal cancers; gastric, pulmonary, uterine account for more than 50%	90	External post-scarification; IT; IV; oral; rectal; inhalation of purified human saliva or TC supernatant	37/90 complete regression or decrease >50%; 42/90 decrease <50% or growth suppression; 11/90 unresponsive	7/90 adverse reactions: bleeding, fever

Abbreviations: IA, intra-arterial; IT, intratumoral; IV, intravenous; TC, tissue culture.

Hepatitis viruses were among the first to be used for therapy. As early as 1897 it had been noticed that viral hepatitis had ameliorating effects on a variety of human diseases.[28] Then in 1949, when two patients with Hodgkin›s disease were observed to go into brief remission after contracting viral hepatitis, clinical trials were undertaken in which 22 patients suffering from Hodgkin›s disease were treated by administration of a total of 35 sera or tissue extracts containing «the hepatitis virus» (Table 1).[32] The patients who contributed sera and tissue samples for these studies were suffering either from infectious hepatitis, a self-limited picornavirus infection, or from serum hepatitis, most likely as a result of hepatitis B. Undissuaded by the death of the first patient being treated with the regimen of multiple inoculations of hepatitis serum, Hoster *et al.* expanded the experimental group to a further 21 patients. Of these, 13 developed hepatitis and 7 were reported to show improvement lasting at least one month in one or more aspects of their disease. Death directly attributed to viral hepatitis was reported to occur on occasion, yet the number of patients to whom this pertains is undisclosed.

Owing perhaps to the suggestion of success given by Hoster *et al.* and additional evidence pointing to remission of monocytic leukemia in a patient naturally infected with Epstein-Barr virus, clinical trials were undertaken in the United Kingdom using glandular fever serum for the treatment of acute leukemia.[27] In this case, the results were a little more encouraging, as three of five treated patients who acquired symptoms of glandular fever did indeed go into remission, albeit briefly. Side effects attributed to the treatment were comparatively minor and of limited duration.

After these early clinical studies, many different human pathogens were administered to cancer patients during the next two decades. Also at this time, Alice Moore, who pioneered the testing of oncolytic viruses in animal cancer models (see below), teamed up with Chester Southam, now considered a rather overzealous clinical oncologist,[35] at Memorial Sloan-Kettering hospital. As the dominant players of that era in the field, Southam and Moore[36,37,38,39,40,41,42,43,44] contributed much in the form of both preclinical trials with animal models and clinical trials employing oncolytic viruses, but perhaps even more so in terms of cautionary tales of transitioning from bench to bedside.

With mosquitoes rampant across the United States and abroad, flavivirus infections such as West Nile, Uganda, dengue, and yellow fever were exceedingly common and were thus some of the first employed for virotherapy.[36,37,38,40,41,42,45,46] The Egypt 101 isolate of West Nile virus was used in more than 150 virus therapy trials against a wide range of cancers (Table 1).[36,38] Viremia and intra-tumoral virus replication were confirmed in most patients, but tumor responses were rare. Immunosuppressed patients with leukemia or lymphoma were more likely to respond to therapy, but were also at higher risk of fatal neurotoxicity. Thus, of eight patients with leukemia or lymphoma, five experienced severe encephalitis. With limited success in controlling the neurotoxicity of the aforementioned agents, large-scale screens of putative oncolytics were carried out. Most were abandoned as lacking in efficacy or safety, however, and a gradual shift of emphasis occurred such that adenoviruses, herpes viruses, paramyxoviruses, picornaviruses, and the pox viruses eventually emerged as the new favorites.

Identified as an oncolytic agent in preclinical models in the 1950s, adenoidal-pharyngeal-conjuctival virus (APC, now known to be an adenovirus) quickly progressed to the bedside and was found to have relatively modest side effects: those who were administered APC occasionally developed inflammation of the eye or pharynx, but were encephalitis-free and, better yet, alive after inoculation.[47] Overcoming this first obstacle, APC rapidly made its way to clinical trials for the treatment of cervical cancer.[48] By intravenous, intravascular, or intra-arterial routes, APC was administered to 30 patients with advanced epidermoid carcinoma of the cervix (Table 1). In two-thirds of the cases, areas of necrosis were present in tumors within 10 days and, most remarkably, appeared to be confined to the cancerous tissue (though no biopsies were performed to test for viral recovery in the surrounding tissues). In those who responded to APC administration, cancerous tissue was shed in large amounts. Although the APC oncolytic produced striking effects, liquefying, causing severe hemorrhaging, and frequently a generalized necrosis specifically to the site of the tumor, infections were quickly eradicated by the host immune system and survival was not significantly prolonged. More than half of those who were treated with APC died within a few months of the beginning of the trial, all from the primary disease. As expected, responses were diminished in patients with pre-existing anti-adenovirus antibodies, emphasizing the problem of premature immune-mediated virus elimination. Though investigation into adenovirus for the treatment of cervical cancer continued,[49,50] in the absence of any useful prolongation of survival, it slowed to a trickle.

Picornavirus implementation for oncolysis briefly came into vogue thereafter. In one study, published in 1957, tumor xenografts were established by intraperitoneal inoculation of HeLa cells in rats that had first been irradiated and treated with cortisone. After several adenoviruses, enteroviruses (Coxsackie A, B, and enteric cytopathic human orphan viruses), vaccinia, and vesicular stomatitis virus were tested for intra-tumoral amplification in this model, Coxsackie B3 virus emerged as the winner.[51] Poliovirus was later used and was shown to cause necrosis and regression in guinea pigs carrying HeLa tumors without any appreciable side effects,[52] yet data still were lacking for an oncolytic virus that was efficacious in human trials.

With enthusiasm for adenovirus as an experimental therapeutic starting to ebb, with no data on picornaviruses in clinical trials equal to what had been shown in rodent models, and with use of the neurotropic viruses of antiquity still out of the question to the circumspect a penchant emerged for a virus that was already widespread and with which there was a substantial amount of experience as a pathogen-thus, the paramyxovirus mumps.

Mumps virus was initially used not as an oncolytic agent but, similarly to some other viruses,[53,54] as an agent of immunotherapy to stimulate the immune system of those who did not respond to a combinatorial treatment of surgery plus chemotherapy plus BCG vaccine for metastatic melanoma.[55] Inoculation with killed mumps virus did appear to enhance tumor regression, but the virus garnered much more attention when it was allowed to replicate. In what was certainly a massive clinical trial for the time, Asada used non-attenuated mumps viruses in Japan to treat 18 types of tumors (**Table 1**).[56] Delivery methods ranged from the ubiquitous intravenous and intravascular administration to rather peculiar methods employing bread or pieces of tampon orally administered after soaking in virus-containing supernatant. They were extended to include topical application after scarification of the skin, rectal administration, and inhalation. Lest the results be too straightforward, the mumps virus itself was obtained from various sources, including the saliva of infected patients and infected cultures of monkey or human embryonic kidney cells. Despite the utter lack of controls for these studies, an admitted limited quantity of the virus itself, and the fact that most of the patients undergoing mumps therapy had neutralizing antibodies, Asada›s results as reported were among the most dramatic ever seen. Minimal toxicity was reported in mumps therapy, and in 37 of 90 patients treated, the tumor regressed completely or decreased to less than half of the initial size. Indeed, all save 10% of the treated patients were reported to have some appreciable response to oncolytic mumps virus therapy. The treatment often produced these effects within a few days of the first administration, before the spread of the virus was brought to a halt by a strong anamnestic anti-mumps immune response. Boosting of anti-tumoral immunity was also reported in a number of cases. After the euphoria of Asada›s initial clinical experience with mumps virus therapy, the subsequent performance of mumps seemed lackluster by comparison.[57,58,59,60]

Perhaps because of the regulatory barriers that would have had to be confronted if non-attenuated pathogens were to become a standard anticancer therapy, there followed a slump in reported clinical trials employing oncolytic viruses, with the number decreasing rather dramatically in the 1970s and 1980s. Nonetheless, the dream of effective oncolytic virotherapy persisted, but now with focus shifting to viruses with diminished pathogenic potential.

IN VIVO CANCER MODELS: VIROTHERAPY BECOMES A SCIENCE

Ex vivo culture of human cells had become possible in 1948 (ref. 61), and attempts to implant these cells into laboratory rodents followed, providing the first opportunity to test the *in vivo* anti-tumor activity of an oncolytic virus under controlled laboratory conditions. Moore, the first to investigate oncolytic viruses using newly developed rodent cancer models, used an *in vivo* tumor model to demonstrate conclusively that an oncolytic virus, in this case Russian Far East encephalitis virus, could selectively seek out and destroy cancer cells in a living animal, reporting her initial findings in 1949. She found that, in certain instances, the mouse sarcoma 180 could be completely destroyed,[37] a landmark for virotherapy. In an effort to extend these results to other tumors, Moore tested the virus against five mouse tumors of various origins[40] and found that it was indeed able to cause complete regression, provided the dose was sufficient. In addition, after infection, tumors could no longer be transplanted to other immunocompetent mice. Although infection with the Russian Far East encephalitis virus did eventually cause fatal encephalitis in the mice, proof of principle for its extraordinary oncolytic potency had been given and much interest was spurred in the field.

In the years that followed, many other human pathogens were investigated for oncolytic activity employing rodent models, including Bunyamwera, Ilheus, dengue, yellow fever, West Nile virus[41,42] and its Egypt 101 isolate,[36] Semliki Forest virus, mumps, vaccinia,[62] and adenovirus.[48,63] As previously discussed, many of these were also evaluated in clinical trials that served primarily to demonstrate that complete tumor regression was much more likely to occur in the mouse than in the patient,[41,48,62,63] bringing into question the relevance of the responses seen in heterotransplanted cancer tissues in the mouse model. Nevertheless, proof of activity in rodent models quickly became a necessary step to establish proof of principle for oncolytic activity of newly identified oncolytic viruses before clinical testing.

It is difficult to speculate on the impetus, but perhaps in an attempt to validate the mouse model, Southam and Moore abandoned briefly the use of mice and began transplanting human tumor cell lines into purported human volunteers before treating them with oncolytic viruses.[39,64] Tissue cultures of HeLa, HEp#1, HEp#2, HEp#3, J-111, and HS#1 cell lines were implanted by subcutaneous inoculation in the forearm, both in normal control patients and in 22 cancer patients. Subsequently, virus replication in tumor implants of cancer patients was compared with replication in healthy adults, and antiviral antibody responses were evaluated. The tumor cell lines were rejected by normal recipients, usually within 3 weeks of implantation, but in 20 of the 22 cancer patients they survived and grew at the sites of implantation, although typically regressing completely by week 6. In a small number of cancer patients, regression was not seen and the tumors had to be excised, sometimes more than once as they tended to regrow at the site of excision. In at least one instance there was clear evidence of lymphatic spread to the axillary lymph nodes. The studies attracted quite harsh criticism from some quarters[35] and may have served to tarnish the entire field of research. It is unclear whether any of the results significantly advanced our understanding of the propagation of oncolytic viruses in cancer patients.

CIRCUMVENTING PATHOGENICITY: USING ANIMAL VIRUSES FOR HUMAN THERAPY

In an effort to control virulence and at the same time avoid the problem of rapid virus elimination resulting from pre-existing antiviral immunity, it was initially hypothesized that a non-human animal virus might retain oncolytic activity even in a host not traditionally susceptible to that particular virus. In support of this theory was Moore's early work with Russian Far East encephalitis virus, a human pathogen showing activity against a murine tumor.[40]

In one early study, a high-throughput screen for non-human animal viruses that possessed oncolytic activity was conducted against a panel of human cancer cell lines. Adaptation by *in vitro* passing in this case was still assumed to be advantageous as the parental strain was either non-infectious or non-pathogenic in humans. Therefore, in the absence of normal cells, it was supposed that the viruses would acquire no additional tropism for cells other than those in which they were adapted.[65] Six animal viruses were identified that had the ability to propagate in human cell lines out of 24 candidates.[65] Even with the identification of these putative apathogenic oncolytics, Hammon *et al.* retained an air of despondency toward the use of viruses in cancer treatment. With chemotherapy still in its fledgling stages, however, the authors expressed a sense of obligation to explore virotherapy as it seemed as promising as any experimental therapy.

In a rigorous screening of the six viruses, Yohn *et al.* evaluated oncolytic potential by the ability to inhibit heterotransplanted human tumors from growing or by causing necrosis and regression of the tumors that did. Of these, two herpes viruses non-pathogenic in humans (equine rhinopneumonitis and infectious bovine rhinotracheitis) were identified as oncolytic for one or more human tumors.[66]

Arenaviruses have been little employed as oncolytics after the rather disappointing clinical performance of a virus referred to as the "M-P" virus (after the authors Molomut and Padnos), now identified as a strain of lymphocytic choriomeningitis virus. Par for the course in virotherapy, the M-P virus brought about dramatic tumor regressions in rodent models, increasing survival by more than 60% over controls in certain cases[66] yet offered little therapeutic benefit in human clinical trials, failing to prolong survival.[67]

In addition to herpes and arenaviruses, the use of avian viruses as oncolytic agents excited a great deal of interest in the 1960s, some of which has continued to this day. Among the first of the avian viruses used for cancer therapy was the avian plague virus. Avian plague virus, presumed to be less pathogenic than many of the neurotropic viruses used by Southam and Moore,[42] was inoculated in sarcoma 97 or epithelioma tumors of the mouse. Although the virus brought about a reduction in tumor size and a transient remission of sarcoma 97 tumors, it rapidly localized to the brain in mice with epithelioma MI, ultimately resulting in death. The observation that nearly every virus identified to date had a neurotropism as strong or stronger than any tropism toward the tumor certainly gave reason to question what, if any, ameliorative effects viral therapy could really produce.

Hope for success, however, came with the testing of another avian virus, Newcastle disease virus (NDV). Although some complete tumor regressions had been seen in virotherapy previously, in nearly every instance the experimental animal died of infection caused by the virus.[37,38,41] In this instance, NDV was inoculated intraperitoneally in animals with Ehrlich ascites carcinomas and was found to be curative with no ill effects that could be attributed to the virus.[70]Even upon rechallenge, nearly 90% of mice were found to be immune to subcutaneous injection.[71] Newcastle disease virus continues to be used in cancer therapy and has been reported in follow-up studies to provide remissions lasting at least 10 years.[72,73]

Whereas much attention had been given to viral adaptation being advantageous for targeting, not much focus had yet been placed on adaptations that a non-human pathogen might acquire that could increase its virulence in a host not normally susceptible. Introduction of wild-type viruses in a traditionally naïve host where populations have not evolved any resistance to the virus would today be considered quite risky. Indeed, one virus that had been used in virotherapy, feline panleukopenia virus,[5] later evolved independently to be transmissible to dogs, resulting in the pandemic canine parvovirus that is believed to have infected more than 80% of wild and domestic dogs between 1978 and 1979 across the world.[74]

For these and other reasons, few animal viruses are still pursued as oncolytic agents today. However, there are exceptions, such as vesicular stomatitis virus, an impotent pathogen of domestic cattle belonging to the Rhabdovirus family. This virus is selectively destructive to human tumor cells with interferon pathway defects[75] and centuries of human exposure to vesicular stomatitis virus-infected cattle has led to little more than occasional cases of conjunctivitis, implying it is quite unlikely to evolve to human pathogenicity when administered as an oncolytic. For similar reasons, NDV is also considered safe for human applications.

ADAPTATION, ATTENUATION, AND ENGINEERING: BUILDING A BETTER VIRUS

Viruses appeared to have tremendous potential but needed manipulation to be targeted more specifically to cancerous cells. Thus began the era of adaptation and, ultimately, genetic engineering of viruses. It was recognized early that viruses were capable of adapting themselves for replication in specific tissues. Because direct manipulation of the viral genome was not yet possible, Southam, Moore, and others[41,65,66] tried to apply this property to a targeted evolution of sorts-making the virus more oncolytic or more tumor-specific. The technical foundation for this concept came from the observation of Levaditi *et al.* in 1922 that a smallpox vaccine (vaccinia virus) was able to inhibit various tumors in rats and mice.[30] Subsequently, Pack reported in 1950 that he had observed a long remission of metastatic melanoma in a patient who was vaccinated against rabies after a dog bite.[53] He therefore went on to treat an additional 12 patients with metastatic melanoma by repeated intramuscular injection of rabies vaccine and saw two responses. The closing paragraph of his report states, «I am motivated in publishing this incomplete and unsuccessful story at this premature time by the unfortunate fact that news of this experiment has become widely disseminated. I have been besieged by innumerable telegrams, letters and telephone calls begging for more specific information.» Clearly, the need for effective cancer therapy in 1950 was dire.

By 1952, Moore was already utilizing the adaptive capacity of viruses to enhance oncolytic activity. Perturbed, perhaps, by the differing oncolytic activity of a given virus in tumors of various origins, Moore hypothesized that successive passage of a virus with known oncolytic activity in a mouse tumor might increase its destructive capacity for that tumor. Indeed, this was found to be the case, as Russian encephalitis virus had greater oncolytic activity after 20-30 passages in sarcoma 180 tumors than the parental strain.[41] Moore had surmised that some progeny of the parental virus would acquire mutations that could be beneficial to replication specifically in those cells in which they were propagated. Thus, the era of viral manipulation had begun, even if at this time viruses themselves were doing the majority of it.

Even before recombinant DNA techniques became available, it was suggested that the alteration of the viral genome could provide improved targeting of oncolytic viruses.[44] Although Moore had developed a form of targeted evolution by propagation in a tumor, direct engineering of viruses was not yet possible. As there had been little success in excluding viruses from specific tissues, efforts began to interfere with their tropism for sites other than tumors-especially to the brain, which caused death in a number of cases. Chemical interference with neurotropism had been suggested previously,[44] and although there was some evidence to suggest that drugs and small molecules were somewhat efficacious in the treatment of viral disease,[76,77,78,79] exclusion of viral replication in only specific tissues was regarded as quite difficult and was little investigated.

In an effort to block the neurotropism yet retain the anti-tumor potency of a primary oncolytic virus, a second virus, NDV, was employed as an agent of interference.[80] The concept was that the brain, after infection with a non-pathogenic virus, would be resistant to invasion by an oncolytic virus administered concurrently. Although the data seemed to suggest that inoculation of NDV intra-cerebrally before infection with the active oncolytic agent did prolong survival by a short time, it offered little total protection. More disappointing, the virus that had been previously shown to be most efficacious in cancer therapy, the Egypt 101 isolate of West Nile virus, was negligibly attenuated by NDV.

With limited success in virus adaptation or with viral interference to reduce the pathogenicity of oncolytic viruses, focus shifted to the manipulation of the viral genome. Before the recombinant DNA revolution, it was first demonstrated in 1968 that genetic alteration of a viral genome was possible when polynucleotides were added to the tobacco mosaic virus genome,[81] leading to expression of polylysine. However, it was not until the early 1990s, when recombinant DNA technology became standard, that virus engineering could provide any scientific furtherance of virotherapy. Not until more than thirty years after the suggestion by Southam of viral attenuation by genomic alteration of an oncolytic virus[44] did the first studies take place in which an engineered virus was employed for cancer therapy.

At the start of the 1990s, the engineering of viruses was in full swing, and virus-based gene therapy was underway for severe combined immunodeficiency,[82,83] liver failure, hemophilia,[84] and other diseases. Also, the specter of ethically questionable clinical trials employing oncolytic viruses was by now in the distant past. Enter Martuza and the treatment of malignant glioma using a mutant herpes simplex virus (HSV).

In contrast to «suicide» gene therapy that employed the thymidine kinase gene to render tumor cells sensitive to the prodrug ganciclovir,[85] Martuza›s approach was to completely remove it from the HSV genome. Although a mutant HSV lacking the enzymatic activity of thymidine kinase had been isolated almost twenty years previously,[86] not until Martuza et al.'s work was its potential as a cancer therapeutic appreciated. By extending on the observation that a thymidine kinase-negative HSV replicated in dividing cells but was crippled in non-dividing cells, Martuza was able to apply HSV-thymidine kinase to treat malignant gliomas by intra-cerebral inoculation in mice, thereby eradicating tumors completely.[87] Although the historical problem of encephalitis still lingered to some degree, nearly one-third of treated mice were spared.

Recombinant technology, when it arrived, focused predominantly on engineering adeno-,[88,89] paramyxo-,[90,91] herpes,[92] picorna-, and poxviruses.[93,94] Yet, even with the newfound ability to engineer viral genomes to produce a new generation of safer, specific oncolytics, a true therapeutic frontrunner has yet to emerge. This is likely due not to inherent problems with the viruses now in circulation, but rather to their rapid clearance by the host immune system.

CONCLUSIONS

Under the right circumstances, viruses are capable of destroying tumor tissue in human cancer patients. Remarkably, for certain human pathogens, the damage inflicted on tumors is far more significant than the damage inflicted on normal host tissues. However, because of their pathogenicity, most human viruses cannot be considered suitable as drugs. Possible exceptions include Coxsackie virus A21, an oncolytic human picornavirus whose pathogenicity is limited to mild upper respiratory tract infections,[95] and human reovirus, which is oncolytic in preclinical models but apparently apathogenic in infected humans.[96]

Fortunately, most viruses can now be adapted or engineered to eliminate their pathogenicity without destroying their oncolytic potency. Some animal viruses lack pathogenicity in humans but are nevertheless capable of destroying human tumor tissue. The main impediment to using animal viruses for human cancer therapy is the ever-present risk of virus evolution giving rise to a new human pathogen able to spread from the patient to non-immune contacts. This risk is not easy to quantify, and it is noteworthy that certain viruses of animal origin (*e.g.,* NDV) have been administered so frequently to humans without adverse consequences that they are now considered to be safe platforms for the development of oncolytics.[70,71,72,73,80,97]

Most of the oncolytic viruses currently in clinical testing are attenuated derivatives of prevalent human pathogens. Typically, in recent years, they have been genetically engineered to further attenuate their pathogenicity, increase their oncolytic potency, or enhance their specificity for cancer tissue. Measles virus, for example, was first attenuated by serial passage in cultured cells, then genetically engineered to enhance its oncolytic potency and tumor specificity.[98,99,100,101,102] By and large, through a variety of mechanisms that have recently been reviewed[103,104,105,106,107,108,109,110] these clinically tested oncolytic viruses have shown strong specificity for neoplastic tissue.

With specificity addressed, a major remaining weakness of current oncolytics is their vulnerability to antiviral host defenses, in that they are usually eliminated from the body before they have had a chance to cause substantial damage to the tumor. Thus, the viruses run into resistance from innate and acquired antiviral immune responses, including the release of interferons, activation of natural killer cells, amplification of antigen-specific cytotoxic T cells, and the secretion of high-affinity antiviral antibodies. These host responses to virus invasion have long been recognized as major impediments to virus-mediated tumor destruction, but clinical trials of old suggest that oncolytic viruses can be effective provided the immune response is suppressed. Current scientific endeavors to develop strategies to control or circumvent these antiviral responses, therefore, seem highly prudent. With studies well underway to address specificity, delivery, and potency, there is a justified sense of optimism in the field and a feeling that the challenges of antiquity are finally (or soon to be) history.

REFERENCES

1. Dock, G (1904). The influence of complicating diseases upon leukemia. *Am J Med Sci* **127**: 563–592.

2. Pelner, L, Fowler, GA and Nauts, HC (1958). Effects of concurrent infections and their toxins on the course of leukemia. *Acta Med Scand Suppl* **338**: 1–47.

3. Garber, K (2006). China approves world's first oncolytic virus therapy for cancer treatment. *J Natl Cancer Inst* **98**: 298–300.

4. Brill, S, Gurman, GM and Fisher, A (2003). A history of neuraxial administration of local analgesics and opioids. *Eur J Anaesthesiol* **20**: 682–689.

5. Bierman, HR, Crile, DM, Dod, KS, Kelly, KH, Petrakis, NL, White, LP *et al.* (1953). Remissions in leukemia of childhood following acute infectious disease: staphylococcus and streptococcus, varicella, and feline panleukopenia. *Cancer* **6**: 591–605.

6. Bernier, J, Hall, EJ and Giaccia, A (2004). Radiation oncology: a century of achievements. *Nat Rev Cancer* **4**: 737–747.

7. McGurk, M and Goodger, NM (2000). Head and neck cancer and its treatment: historical review. *Br J Oral Maxillofac Surg* **38**: 209–220.

8. Farber, S, Diamond, LK, Mercer, RD, Sylvester, RF Jr. and Wolff, JA (1948). Temporary remissions in acute leukemia in children produced by folic acid antagonist, 4-aminopteroyl-glutamic acid (aminopterin). *N*

Engl J Med **238**: 787–793. |

9. Pearson, OD, Eliel, LP, Rawson, RW, Dobriner, K and Rhoads, CP (1949). Adrenocorticotropic hormone- and cortisone-induced regression of lymphoid tumors in man: a preliminary report. *Cancer* **2**: 943–945. |

10. Hill, JM, Marshall, GJ and Falco, DJ (1956). Massive prednisone and prednisolone therapy in leukemia and lymphomas in the adult. *J Am Geriatr Soc* **4**: 627–641.

11. Burchenal, JH (1977). The historical development of cancer chemotherapy. *Semin Oncol* **4**: 135–146.

12. Beijerinck, MW (1898). Concerning a contagium vivum fluidum as a cause of the spot-disease of tobacco leaves. *Verh Akad Wetensch* **6**: 3–21.

13. Ivanofsky, D (1892). Concerning the mosaic disease of the tobacco plant. *St. Petersburg Acad Imp Sci Bul* **35**: 67–70.

14. Mayer, A (1886). On the mosaic disease of tobacco. *Landwyn VerSStnen* **32**: 451–467.

15. Loeffler, F and Frosch, P (1898). *Zentralbl Bakteriol 1 Orig* **28**: 371.

16. Reed, W, Carroll, J, Agramonte, A and Lazear, J (1901). *Senate Documents* **66**: 156.

17. d'Herelle, FH (1917). Sur un microbe invisible antagoniste des bacilles dysenteriques. *C R Hebd Seances Acad Sci Paris* **165**: 373–390.

18. Kausche, G, Ankuch, P and Ruska, H (1939). Die Sichtbarmachung von PF lanzlichem Virus in Ubermikroskop. *Naturwissenschaften* **27**: 292–299. |

19. Gey, C, Coffman, W and Kubicek, M (1952). Tissue culture studies of the proliferative capacity of cervical carcinoma and normal epithelium. *Cancer Res* **12**: 264–265.

20. Weller, TH, Robbins, FC and Enders, JF (1949). Cultivation of poliomyelitis virus in cultures of human foreskin and embryonic tissues. *Proc Soc Exp Biol Med* **72**: 153–155.

21. Sinkovics, J and Horvath, J (1993). New developments in the virus therapy of cancer: a historical review. *Intervirology* **36**: 193–214. |

22. Pasquinucci, G (1971). Possible effect of measles on leukaemia. *Lancet* **1**: 136.

23. Gross, S (1971) Measles and leukaemia. *Lancet* **1**: 397–398.

24. Zygiert, Z (1971). Hodgkin's disease: remissions after measles. *Lancet* **1**: 593.

25. Taqi, AM, Abdurrahman, MB, Yakubu, AM and Fleming, AF (1981).

Regression of Hodgkin's disease after measles. *Lancet* 1: 1112. |

26. Bluming, AZ and Ziegler, JL (1971). Regression of Burkitt›s lymphoma in association with measles infection. *Lancet* 2: 105–106. |

27. Taylor, AW (1953). Effects of glandular fever infection in acute leukemia. *Br Med J* 1: 589–593.

28. Still, GF (1897). On a form of chronic joint disease in children. *Med Chir Soc* 80: 52.

29. DePace, N (1912). Sulla scomparsa di un enome canco vegetante del collo dell'utero senza cura chirurgica. *Ginecologia* 9: 82–89.

30. Levaditi, C and Nicolau, S (1922). Sur le culture du virus vaccinal dans les neoplasmes epithelieux. *CR Soc Biol* 86: 928.

31. Levaditi, C and Nicolau, S (1936). Affinite u virus de la peste aviare pour les cellules neoplastiques (epithelioma) de la souris. *CR Soc Biol* 202: 218–220.

32. Hoster, H, Zanes R and von Haam E (1949). The association of "viral" hepatitis and Hodgkin's disease. *Cancer Res* 9: 473–480.

33. Carter, SK (1976). Immunotherapy of cancer in man: current status and prospectus. *Ann N Y Acad Sci* 277: 722–740.

34. Onuigbo, WI (1962). Historical trends in cancer surgery. *Med Hist* 6: 154–161.

35. Lerner, BH (2004). Sins of omission-cancer research without informed consent. *N Engl J Med* 351: 628–630.

36. Southam CM, and Moore, AE (1952). Clinical studies of viruses as antineoplastic agents, with particular reference to Egypt 101 virus. *Cancer* 5: 1025–1034.

37. Moore, AE (1949). The destructive effects of the virus of Russian Far East encephalitis on the transplantable mouse sarcoma 180. *Cancer* 2: 525–534.

38. Moore, AE (1954). Effects of viruses on tumors. *Annu Rev Microbiol* 8: 393–410.

39. Moore, AE, Rhoads, CP and Southam, CM (1957). Homotransplantation of human cell lines. *Science* 125: 158–160.

40. Moore, A (1951). Inhibition of growth of five transplantable mouse tumours by the virus of Russian far east encephalitis. *Cancer* 4: 375–382.

41. Moore, A (1952). Viruses with oncolytic properties and their adaptation in tumours. *Ann N Y Acad Sci USA* 54: 945–952. | |

42. Southam, CM and Moore, AE (1951). West Nile, Ilheus, and Bunyamwera virus infections in man. *Am J Trop Med Hyg* **31**: 724–741.

43. Southam, CM (1976). History and prospects of immunotherapy of cancer: an introduction. *Ann N Y Acad Sci* **277**: 1–6.

44. Southam, CM (1960). Present status of oncolytic virus studies. *Trans N Y Acad Sci* **22**: 657–673.

45. Mettler, NE, Clarke, DH and Casals, J (1982). Virus inoculation in mice bearing Ehrlich ascitic tumors: antigen production and tumor regression. *Infect Immun* **37**: 23–27.

46. Webb, HE, Wetherley-Mein, G, Smith, CE and McMahon, D (1966). Leukaemia and neoplastic processes treated with Langat and Kyasanur Forest disease viruses: a clinical and laboratory study of 28 patients. *Br Med J* **1**: 258–266.

47. Huebner, RJ, Bell, JA, Rowe, WP, Ward, TG, Suskind, RG, Hartley, JW *et al.* (1955). Studies of adenoidal-pharyngeal-conjunctival vaccines in volunteers. *J Am Med Assoc* **159**: 986–989.

48. Huebner, RJ, Rowe, WP, Schatten, WE, Smith, RR and Thomas, LB (1956). Studies on the use of viruses in the treatment of carcinoma of the cervix. *Cancer* **9**: 1211–1218. |

49. Zielinski, T and Jordan, E (1969). [Remote results of clinical observation of the oncolytic action of adenoviruses on cervix cancer]. *Nowotwory* **19**: 217–221.

50. Georgiades, J, Zielinski, T, Cicholska, A and Jordan, E (1959). Research on the oncolytic effect of APC viruses in cancer of the cervix uteri; preliminary report. *Biul Inst Med Morsk Gdansk* **10**: 49–57.

51. Suskind, RG, Huebner, RJ, Rowe, WP and Love, R (1957). Viral agents oncolytic for human tumours in heterologous host; oncolytic effect of Coxsackie B viruses. *Proc Soc Exp Biol Med* **94**: 309–318. |

52. Pond, AR and Manuelidis, EE (1964). Oncolytic effect of poliomyelitis virus on human epidermoid carcinoma (Hela Tumor) heterologously transplanted to guinea pigs. *Am J Pathol* **45**: 233–249.

53. Pack, GT (1950). Note of the experimental use of rabies vaccine for melanomatosis. *Arch Dermatol* **62**: 694–695. | |

54. Milton, GW and Brown, MM (1966). The limited role of attenuated smallpox virus in the management of advanced malignant melanoma. *Aust N Z J Surg* **35**: 286–290.

55. Minton, JP (1973). Mumps virus and BCG vaccine in metastatic melanoma. *Arch Surg* **106**: 503–506.

56. Asada, T (1974). Treatment of human cancer with mumps virus. *Cancer* **34**: 1907–1928. |

57. Sato, M, Urade, M, Sakuda, M, Shirasuna, K, Yoshida, H, Maeda, N *et al.* (1979). Attenuated mumps virus therapy of carcinoma of the maxillary sinus. *Int J Oral Surg* **8**: 205–211.

58. Okuno, Y, Asada, T, Yamanishi, K, Otsuka, T, Takahashi, M, Tanioka, T *et al.* (1978). Studies on the use of mumps virus for treatment of human cancer. *Biken J* **21**: 37–49.

59. McCarthy, M, Jubelt, B, Fay, DB and Johnson, RT (1980). Comparative studies of five strains of mumps virus *in vitro* and in neonatal hamsters: evaluation of growth, cytopathogenicity, and neurovirulence. *J Med Virol* **5**: 1–15.

60. Shimizu, Y, Hasumi, K, Okudaira, Y, Yamanishi, K and Takahashi, M (1988). Immunotherapy of advanced gynecologic cancer patients utilizing mumps virus. *Cancer Detect Prev* **12**: 487–495.

61. Sanford, K, Earle, W and Likely, G (1948). The growth *in vitro* of single isolated tissue cells. *J Natl Cancer Inst* **23**: 1035–1069.

62. Newman, W and Southam, CM (1954). Virus treatment in advanced cancer; a pathological study of fifty-seven cases. *Cancer* **7**: 106–118.

63. (1957). Viruses in the treatment of cancer. *Br Med J* **2**: 1481–1482.

64. Southam, CM (1958). Homotransplantation of human cell lines. *Bull N Y Acad Med* **34**: 416–423.

65. Hammon, WM, Yohn, DS, Casto, BC and Atchison, RW (1963). Oncolytic potentials of nonhuman viruses for human cancer. I. Effects of twenty-four viruses on human cancer cell lines. *J Natl Cancer Inst* **31**: 329–345.

66. Yohn, DS, Hammon, WM, Atchison, RW and Casto, BC (1968). Oncolytic potentials of nonhuman viruses for human cancer. II. Effects of five viruses on heterotransplantable human tumors. *J Natl Cancer Inst* **41**: 523–529.

67. Molomut, N and Padnos, M (1965). Inhibition of transplantable and spontaneous murine tumours by the M-P virus. *Nature* **208**: 948–950.

68. Webb, HE, Molomut, N, Padnos, M and Wetherley-Mein, G (1975). The treatment of 18 cases of malignant disease with an arenavirus. *Clin Oncol* **1**: 157–169.

69. Nastac, E and Anagnoste, B (1963). Experimental investigations on the oncolytic action of certain viruses. *Neoplasma* **10**: 65–74.

70. Cassel, WA and Garrett, RE (1965). Newcastle disease virus as an antineoplastic agent. *Cancer* **18**: 863–868.

71. Cassel, WA and Garrett, RE (1966). Tumor immunity after viral oncolysis. *J Bacteriol* **92**: 792.

72. Murray, DR, Cassel, WA, Torbin, AH, Olkowski, ZL and Moore, ME (1977). Viral oncolysate in the management of malignant melanoma. II. Clinical studies. *Cancer* **40**: 680–686.

73. Cassel, WA and Murray, DR (1992). A ten-year follow-up on stage II malignant melanoma patients treated postsurgically with Newcastle disease virus oncolysate. *Med Oncol Tumor Pharmacother* **9**: 169–171.

74. Parrish, CR and Kawaoka, Y (2005). The origins of new pandemic viruses: the acquisition of new host ranges by canine parvovirus and influenza A viruses. *Annu Rev Microbiol* **59**: 553–586.

75. Stojdl, DF, Lichty, B, Knowles, S, Marius, R, Atkins, H, Sonenberg, N *et al.* (2000). Exploiting tumor-specific defects in the interferon pathway with a previously unknown oncolytic virus. *Nat Med* **6**: 821–825.

76. Hurst, EW (1953). Chemotherapy of virus diseases. *Br Med Bull* **9**: 180–185.

77. Hurst, EW (1957). Approaches to the chemotherapy of virus diseases. *J Pharm Pharmacol* **9**: 273–292.

78. Brownlee, KA and Hamre, D (1951). Studies on chemotherapy of vaccinia virus. I. An experimental design for testing antiviral agents. *J Bacteriol* **61**: 127–134.

79. (1958). ANTIVIRAL chemotherapy. *Br Med J* **1**: 1290–1291.

80. Speir, RW and Southam, CM (1960). Interference of Newcastle disease virus with neuropathogenicity of oncolytic viruses in mice. *Ann N Y Acad Sci* **83**: 551–563.

81. Rogers, S (1968). Use of viruses as carriers of added genetic information. *Nature* **219**: 749–751.

82. (1990). The ADA human gene therapy clinical protocol. *Hum Gene Ther* **1**: 327–362.

83. Anderson, WF (1984). Prospects for human gene therapy. *Science* **226**: 401–409.

84. Anderson, WF (1992). Human gene therapy. *Science* **256**: 808–813.

85. Moolten, FL, Wells, JM, Heyman, RA and Evans, RM (1990). Lymphoma regression induced by ganciclovir in mice bearing a herpes thymidine kinase transgene. *Hum Gene Ther* **1**: 125–134.

86. Jamieson, AT (1974). Induction of both thymidine and deoxycytidine kinase activity by herpes viruses. *J Gen Virol* **24**: 465–480.

87. Martuza, RL, Malick, A, Markert, JM, Ruffner, KL and Coen, DM (1991). Experimental therapy of human glioma by means of a genetically engineered virus mutant. *Science* **252**: 854–856.

88. Alemany, R, Balague, C and Curiel, DT (2000). Replicative adenoviruses for cancer therapy. *Nat Biotechnol* **18**: 723–727.

89. Mathis, JM, Stoff-Khalili, MA and Curiel, DT (2005). Oncolytic adenoviruses-selective retargeting to tumor cells. *Oncogene* **24**: 7775–7791.

90. Nakamura, T and Russell, SJ (2004). Oncolytic measles viruses for cancer therapy. *Expert Opin Biol Ther* **4**: 1685–1692.

91. Fielding, AK (2005). Measles as a potential oncolytic virus. *Rev Med Virol* **15**: 135–142.

92. Latchman, DS (2005). Herpes simplex virus-based vectors for the treatment of cancer and neurodegenerative disease. *Curr Opin Mol Ther* **7**: 415–418.

93. Shen, Y and Nemunaitis, J (2005). Fighting cancer with vaccinia virus: teaching new tricks to an old dog. *Mol Ther* **11**: 180–195.

94. McFadden, G (2005). Poxvirus tropism. *Nat Rev Microbiol* **3**: 201–213.

95. Shafren, DR, Au, GG, Nguyen, T, Newcombe, NG, Haley, ES, Beagley, L *et al.* (2004). Systemic therapy of malignant human melanoma tumors by a common cold-producing enterovirus, coxsackievirus a21. *Clin Cancer Res* **10**: 53–60.

96. Norman, KL and Lee, PW (2000). Reovirus as a novel oncolytic agent. *J Clin Invest* **105**: 1035–1038.

97. Sinkovics, JG and Horvath, JC (2000). Newcastle disease virus (NDV): brief history of its oncolytic strains. *J Clin Virol* **16**: 1–15.

98. Nakamura, T, Peng, KW, Vongpunsawad, S, Harvey, M, Mizuguchi, H, Hayakawa, T *et al.* (2004). Antibody-targeted cell fusion. *Nat Biotechnol* **22**: 331–336.

99. Hadac, EM, Peng, KW, Nakamura, T and Russell, SJ (2004). Reengineering paramyxovirus tropism. *Virology* **329**: 217–225.

100. Nakamura, T, Peng, KW, Harvey, M, Greiner, S, Lorimer, IA, James, CD *et al.* (2005). Rescue and propagation of fully retargeted oncolytic measles viruses. *Nat Biotechnol* **23**: 209–214.

101. Peng, KW, Facteau, S, Wegman, T, O›Kane, D and Russell, SJ (2002). Non-invasive in vivo monitoring of trackable viruses expressing soluble marker peptides. *Nat Med* **8**: 527–531.

102. Dingli, D, Peng, KW, Harvey, ME, Greipp, PR, O›Connor, MK, Cattaneo,

R *et al.* (2004). Image-guided radiovirotherapy for multiple myeloma using a recombinant measles virus expressing the thyroidal sodium iodide symporter. *Blood* **103**: 1641–1646.

103. Russell, SJ (1994). Replicating vectors for cancer therapy: a question of strategy. *Semin Cancer Biol* **5**: 437–443.

104. Russell, SJ (1994). Replicating vectors for gene therapy of cancer: risks, limitations and prospects. *Eur J Cancer* **30A**: 1165–1171.

105. Parato, KA, Senger, D, Forsyth, PA and Bell, JC (2005). Recent progress in the battle between oncolytic viruses and tumours. *Nat Rev Cancer* **5**: 965–976.

106. Campbell, SA and Gromeier, M (2005). Oncolytic viruses for cancer therapy I. Cell-external factors: virus entry and receptor interaction. *Onkologie* **28**: 144–149.

107. Campbell, SA and Gromeier, M (2005). Oncolytic viruses for cancer therapy II. Cell-internal factors for conditional growth in neoplastic cells. *Onkologie***28**: 209–215.

108. Davis, JJ and Fang, B (2005). Oncolytic virotherapy for cancer treatment: challenges and solutions. *J Gene Med* **7**: 1380–1389.

109. Aghi, M and Martuza, RL (2005). Oncolytic viral therapies-the clinical experience. *Oncogene* **24**: 7802–7816.

110. Chernajovsky, Y, Layward, L and Lemoine, N (2006). Fighting cancer with oncolytic viruses. *BMJ* **332**: 170–172.

111. Hammonds, WD and Steinhaus, JE (1993). Crawford W. Long: pioneer physician in anesthesia. *J Clin Anesth* **5**: 163–167.

112. Freireich, E (1984). Landmark perspective: nitrogen mustard therapy. *JAMA***251**: 2262–2263.

113. Li, MC, Hertz, R and Bergenstal, DM (1958). Therapy of choriocarcinoma and related trophoblastic tumors with folic acid and purine antagonists. *N Engl J Med* **259**: 66–74.

114. Frei, E 3rd, Karon, M, Levin, RH, Freireich, EJ, Taylor, RJ, Hananian, J *et al.* (1965). The effectiveness of combinations of antileukemic agents in inducing and maintaining remission in children with acute leukemia. *Blood***26**: 642–656.

115. Riedel, S (2005). Edward Jenner and the history of smallpox and vaccination. *Proc (Bayl Univ Med Cent)* **18**: 21–25.

116. Pasteur, L (1885). Methode pour prevenir la rage apres morsure. *C R Acad Sci* **101**: 765–772.

117. Landsteiner, K and Popper, E (1908). Mikroskopische Praparate

von einem menschlichen und zwei Affrenuckenmarken. *Wien Klin Wochenschr* **21**: 1830.

118. Salk, JE, Krech U, Youngner JS, Bennett BL, Lewis LJ, Bazeley PL *et al*. (1954). Formaldehyde treatment and safety testing of experimental poliomyelitis vaccines. *Am J Public Health* **44**: 563–570.

119. Baltimore, C (1970). RNA-dependent DNA polymerase in virions of RNA tumour viruses. *Nature* **226**: 1209–1222. |

120. Racaniello, V and Baltimore, D (1981). Cloned poliovirus complementar DNA is infectious in mammalian cells. *Science* **214**: 916–919.

121. Hogle, JM, Chow, M and Filman, DJ (1985). Three-dimensional structure of poliovirus at 2.9 A resolution. *Science* **229**: 1358–1365.

Chapter 11

USE OF THE Λ RED-RECOMBINEERING METHOD FOR GENETIC ENGINEERING OF PANTOEA ANANATIS

Joanna I Katashkina[1], Yoshihiko Hara[2], Lyubov I Golubeva[1], Irina G Andreeva[1], Tatiana M Kuvaeva[1] and Sergey V Mashko[1]

[1]Closed Joint-Stock Company "Ajinomoto-Genetika Research Institute"

[2]Fermentation and Biotechnology Laboratories, Ajinomoto Co, Inc

ABSTRACT

Background

Pantoea ananatis, a member of the Enterobacteriacea family, is a new and promising subject for biotechnological research. Over recent years, impressive progress in its application to L-glutamate production has been achieved. Nevertheless, genetic and biotechnological studies of Pantoea ananatis have been impeded because of the absence of genetic tools for rapid construction of direct mutations in this bacterium. The λ Red-recombineering technique previously developed in E. coli and used for gene inactivation in several other bacteria is a high-performance tool for rapid construction of precise genome modifications.

Results

In this study, the expression of λ Red genes in P. ananatis was found to be highly toxic. A screening was performed to select mutants of P. ananatis that were resistant to the toxic affects of λ Red. A mutant strain, SC17(0) was identified that grew well under conditions of simultaneous expression of λ gam, bet, and exo genes. Using this strain, procedures for fast introduction of multiple rearrangements to the Pantoea ananatis genome based on the λ Red-dependent integration of the PCR-generated DNA fragments with as short as 40 bp flanking homologies have been demonstrated.

Conclusion

The λ Red-recombineering technology was successfully used for rapid generation of chromosomal modifications in the specially selected P. ananatis recipient strain. The procedure of electro-transformation with chromosomal DNA has been developed for transfer of the marked mutation between different P. ananatis strains. Combination of these techniques with λ Int/Xis-dependent excision of selective markers significantly accelerates basic research and construction of producing strains.

BACKGROUND

Pantoea ananatis belongs to the Enterobacteriacea family. The P. ananatis strain AJ13355 (SC17) was isolated from soil in Iwata-shi (Shizuoka, Japan) as a bacterium able to grow at acidic pH and showing resistance to high concentrations of glutamic acid [1]. These physiological features made this organism an interesting object for biotechnological studies, and for this reason its genome has been sequenced by Ajinomoto Co. (unpublished results). Nevertheless, up to the recent past, the absence of efficient genetic tools has hampered manipulations of this bacterium and retarded both basic research and applied investigations.

Over the last decade [2–7], the most powerful method for generating a wide variety of DNA rearrangements in E. coli has been termed "recombineering" (recombination-mediated genetic engineering) [8]. The term generally refers to in vivo genetic engineering with DNA fragments carrying short homologies with a bacterial chromosome, using the proteins of a homologous recombination system of the bacteriophage λ (λ Red system). Lambda red operon includes only three genes encoding Exo, Beta and Gam proteins. Gam inhibits the host nucleases, RecBCD and SbcCD, thereby protecting the dsDNA substrate for recombination [9, 10]; Exo degrades linear dsDNA from each end in a 5ʹ→3ʹ direction, creating dsDNA with 3ʹ single-stranded DNA tails [11–14]; and Beta stably binds a ssDNA greater than 35 nucleotides in length and mediates pairing the one with a complementary target [15–17].

The λ Red-mediated recombineering technology developed initially for modification of the genome of Escherichia coli K12 [2, 4–7], was later broadened to other E. coli strains including enteropathogenic ones [18], and to Salmonella [19, 20], Shigella [21, 22], Yersinia [23, 24], Pseudomonas [25], as well. Probably, one of the factors impeding application of this system in other hosts is the toxicity of expression of the λ Red genes for the cells.

In the present study, to overcome this toxicity for Pantoea ananatis, the special strain resistant to simultaneous expression of the λ Red genes was selected. Using this mutant, construction of all types of chromosomal rearrangements previously obtained in E. coli by recombineering was reproduced. The approach described may be used for adjusting the technology to other hosts.

RESULTS

Construction of the New Broad-Host-Range λ Red-Expressing Plasmid

To provide regulated expression of the λ Red genes in different bacteria, the plasmid pRSFRedTER [GenBank:FJ347161] based on the broad-host-range replicon of RSF1010 [26] has been constructed (see Additional file 1). This plasmid is useful for λ Red-mediated recombineering because: 1) the replicon is stably maintained in many Gram-negative [27] and some Gram-positive bacteria [28]; 2) the λ Red genes are placed under the control of the $P_{lac\ UV5}$ promoter recognized by different bacterial RNA polymerases [29, 30]; 3) the auto-regulated element $P_{lac\ UV5}$-lac I provides IPTG-inducible expression of the λ Red genes with low basal level [31]; 4) the plasmid contains the levansucrase gene from B. subtilis allowing rapid and efficient recovery of this plasmid from the cells in a medium containing sucrose.

To test recombineering efficiency, pRSFRedTER mediated disruption of galK gene in E. coli MG1655 chromosome by integration of the PCR-generated DNA fragment carrying the Km^R gene from pUC4K flanked by $attL_{λ}$ and $attR_{λ}$ sites ($attL_{λ}$-Km^R-$attR_{λ}$) was performed. The constructed plasmid provided about 300 transformants per 10^8 survivors following electroporation. A similar frequency was obtained when pKD46 plasmid [4] was used as λ Red-expressing plasmid. In each case a chromosome structure of ten Km^R colonies was confirmed with PCR analysis.

Another plasmid carrying λ Red genes and named pRSFRedkan [GenBank:FJ347162], has been constructed via substitution of Cm^R and B. subtilis sacB genes by the Km^R gene from pUC4K [32].

Concerted Expression of λ Red Genes is Highly Toxic for the *P. ananatis* wild-type Cells

Clones of P. ananatis SC17 strain [1] obtained after electroporation by pRSFRedTER and plated on a solid LB-medium supplemented with Cm (50 μg/ml) were of very small size. Bacteria from these colonies grew very slowly in comparison with bacteria carrying pRSFsacB plasmid (see Additional file 1), which served as a vector for cloning of λ Red genes. Addition of IPTG (1 mM) for induction of expression of λ Red genes, led to complete cessation of the growth of pRSFRedTER containing cells. This effect was not detected for the cells carrying pRSFsacB and was based on the toxicity of the expression of λ Red genes.

To establish which component of the λ Red system caused the toxic effect, we constructed the pRSFGamBet [GenBank;FJ347163] plasmid lacking the exo gene encoding 5'→3' exonuclease.

It is well known that exonuclease activity is necessary for integration of dsDNAs only; integration of ssDNAs, such as chemically synthesized oligos, requires only recombinase (product of the bet gene, see [8]). To test the functional activity of the constructed pRSFGamBet, the plasmid was used to promote recombination between the artificial ss-oligos comprising two 36-nt homologies to the galK sequences and chromosome of the E. coli MG1655galK::(attL$_\lambda$- Km^R-attR$_\lambda$) strain. As a result of recombination a native structure of galK gene was restored. Between $(1.5–2.5) \times 10^4$ Gal$^+$ integrants per 10^8survivors following electroporation were obtained in three independent experiments.

When introduced into P. ananatis SC17 strain, the pRSFGamBet plasmid did not inhibit cell growth even under the induced conditions (in the presence of 1 mM IPTG). Hence, the detected toxicity of pRSFRedTER was apparently caused by exoexpression, or by simultaneous expression of all λ Red genes in P. ananatis cells.

Selection of the Recipient Strain for λ Red-Mediated Recombineering in *P. Ananatis*

A mutant P. ananatis strain, SC17(0), resistant to concerted expression of all λ Red genes, and thus manifesting the properties of a suitable recipient strain for λ Red-mediated integration of the linear DNAs into the chromosome, was obtained as follows. About 10 clones from 10^6 transformants obtained after electroporation of P. ananatis SC17 strain by pRSFRedTER, were of larger size after being plated on LB-agar with Cm. In LB-broth, bacteria from the "large" clones had a growth rate similar to the control strain with the pRSFsacB plasmid, and induction of the λ Red genes by IPTG caused only slight retardation in the growth of these cells.

Several of the selected «large» clones were cured from the plasmid on LB-agar containing sucrose, and re-transformed with pRSFRedTER. All clones grown after this re-transformation were of large size, similar to the parental clones. Three of the pRSFRedTER transformants that grew well were used as recipient strains for λ Red-mediated disruption of hisD gene. A PCR substrate, containing (attL$_\lambda$ - Km^R-attR$_\lambda$)-marker flanked by 40-bp homologous to the hisD gene, was electroporated into these strains. From 100 to 150 Km^{R}His$^-$ clones per 10^8 survivors following electroporation were obtained for each tested recipient strain. The insertion of the marker in the desired point of hisD gene was confirmed by PCR-analysis of 10 independent Km^{R}His$^-$ clones in each case. The observed integration frequency was similar to that obtained in the corresponding experiments with E. coli [4–6]. One of the plasmid-less strain used as a recipient in this experiment was named as SC17(0), and the obtained hisD strain constructed on its basis – as SC17(0)hisD::(attL$_\lambda$ - Km^R-attR$_\lambda$).

We tried to determine the nature of the mutation/mutations that provide resistance to concerted expression of λ Red genes to P. ananatis. There were no auxotrophic properties for SC17(0) strain growing on the M9 minimal media, supplemented with different carbon sources, in comparison with initial SC17 strain [1]. One of the possible explanations for the SC17(0) resistance is the reduced level of accumulation of λ Red proteins in this strain. To test the level of accumulation of λ Red proteins in SC17 and SC17(0), we performed SDS-PAGE of extracts of both strains carrying pRSFRedTER plasmid. Unfortunately, bands of λ Red proteins were not detected among the total cellular proteins even in conditions of IPTG induction for the both plasmid-carrier strains. The reduction in level of accumulation of λ Red proteins in SC17(0) strain, also, can be caused by decreased copy-number of RSF1010-replicon carrying plasmids in this strain. However, no reliable change in copy-number of pRSFsacB plasmid extracted from the cells of P. ananatis SC17 or SC17(0) strains could be experimentally found. On the other hand, the toxicity of the expression of λ Red genes has been detected for the plasmid-carrier cells of SC17 strain grown even without IPTG addition to the medium. In this case the transcription of the operon mediated by auto-regulated P$_{lac\ UV5}$-lacI genetic element has to be tightly repressed. According to Skorokhodova et al. addition of 1 mM IPTG to the culture medium provides an increase of transcriptional level up to 10–20 fold [31]. But, the transcription level of λ Red genes under such conditions is not toxic for SC17(0). Hence, it is unlikely, that the synthesis of λ Red proteins in SC17 under the repressed conditions could be significantly higher than in SC17(0) after induction.

Perhaps, mutation/mutations that are present in the genome of the SC17(0) strain does not affect most of the genes encoding factors of global cellular regulation. At least, the patterns of total cellular proteins separated by

2D-PAGE in such a fashion that about 350 of individual polypeptides could be quantitatively analyzed [33], were not different for the SC17 and SC17(0) strains (data not shown). Thus, the molecular mechanism of the resistance of the mutant SC17(0) strain to concerted expression of all λ Red genes is unknown as yet. Nevertheless, as will be shown below, various λ Red-driven modifications of bacterial chromosome could be provided on the basis of this selected strain.

Use of the Combined λ Red-Int/Xis System for Introduction of Multiple Modifications

Earlier, a λ Int/Xis-driven system for removing the markers from E. coli chromosome was constructed, similar to one developed by Peredelchuk & Bennet [34, 35]. It comprised of the plasmids carrying removable markers of Km^R or Cm^R flanked by $attL_λ$ and $attR_λ$ sites and the plasmid pMW-intxis-ts with thermo-sensitive pSC101-like replicon. This plasmid provided thermo-inducible expression of the xis-int genes under the control of λ P_R promoter regulated by the temperature sensitive λ CIts857 repressor. Even being partially induced at 37°C, this system provided a high frequency (about 30%) of marker excision in E. coli. The high frequency of marker eviction at +37°C was very important for use of the system in P. ananatis because this bacterium has a lower temperature optimum than E. coli and cannot grow at 42°C (the standard temperature for temperature sensitive λ CIts857 repressor inactivation).

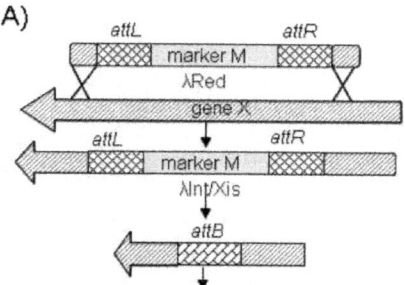

Selective marker M can be used for next type of the chromosome modification

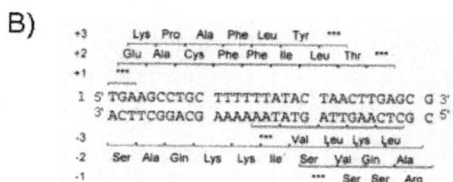

Figure 1: Scheme of a construction of multiple chromosomal modifications, using the

combined λ Red-Int/Xis system. A) Selective marker M flanked by attL$_\lambda$/attR$_\lambda$ is used for introduction of an appropriate mutation into the chromosome by λ Red-dependent recombination. Then the marker is eliminated from the chromosome by λ Int/Xis site-specific recombination. As a result, only the 31 bp long attB$_\lambda$ sequence linked to the mutation remains in the chromosome. Selective marker M can then be used in the next step of the introduction of multiple chromosomal modifications. B) The sequence of the attB$_\lambda$ site. One of the six ORFs provided by this sequence does not contain stop codons. Hence, it is possible to design an «in-frame» deletion of a gene. Asterisks mark stop codons.

The pMW-intxis-ts plasmid contains Ap^R gene as a selective marker that was not practicable for P. ananatis because of its high natural resistance to Ap. For this reason, we have substituted this marker with the Cm^R gene. The resulting plasmid (pMW-intxis-cat) was introduced to the SC17(0)hisD::(attL$_\lambda$ - Km^R-attR$_\lambda$) strain described above by electroporation. More than 30% of the transformants grown at 37°C on the plates containing LB-agar with Cm had lost the Km resistance. Loss of the Km^R cassette in the kanamycin-sensitive colonies was verified by PCR. Thus, the Km^R cassette can be used in the next step of the chromosomal modifications of this strain (Fig. 1A). Curing of the Km^S clones from the pMW-intxis-cat plasmid (with a frequency of 10%) was performed by re-streaking bacteria on LB-agar without antibiotic followed by cultivation at 37°C.

The attB$_\lambda$ site (31 bp in length), remaining in the chromosome after marker excision, contains six possible reading frames. One of these reading frames does not contain stop codons (Fig. 1B). Therefore, usage of the removable markers flanked with attL$_\lambda$ and attR$_\lambda$ sites allows design and construction of «in frame» deletions.

Using the λ Red-driven chromosomal modification followed by λ Int/Xis-mediated excision of the selective marker, it was possible to provide, step-by-step, the multiple chromosomal modifications in P. ananatis SC17(0) strain. This approach was repeatedly used for different modifications. Among them were 1) combinations of the simple or «in frame» deletions of several genes/operons; 2) integration of the marked heterologous genes into the chromosome of P. ananatis and 3) modification of the regulatory regions of the genes of interest [36]. Up to the present, several P. ananatis strains carrying more than 10 different modifications have been constructed using this strategy; the presence of multiple attB$_\lambda$ sites in their chromosomes did not hamper the repeated exploitation of this system.

Transfer of Marked Mutations by Electroporation of Chromosomal DNA

General transduction is the most efficient and popular method for transfer of mutations between different E. coli strains. Although P. ananatis and E. coli are close relatives, the known E. coli transducing phages cannot infect P. ananatis cells. Therefore, development of another method for transfer of mutations between P. ananatis strains was necessary.

The electroporation of genomic DNA has been described for the transfer of genetic markers between different backgrounds of E. coli and Pseudomonas [37, 38]. We tried to apply this technique to P. ananatis. The SC17(0)hisD::(attL$_\lambda$ -Km^R-attR$_\lambda$) strain was used as a donor of Km^R marker. Wild type strain SC17 was used as a recipient for electro-transformation of chromosomal DNA.

Previously, it was shown that special electroporation conditions are needed for the transformation of E. coli cells with large DNA molecules (see [39] for details). Different cultivation conditions of recipient strain and parameters of electroporation (electric field strength – E, time constant – τ) were tested. For P. ananatis, the highest yield of integrants (about 100 Km^{R}His$^-$ integrants per 10^8 survivors following electroporation) was obtained under the following conditions. Recipient strain was grown up to absorbance of 0.8 – 1.0. Then electro-competent cells were prepared using 10 ml of culture as described in the "Plasmid electro-transformation" section (see "Methods"). Electroporation was performed at E = 12.5 kV/cm and τ = 10 msec (resistance of 400 Ω and capacity of 25 μF). Electro-transformation with chromosomal DNA is very fast method: all procedures, including DNA isolation and electroporation, can be performed in one day.

We found that marked chromosomal modification, obtained in P. ananatis SC17(0) strain via λ Red-recombineering method, could be easily transferred into the wild type SC17 strain by electroporation of chromosomal DNA. At the time of writing up to ten different mutations had been combined in the chromosome of the SC17 strain by repeated electro-transformation with chromosomal DNA followed by λ Int/Xis-driven excision of selective marker. The frequency of marked mutation transfer varied from several up to several hundreds of integrants per trial.

Two-step λ Red-Mediated Introduction of Unmarked Mutations into *P. Ananatis* Chromosome

A two-step λ Red-mediated procedure for the introduction of unmarked mutations was elaborated for P. ananatis SC17(0). It comprises: 1) λ Red-driven insertion of dual selective/counter-selective marker into the desired point; 2) elimination of the marker via λ Red-mediated integration of the short dsDNA fragment containing the mutation of interest flanked with sites homologous to the appropriate target. Such an approach based on ET-recombination or λ Red-recombination was previously exploited for introduction of the unmarked mutations in E. coli [40, 41]. One of the most popular counter-selective markers used for this purpose is the sacB gene from B. subtilis, whose introduction imparts sucrose sensitivity to the bacterium.

To create a template for the PCR amplification of the dual selective/counter-selective marker, cat/sacB, the pRSFPlacsacB plasmid was constructed. The cassette $P_{lac\ UV5}$-sacB-cat was amplified with primers containing 36-nt homologies to the target on their 5'-ends, and integrated into Sma I recognition site located in hisD gene using SC17(0) harboring pRSFRedkan as a recipient and the cat gene in the cassette, as the selective marker. A short 170-bp long dsDNA fragment harboring an appropriate mutation and 82-bp long flanks homologous to the target region has been constructed (Fig. 2A). The desired modification of bacterial chromosome (substitution of the artificial Xho I for the native Sma I site) was finally achieved via the λ Red-mediated integration of the obtained DNA fragment accompanied by elimination of $P_{lac\ UV5}$-sacB-cat, using sacBgene for counter-selection (Fig. 2B). The integrants of interest were found with a rather high frequency (25%) among the clones grown on LB-agar containing 30% sucrose.

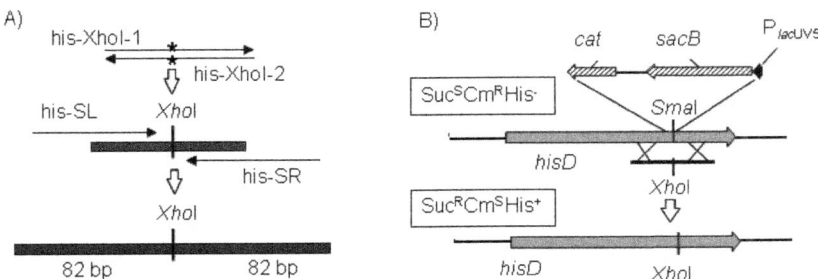

Figure 2: Construction of the unmarked nucleotide exchange in the P. ananatis hisD gene. A) Construction of a dsDNA fragment with appropriate mutation in the center. First, his-XhoI-1 and his-XhoI-2 oligos are annealed to each other. The nucleotide exchange of interest included in the sequence of his-XhoI-1/his-Xho2 olig is indicated by asterisks. The resulting dsDNA fragment is then used as DNA-template for PCR-amplification with his-SL and his-SR oligos. As a result a linear dsDNA fragment, harboring a Xho I restriction site and 82 bp long arms homologous to the target region of hisD gene, was obtained. B) The cassette containing dual selective/contra-selective marker is integrated into the target point of hisD gene. The constructed in vitro linear dsDNA fragment or ssDNA harboring appropriate mutation in the center is then integrated

into the chromosome of this strain by the λ Red recombination system. As a result, the dual selective/contra-selective marker is eliminated from the chromosome with simultaneous introduction of the desired mutation into the hisD gene. This mutation leads to substitution of the native Sma I restriction site by the Xho I restriction site and restoration of the amino-acid sequence of HisD protein. Integrants are selected as colonies resistant to sucrose. Such colonies are subsequently tested for Cm sensitivity, ability for growth without histidine and presence of Xho I restriction site in the target chromosome point.

This procedure could be applied for introduction of unmarked mutations to the genome of SC17(0) strain. Transfer of the obtained unmarked mutations to other strains by electro-transformation with chromosomal DNA is difficult because of impossibility of direct selection of integrants. Hence, to provide construction of unmarked mutations in the genome of theP. ananatis strains, sensitive to expression of all λ Red genes, we performed the method of λ Beta-driven integration of single-stranded oligos into the chromosome of SC17 strain.

λ Beta-Driven Integration of Single-Stranded Oligos

As mentioned above, ssDNAs, e.g. oligos containing short flanks homologous to the target site, can be integrated into the E. coli chromosome using only the product of the λ bet gene [8]. As the expression of gam and bet genes from pRSFGamBet plasmid was not toxic towards P. ananatis SC17 even under induced conditions, we tried to perform λ Beta-driven integration of oligos into the chromosome of this strain.

The hisD::P$_{lac\ UV5}$-sacB-cat chromosomal modification was transferred into the SC17 strain by electro-transformation with genomic DNA isolated from SC17(0)hisD::P$_{lac\ UV5}$-sacB-cat strain. The resulting strain, selected on LB medium supplemented with Cm, was named SC17hisD::P$_{lac\ UV5}$-sacB-cat. The pRSFGamBet-kan [GenBank:FJ347164] plasmid was constructed by substitution of sacB and cat genes of pRSFGamBet for the Km^R gene from pUC4K. This plasmid was introduced to the obtained SC17hisD::P$_{lac\ UV5}$-sacB-cat strain. It is known from the literature [8, 42] that the efficiency of λ Red-dependent integration of the oligos depends on a direction of replication through the recombination site. The direction of replication at the hisD locus in the P. ananatis chromosome is not known. Hence, the plasmid-carrier cells were independently electroporated with hisD-XhoI-1 and hisD-XhoI-2 ss-oligos complementary to each other. Cells were plated on LB medium containing 30% sucrose. No colonies were observed after 24-hour cultivation when hisD-XhoI-2 olig was used for electroporation. About 200 clones were obtained after electro-transformation with hisD-XhoI-1 olig. To test the phenotype of the obtained transformants, 100 clones were replicated on the following solid mediums: LB medium supplemented with 30% sucrose, LB medium supplemented with Cm and M9 minimal medium. Seven of the tested clones had $Suc^R Cm^S His^+$ phenotype. These colonies were further verified by PCR and restriction of the amplified product. The presence of the expected mutation (substitution of the Sma I by Xho I recognition site) was confirmed in all of the seven $Suc^R Cm^S His^+$ clones.

Although the frequency of selection of the desired mutation was not high, in principle, λ Beta-dependent integration of ssDNAs allowed the construction of the unmarked mutations in the P. ananatis.

Discussion

It is currently well established that high-frequency recombination between short homologies can be catalyzed in E. coli cells by the λ Red functions [4–7, 43–45]. Unfortunately, up to the present, the range of bacteria for which this system has been utilized is limited. The main goal of the present study was to widen use of the λ Red-recombineering technology to P. ananatis, a bacterium of interest in the field of metabolic engineering. A broad-host-range λ Red-expressing plasmid useful for P. ananatis was constructed. The observed general toxicity of simultaneous expression of the λ gam, bet, and exo genes for P. ananatis SC17 cells was overcome by selection of the special recipient, SC17(0).

It is known that expression of λ Red functions in E. coli cells can lead to toxic effects [46, 47]. Certainly, expression of λ Red genes has been investigated for E. coli in detail. In addition to its direct influence on traditional recombination pathways due to inhibition of RecBCD and SbcCD nucleases by λ Gam, the activity of λ Red proteins can interfere with the processes of replication and repair (see, [48, 49] for reviews). For example, prolonged expression of gam gene could lead to formation of linear multimers of high, medium and low copy-number plasmids, and, even, of minichromosomes [50–53]. Expression of exo gene in addition to gam enhances this effect [50, 52]. The plasmid linear multimers, also, may interfere with λ Red-recombination [18]. Murphy and co-workers showed that extended expression of λ Red-recombination functions could significantly induce a spontaneous mutagenesis, probably caused by interfering with mismatch repair UvrD-dependent pathway of E. coli [18].

Toxicity of expression of λ Red genes for the bacteria closely related to E. coli, but differing in the enzymes of replication, recombination and reparation, could not be predicted in advanced, as in the case of P. ananatis SC17. It is difficult, as well, to give an univocal explanation of how λ Red-mediated toxicity could be overcome. It seems that the decrease of this toxicity could be based on (i) lower intracellular level of λ Red proteins in the mutant cell, caused by reduced level in the synthesis of λ Red proteins or by increased efficiency of the specific proteolysis of λ Red proteins, (ii) decreased affinity of specific targets for interaction with λ Red proteins or increased level of biosynthesis of these targets. As for the P. ananatismutant strain obtained, both variants of the explanation may be possible. As mentioned in the Results section, it is not very probable that SC17(0) strain possesses reduced level of the synthesis of λ Red proteins. But this possibility can not be completely rejected. Nevertheless, even without information concerning the nature of the SC17(0) mutation, it is possible to use the corresponding strain for the desired λ Red-mediated rearrangements of P. ananatis chtomosome. All types of chromosome modifications constructed in E. coli by the λ Red-mediated recombineering, have been successfully reproduced in SC17(0) with pRSFRedTER as λ Red-expressing plasmid. Typically, the yield of recombinants varied from several tens to several hundreds per trial.

Using SC17(0) as the initial recipient for λ Red-promoting modifications it is subsequently possible to transfer the marked mutation to other P. ananatis strains of interest using the method of electro-transformation with chromosomal DNA. This method is a unique way to combine the set of marked mutations constructed in different P. ananatis strains into a single strain. In addition, potentially, co-transfer of rather closed mutations could be prevented by digestion of the chromosomal DNA by the appropriate restriction endonucleases, whose recognition sites are located between the mutations.

As the number of combined mutations of interest exceeds the number of available antibiotic resistance markers, curing of the intermediate strains from the used selective markers is necessary. Moreover, the presence of antibiotic resistance genes in the genomes of the industrial strains is rigorously restricted by the legislations of different countries. A wide variety of systems for marker curing based on the site-specific recombination systems (Cre/lox, Flp/FRT) are well-known [54–57]. These systems provide "symmetrical" recombination reaction between two identical sites flanking the removing marker, e.g. FRTxFRT = 2FRT. In this case, after removing the marker, the active site remains in the chromosome. Repeated action of such systems can lead to inversion or deletion of extended chromosomal fragments caused by site-specific recombination between the sites remaining at different points in the chromosome. Therefore, use of systems providing "asymmetrical" recombination reaction would be preferable. One such system is the site-specific recombination system of λ phage. It includes the Int and Xis proteins encoded by int and xis genes that, together with the host factors (IHF, RecA and Fis), provide the following reaction: $attL_\lambda \times attR_\lambda = attP_\lambda + attB_\lambda$. Peredelchuk

& Bennet [34] were the first who used this system for removal of the selective markers flanked by attL$_\lambda$ and attR$_\lambda$ sites. The attB$_\lambda$ site remaining in the chromosome after marker excision cannot recombine with attL$_\lambda$ or attR$_\lambda$ site in the next steps of strain construction. Thus, repeated action of the system would not influence the strain stability. Certainly, residual attB$_\lambda$ sites could recombine with each other via host general recombination system or provoke replication errors, especially if many left over scars (attB$_\lambda$) are presented in chromosome and their positions were rather close to each other. Such events would lead to deletions of chromosomal fragments.

We adjusted the λ Int/Xis-dependent system for use in high-efficient marker excision in P. ananatis. Note that it is also possible, in particular, to design marker-less strains, carrying "in-frame" deletions.

Finally, the two-step procedure for introducing the unmarked mutations into the P. ananatis genome was demonstrated using B. subtilis sacB gene as a counter-selective marker. Desirable mutants were achieved at the second stage via λ Red-mediated integration of the short dsDNA in SC17(0) or ssDNA in any other P. ananatis strain carrying the dual selective/counter-selective marker in the target point of the genome.

Up to the present, the developed λ Red-mediated method has been used for deletion of more than 50 P. ananatis genes whose products were involved in central metabolism, respiration, transcription regulation, etc. Several marker-less P. ananatis strains carrying multiple (> 10) different chromosomal modifications including "in-frame" deletions, point mutations, rearrangements of regulatory regions, in particular, were constructed for basic research and applied purposes using combined application of the λ Red-Int/Xis systems and electro-transformation with chromosomal DNA.

Conclusion

The λ Red-mediated recombineering has been adjusted for rapid and efficient construction of genome rearrangements inP. ananatis. In combination with the established procedures of λ Int/Xis-dependent marker elimination and electro-transformation with chromosomal DNA, this method provides a simple route to obtaining marker-less strains carrying multiple mutations of different types (deletions, substitutions of regulatory regions, integration of heterologous genes, and point mutations). The described approach of selection of the recipient strain resistant to expression of λ Red genes could be useful in exploiting λ Red-recombineering in other bacteria.

METHODS

Strains and Plasmids

Strains and plasmids used or generated in this study are listed in Table 1. A detailed description of plasmids obtained in this work is in Additional file 1.

Table 1: Strains and plasmids used or generated in this study

Name	Main characteristics/accession number	Source or reference
Strains		
Pantoea anana-tis SC17 (AJ13355)	mutant with decreased secretion of mucus	[1]
Pantoea ananatis SC17(0)	Derivative of SC17 resistant to expression of λ Red genes	This work
E. coli K12 MG1655	Wild type	VKPM
Plasmids		
RSF1010	GenBank accession number NC_001740	[26]
pUC4K	GenBank accession number X06404	[32]
pMW-attL$_\lambda$-Km^R-attR$_\lambda$	Donor attL$_\lambda$- Km^R-attR$_\lambda$ cassette; Ap^R; Km^R	[35]
pKD46	pINT-ts; λ gam, bet, and exo genes are under P$_{araB}$ promoter; Ap^R	[4]
pRS-FRedTER	λ gam, bet, and exo genes are under control of P-element; sacB gene; Cm^R	This work
pRSFRed-kan	λ gam, bet, and exo genes are under control of P-element; Km^R	This work
pRSFGam-Bet	λ gam and bet genes are under control of P-element; sacB gene; Cm^R	This work
pRSFGam-Betkan	λ gam and bet genes are under control of P-element; Km^R	This work
pRSFPlac-sacB	P$_{lacUV5}$-sacB-cat cassette; Cm^R	This work
pMW-intxis-cat	pSC101-ts; λ xis-int genes transcribed from λ P$_R$ promoter under CIts857 control; Cm^R	This work

A P-element marks the auto-regulated element P$_{lacUV5}$-lac I [see [31] for details].

Media and Growth Conditions

E. coli and P. ananatis strains were cultivated with aeration in LB medium at 37°C and 34°C, respectively. The following antibiotic concentrations were used to select transformants and to maintain the plasmids: Km – 40 mg/l, Cm – 50 mg/l. The M9 salt medium supplemented with galactose (1 g/l) or glucose (1 g/l) was used to select Gal$^+$ or His$^+$ cells.

Recombinant DNA Techniques

DNA manipulations were performed according to standard methods [58]. Restrictases were provided by "Fermentas" (Lithuania). T4-DNA ligase was from Promega (USA). All reactions were performed according to the manufacturer's instructions. PCR was carried out with Taq-polymerase ("Fermentas"). Primers were purchased from "Syntol" (Russia). All primers used in this work are listed in Additional file 2.

Construction of Integrative Cassettes

To provide cassettes for λ Red-dependent integration, the appropriate selective marker was amplified by PCR with oligos containing on their 5'-ends 36-nt sequences homologous to the target region. To disrupt E. coli galK and P. ananatis hisDgenes, a removable Km^R marker flanked by attL$_\lambda$ and attR$_\lambda$ was amplified with galK-5/galK-3 and hisD-5/hisD-3 primers, respectively. The pMW-attL$_\lambda$ - Km^R-attR$_\lambda$ plasmid was used as DNA template. To obtain insertion into the P. ananatis hisDgene, the P$_{lac}$-sacB-cat cassette was amplified in PCR with his-Plac-5/his-cat-3 primers using pRSFPlacsacB plasmid as template.

Construction of the Linear dsDNA Fragment to Exchange Native *Sma* I Recognition Site in the *P. Ananatis hisD* gene by *Xho* I Site

First, his-XhoI-1 and his-XhoI-2 oligos complementary to each other were annealed. Both oligos contained sequences corresponding to the Xho I restriction site in the center and arms homologous to the sequence surrounding native Sma I site of the P. ananatis hisD. As a result, the short dsDNA fragment containing the Xho I recognition site in its center and 33 bp long arms homologous to the target region, was obtained. The obtained fragment was amplified and extended by PCR with the primers his-SL and his-SR. The resultant DNA fragment generated by PCR contained Xho I recognition site in its center flanked with 82 bp arms homologous to the appropriate site in P. ananatis hisD gene.

Plasmid Electro-Transformation

An overnight culture of P. ananatis strain grown at 34°C with aeration was diluted with fresh LB broth 100 times and the cultivation was continued up to the OD_{600} = 0.5–0.8. Cells from ten millilitres were washed three times with an equal volume of deionized ice water followed by washing with 1 ml of 10% cold glycerol and resuspended in 35 μl of 10% cold glycerol. Just before electroporation, 10–100 ng of the plasmid DNA dissolved in 2 μl of deionized water was added to the cell suspension. The procedure of plasmid electro-transformation was performed using the GenePulser and Pulse Controller («BioRad», USA). The applied pulse parameters were: electric field strength of 20 kV/cm, time constant of 5 msec. After electroporation, 1 ml of LB medium enriched with glucose (5 g/l) was immediately added to the cell suspension. Then the cells were cultivated under aeration at 34°C for 2 h and plated on LB-agar containing the appropriate antibiotic. This was followed by an overnight incubation at 34°C. A competence of P. ananatis cells, determined for RSF1010 plasmid, was 10^6 CFU per μg of DNA. Typically, 10^5–10^6 antibiotic resistant colonies were obtained per 10^8 survivors following electroporation.

Gene Rearrangement

Overnight cultures of P. ananatis or E. coli strains harbouring the plasmid, expressing appropriate λ Red genes, grown in LB broth with Cm (for pRSFRedTER, pRSFGamBet plasmids) or Km (for pRSFRedkan, pRSFGamBetkan plasmids) were diluted 100 times with the same fresh medium supplemented with 1 mM IPTG for induction of the λ Red genes. At culture density of 0.5–0.6, electro-competent cells were prepared as described above. From 200 to 500 ng of a PCR-generated linear dsDNA or 100 ng of a ss-oligos were used for transformation. Electroporation was carried out at electric field strength of 25 kV/cm and time constant of 5 msec for both types of DNA substrates. The chromosome structure of the obtained transformants was verified in PCR with galK-t1/galK-t2 primers for E. coli galK gene disruption, hisD-t1/hisD-t2 for P. ananatis hisD gene disruption and for insertion of the double selective/contra-selective marker into the P. ananatis hisD.

Gene disruption provided with pKD46 plasmid was performed as described in (4).

P. ananatis electro-Transformation with Chromosomal DNA

Cells were grown in LB medium up to OD_{600} = 0.8–1.0. Electro-competent cells were prepared as described above. From 1 to 2 mg of a chromosomal

DNA, isolated using a Genomic DNA Isolation Kit (Sigma), was used for transformation. Electroporation was carried out with an electric field strength of 12.5 kV/cm and time constant of 10 msec.

ACKNOWLEDGEMENTS

Authors acknowledge Prof. B.L. Wanner kindly provided us with BW25113/pKD46 strain. This study was carried out at the request of and with financial support from the Ajinomoto Co.

AUTHORS' ORIGINAL SUBMITTED FILES FOR IMAGES

Below are the links to the authors' original submitted files for images.

12867_2008_402_MOESM3_ESM.pdf Authors' original file for figure 1

12867_2008_402_MOESM4_ESM.jpeg Authors' original file for figure 2

AUTHORS› CONTRIBUTIONS

JIK and YH are the project leaders in AGRI and in Ajinomoto, respectively. JIK obtained SC17(0), designed the main experiments concerned with λ Red-driven modifications, and drafted the manuscript. YH developed the transfer of λ Red-driven mutations to P. ananatis strains differed from SC17(0). LIG constructed all recombinant plasmids, expressing λ Red genes, tested their function initially in E. coli, and edited the manuscript. TMK performed the λ Red-driven modifications of the chromosome of P. ananatis SC17(0) strain. IGA performed λ Red-driven oligos integration into chromosome of P. ananatis SC17 strain. SVM supervised and coordinated the work and edited the manuscript. All authors have read and approved the final version of the manuscript.

REFERENCES

1. Izui H, Hara Y, Sato M, Akiyoshi N: Method for producing L-glutamic acid. United States Patent.

2. Murphy KC: Use of bacteriophage λ recombination functions to promote gene replacement in Escherichia coli. J Bacteriol. 1998, 180: 2063-2071.

3. Zhang Y, Buchholtz F, Muyrers JPP, Stewart AF: A new logic for DNA engineering using recombination in Escherichia coli. Nature Genetics. 1998, 20: 123-128.

4. Datsenko KA, Wanner BL: One-step inactivation of chromosomal genes in Escherichia coli K12 using PCR products. Proc Natl Acad Sci USA. 2000, 97: 6640-6645.

5. Murphy KC, Campellone KG, Poteete AR: PCR-mediated gene replacement in Escherichia coli. Gene. 2000, 246: 321-330.

6. Yu D, Ellis HM, Lee EC, Jenkins NA, Copeland NG, Court DL: An efficient recombination system for chromosome engineering inEscherichia coli. Proc Natl Acad Sci USA. 2000, 97: 5978-5983.

7. Zhang Y, Muyrers JP, Testa G, Stewart AF: DNA cloning by homologous recombination in Escherichia coli. Nat Biotechnol. 2000, 18: 1314-1317.

8. Ellis HM, Yu D, DiTizio T, Court LD: High efficiency mutagenesis, repair, and engineering of chromosomal DNA using single-stranded oligonucleotides. Proc Natl Acad Sci USA. 2001, 98: 6742-6746.

9. Karu AE, Sakaki Y, Echols H, Linn S: The gamma protein specified by bacteriophage λ. Structure and inhibitory activity for the RecBC enzyme of Escherichia coli. J Biol Chem. 1975, 250: 7377-7387.

10. Murphy KC: λ Gam protein inhibits the helicase and chi-stimulated recombination activities of Escherichia coli recBCD enzyme. J Bacteriol. 1991, 173: 5808-5821.

11. Little JW: An exonuclease induced by bacteriophage λ. II. Nature of the enzymatic reaction. J Biol Chem. 1967, 242: 679-686.

12. Carter DM, Radding CM: The role of exonuclease and β protein of phage λ in genetic recombination. II. Substrate specificity and the mode of action of lambda exonuclease. J Biol Chem. 1971, 246: 2502-2512.

13. Cassuto E, Lash T, Sriprakash KS, Radding CM: The role of exonuclease and β protein of phage λ in genetic recombination. V. Recombination of λ DNA in vitro. Proc Natl Acad Sci USA. 1971, 68: 1639-1643.

14. Hill SA, Stahl MM, Stahl FM: Single-strand DNA intermediates in phage λ>s Red recombination pathway. Proc Natl Acad Sci USA. 1997, 94: 2951-2956.

15. Muniyappa K, Radding CM: The homologous recombination system of phage λ. Pairing activities of β protein. J Biol Chem. 1986, 261: 7472-7478.

16. Mythili E, Kumar KA, Muniyappa K: Characterization of the DNA-binding domain of β protein, a component of λ Red-pathway, by UV catalyzed cross-linking. Gene. 1996, 182: 81-87.

17. Li Z, Karakousis G, Chiu SK, Reddy G, Radding CM: The β protein of phage λ promotes strand exchange. J Mol Biol. 1998, 276: 733-744.

18. Murphy KC, Campellone KG: Lambda Red-mediated recombinogenic engineering of enterohemorrhagic and enteropathogenic E. coli. BMC

Mol Biol. 2003, 4: 11-

19. Husseiny MI, Hensel M: Rapid method for the construction of Salmonella enterica serovar Typhimurium vaccine carrier strains. Infect Immun. 2005, 73 (3): 1598-1605.

20. Karlinsey JE: Lambda-Red genetic engineering in Salmonella enterica serovar Typhimurium. Methods Enzymol. 2007, 421: 199-209.

21. Shi Z-X, Wang H-L, Hu K, Feng E-L, Yao X, Su G-F, Huang P-T, Huang L-Y: Identification of alkA gene related to virulence of Shigella flexneri 2a by mutational analysis. World J Gastroenterol. 2003, 9: 2720-2725.

22. Ranallo RT, Barnoy S, Thakkar S, Urick T, Venkatesan MM: Developing live Shigella vaccines using lambda red recombineering. FEMS Immunol Med Microbiol. 2006, 47: 462-469.

23. Derbise A, Lesic B, Dacheux D, Ghigo JM, Carniel E: A rapid and simple method for inactivating chromosomal genes in Yersinia. FEMS Immunol Med Microbiol. 2003, 38: 113-116.

24. Lesic B, Bach S, Ghigo JM, Dobrindt U, Hacker J, Carniel E: Excision of the high-pathogenicity island of Yersinia pseudotuberculosisrequires the combined actions of its cognate integrase and Hef, a new recombination directionality factor. Mol Microbiol. 2004, 52: 1337-1348.

25. Lesic B, Rahme LG: Use of the lambda Red recombinase system to rapidly generate mutants in Pseudomonas aeruginosa. BMC Mol Biol. 2008, 9: 20-

26. Scholz P, Haring V, Wittmann-Liebold B, Ashman K, Bagdasarian M, Scherzinger E: Complete nucleotide sequence and gene organization of the broad-host-range plasmid RSF1010. Gene. 1989, 75: 271-288.

27. Frey J, Bagdasarian M: The molecular biology of IncQ plasmids. Promiscuous plasmids of Gram-negative bacteria. Edited by: Thomas CM. 1989, 79-94. New York: Academic Press, Inc,

28. Gormley EP, Davies J: Transfer of plasmid RSF1010 by conjugation from Escherichia coli to Streptomyces lividans and Mycabacterium smegmatis. J Bacteriol. 1991, 173: 6705-6708.

29. Brunschwig E, Darzins A: A two-component T7 system for the overexpression of genes in Pseudomonas aeruginosa. Gene. 1992, 111: 35-41.

30. Dehio M, Knorre A, Lanz C, Dehio C: Construction of versatile high-level expression vectors for Bartonella henselae and the use of green fluorescent protein as a new expression marker. Gene. 1998, 215: 223-229.

31. Skorokhodova AYu, Katashkina ZhI, Zimenkov DV, Smirnov SV, Gulevich AYu, Biriukova IV, Mashko SV: Design and study on characteristics of auto- and smoothly regulated genetic element O3/P lac UV5/O lac→lacI. Biotechnologiya (Russian). 2004, 5: 3-21.

32. Taylor LA, Rose RE: A correction in the nucleotide sequence of the Tn903 kanamycin resistance determinant in pUC4K. Nucleic Acids Res. 1988, 16: 358-

33. Rabilloud T: Proteome research: two-dimensional gel electrophoresis and identification methods. 2000, Berlin: Springer,**View Article**

34. Peredelchuk MY, Bennett GN: A method for construction of E. coli strains with multiple DNA insertion in the chromosome. Gene. 1997, 187: 231-238.

35. Minaeva NI, Gak ER, Zimenkov DV, Skorokhodova AYu, Biryukova IV, Mashko SV: Dual-In/Out strategy for genes integration into bacterial chromosome: a novel approach to step-by-step construction of plasmid-less marker-less recombinant E. coli strains with predesigned genome structure. BMC Biotechnology. 2008, 8: 63-

36. Katashkina JI, Golubeva LI, Kuvaeva TM, Gaidenko TA, Gak ER, Mashko SV: Method for constructing recombinant bacterium belonging to the genus Pantoea and method for producing L-amino acids using bacterium belonging to the genus Pantoea. Russian Federation Patent application. 2006134574.,

37. Kilbane JJ, Bielaga BA: Instantaneous gene transfer from donor to recipient microorganism via electroporation. Biotechniques. 1991, 10: 354-365.

38. Choi KH, Kumar A, Schweizer HP: A 10-min method for preparation of highly electrocompetent Pseudomonas aeruginosa cells: application for DNA fragment transfer between chromosomes and plasmid transformation. J Microbiol Methods. 2006, 64: 391-397.

39. Sheng YuL, Mancino V, Birren B: Transformation of Escherichia coli with large DNA molecules by electroporation. Nucleic Acids Res. 1995, 23: 1990-1996.

40. Muyrers JJP, Zhang Y, Benes V, Testa G, Ansorge W, Stewart AF: Point mutation of bacterial artificial chromosomes by ET recombination. EMBO reports. 2000, 1: 239-243.

41. Heermann R, Zeppenfeld T, Jung K: Simple generation of site-directed point mutations in the Escherichia coli chromosome using Red®/ET® recombination. Microbial Cell Factories. 2008, 7: 14-

42. Li X, Costantino N, Lu L, Liu D, Watt RM, Cheah KSE, Court DL, Huang J-D: Identification of factors influencing strand bias in oligonucleotide-mediated recombination in Escherichia coli. Nucleic Acids Res. 2003, 31: 6674-6687.

43. Swaminathan S, Ellis HM, Waters LS, Yu D, Lee E-C, Court DL, Sharan SK: Rapid engineering of bacterial artificial chromosomes using oligonucleotides. Genesis. 2001, 29: 14-21.

44. Datta S, Costantino N, Court DL: A set of recombineering plasmids for gram-negative bacteria. Gene. 2006, 379: 109-115.

45. Thomason LC, Costantino N, Shaw DV, Court DL: Multicopy plasmid modification with phage λ Red recombineering. Plasmid. 2007, 58: 148-158.

46. Friedman SA, Hays BH: Selective inhibition of Escherichia coli activities by plasmid-encoded GamS function of phage lambda. Gene. 1986, 43: 255-263.

47. Sergueev K, Yu D, Austin S, Cuort D: Cell toxicity caused by products of the p(L) operon of bacteriophage lambda. Gene. 2001, 227: 227-235. **View Article**

48. Court DL, Sawitzke JA, Thomason LC: Genetic engineering using homologous recombination. Annu Rev Genet. 2002, 36: 361-388.

49. Sawitzke JA, Thomason LC, Costantino N, Bubuenko M, Datta S, Court DL: Recombineering: In vivo genetic engineering in E. coli, S. enterica, and beyond. Methods Enzymol. 2007, 421: 171-199.

50. Enquist LW, Skalka A: Replication of bacteriophage λ DNA dependent on the function of host and viral genes. I. Interaction of red, gamand rec. J Mol Biol. 1973, 75: 185-212.

51. Silberstein Z, Cohen A: Synthesis of linear multimers of OriC and pBR322 derivatives in Escherichia coli K-12: role of recombination and replication functions. J Bacteriol. 1987, 169: 3131-3137.

52. Silberstein Z, Maor S, Berger I, Cohen A: Lambda Red-mediated synthesis of plasmid linear multimers in Escherichia coli K12. Mol Gen Genet. 1990, 223: 496-507.

53. Murphy K: λ Gam protein inhibits the helicase and χ-stimulated activities of Escherichia coli RecBCD enzyme. J Bacteriol. 1991, 173: 5808-5821.

54. Dale EC, Ow DW: Gene transfer with subsequent removal of the selection gene from the host genome. Proc Natl Acad Sci USA. 1991, 88: 10558-10562.

55. Posfai G, Koob M, Hradecna Z, Hasan N, Filutowich M, Szybalski W: In

vivo excision and amplification of large segments of theEscherichia coli genome. Nucleic Acids Res. 1994, 22: 2392-2398.

56. Cherepanov PP, Wackernagel W: Gene disruption in Escherighia coli : TcR and KmR cassettes with the option of FLP-catalyzed excision of the antibiotic-resistance determinant. Gene. 1995, 158: 9-14.

57. Kristensen CS, Eberl L, Sanchez-Romero JM, Givskov M, Molin S, De Lorenzo V: Site-specific deletions of chromosomally located DNA segments with the multimer resolution system of broad-host-range plasmid RP4. J Bacteriol. 1995, 177: 52-58.

58. Sambrook J, Fitsch EF, Maniatis T: Molecular Cloning: A Laboratory Manual. 1989, Cold Spring Harbor: Cold Spring Harbor Press,

59. Laemmli UK: Cleavage of structural proteins during the assembly of the head of bacteriophage T4. Nature. 1970, 227: 680-685.

Chapter 12

NATURAL GENETIC ENGINEERING: INTELLIGENCE & DESIGN IN EVOLUTION?

David W Ussery

Center for Biological Sequence Analysis, Department of Systems Biology, The Technical University of Denmark, Kgs. Lyngby

ABSTRACT

There are many things that I like about James Shapiro's new book "Evolution: A View from the 21st Century" (FT Press Science, 2011). He begins the book by saying that it is the creation of novelty, and not selection, that is important in the history of life. In the presence of heritable traits that vary, selection results in the evolution of a population towards an optimal composition of those traits. But selection can only act on changes - and where does this variation come from? Historically, the creation of novelty has been assumed to be the result of random chance or accident. And yet, organisms seem 'designed'. When one examines the data from sequenced genomes, the changes appear NOT to be random or accidental, but one observes that whole chunks of the genome come and go. These 'chunks' often contain functional units, encoding sets of genes that together can perform some specific function. Shapiro argues that what we see in genomes is 'Natural Genetic Engineering', or designed evolution: "Thinking about genomes from an informatics perspective, it is apparent that systems engineering is a better metaphor for the evolutionary process than the conventional view of evolution as a select-biased random walk through limitless space of possible DNA configurations" (page 6).

In this review, I will have a look at four topics: **1.)** why I think genomics is not the whole story; **2.)** my own perspective of *E. coli* genomics, and how I think it relates to this book; **3.)** a brief discussion on "Intelligence, Design, and Evolution"; and finally, **4.)** a section "in defense of the central dogma".

GENOMICS IS NOT ENOUGH

Merely knowing the DNA sequence of the genome does not give the full picture; knowledge of biological systems can provide a more robust explanation. The emergence of novel functions often comes from the 'retention, duplication, and diversification of evolutionary inventions' (page 133 [1]). For example, the evolution of a novel system of motility has been found in *Myxococcus xanthus*, apparently resulting from an ancient duplication and then diversification of genes originally involved in sporulation [2]. As Shapiro warns the reader, there are many parts of the book that are technical. The case is built up for an analogy between the genome and a read/write storage system, which can, in a sense, program itself. This is in contrast to the traditional view that DNA is for storage only, with occasional change through small incremental mutations. The technical details are meant to present the current views of the subject, and in some ways this section feels a bit like browsing through some of the recent publications in a journal such as Genome Research, with details about genome rearrangements and genomic islands and large regions coming and going from chromosomes. At the time of writing, I read an editorial in Science magazine (7 October, 2011), with the title "Genomics is not enough", about how "Genes and their products almost never act alone, but in networks with other genes and proteins and in context of the environment." This is what is meant by Systems Biology - the subject of evolution as addressed in Shapiro's book. Although Shapiro seems to have difficulty in describing the exact function of a gene, from a bacterial perspective, the concept of a 'gene' is both useful and easy to define - it is just a piece of DNA that encodes a functional RNA. Some of these RNAs encode proteins, others form stable RNAs, and together these products form a complex, which has a particular function. The cell can be thought of as a collection of biopolymer complexes, which can form a cognitive system. The cell can 'think' in that it processes signals from the environment and then acts on those signals, in some cases rearranging the genome to accommodate a new and different environment. This is the whole point of Shapiro's book - that the cells can 'design' their own evolution!

SYSTEMS BIOLOGY OF E. COLI

I work with bacterial genomics, so it is natural for me to think about genome evolution in terms of phages and genomic islands coming and going from a bacterial chromosome. It is a cruel world out there for the poor bacteria - there are more than 10 phages (bacterial viruses) for every one bacterial cell! Many bacterial genomic islands contain sets of genes which can be thought of as encoding a 'system' - for example, a set of proteins which together can form a type three secretion complex, allowing the bacteria to attach to a eukaryotic cell and inject a protein (such as a toxin, in a bad case) into the cell. Figure 1 shows the conservation of a reference genome (this is one of the sequenced chromosomes from an *E. coli* O104 strain) compared to other O104 genomes from the outbreak in Germany this summer (solid blue circles). Each circle represents comparison to a different genome. I have added other pathogenic E. coli genomes (red circles) and non-pathogenic, commensal *E. coli* genomes (turquoise). The three outermost purple circles represent matches to three*Haemophilus influenza* genomes, which is a distant cousin of *E. coli*. The entire reference genome is more than 5 million bp long, so this means that one pixel wide in the innermost circle represents a bit more than 2000 bp, or roughly 2 genes. The white gaps that can be seen scattered throughout are regions with tens or hundreds of genes, are present in the reference genome, but missing in the various other strains.

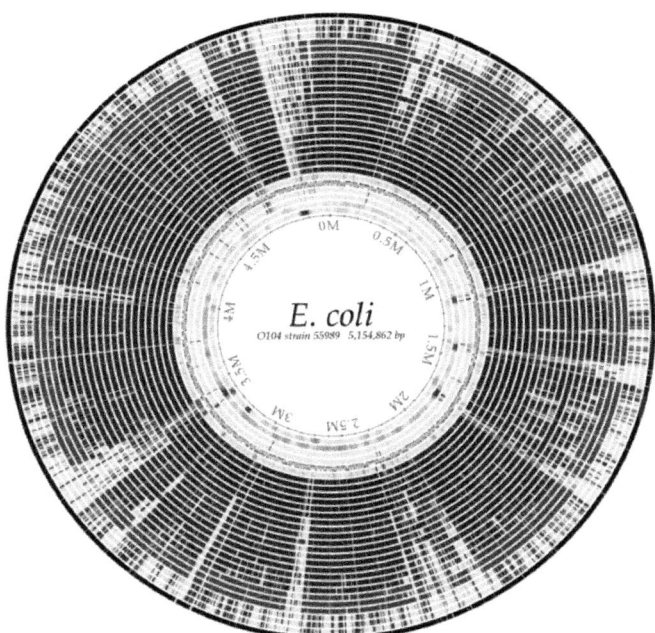

Figure 1: A BLAST atlas [8]of *Escherichia coli* O104 strain 55989, compared to five other *E. coli* O104 isolates (inner blue circles), eleven other pathogenic *E. coli* strains (middle red circles), three commensal, non-pathogenic *E. coli* strains (turquoise circles), and three *H. influenza* genomes (outer violet circles). The outermost black circle represents the reference *E. coli* 55989 compared to itself; since only the protein encoded genes are compared, the gaps shown are due to intergenic and non-protein encoding regions of the chromosome. A full legend, listing all the strains compared is available as Additional File 1.

There are large solid colored regions that are quite similar in all *E. coli* genomes. These conserved regions become smaller and thinner in the outer lanes representing three *H. influenza* genomes. This particular serotype of *E. coli* (O104) was known, but relatively obscure until a major outbreak occurred a few months ago in Germany. Historically, strains of this serotype have not been pathogenic, although they are resistant to many antibiotics. The outbreak strain has an additional virus (bacteriophage) inserted in the genome, near the top of the figure - this virus contains two Shiga toxin genes, which are a source of food-poisoning. From my perspective, this is a good example of the types of changes that Shapiro talks about - in this case the insertion of a new 'system' the phage genes which contain toxins. In experiments with *E. coli* grown continuously in culture over many years, convergent evolution can be seen in the genome (and the results are as expected, affecting DNA topology [3].

Intelligence, Design, and Evolution

Most people don't have problems with the evolution of bacteria, although in the US it seems that many have problems with the evolution of humans. Shapiro points out that there is a fear of teleology within biology, in part due to historical friction between science and religion - many scientists are simply not comfortable with the idea that an organism has a 'purpose' or is 'designed'. "A shift from thinking about gradual selection of localized random changes to sudden genome restructuring by sensor network-influenced cell systems is a major conceptual challenge.... The emphasis is systemic rather than atomistic and information-based rather than stochastic." (pages 145-146) Inspired by Jim Watson's Molecular Biology of the Gene, the title for first chapter in my textbook is "Life Obeys the Laws of Chemistry and Physics", and in my lectures for my course for the past several years, I've used Stephen Meyers' book, Signature in the Cell [4], as an example of a claim that somehow life is 'special' and cannot be explained by the traditional laws of physics and chemistry - something 'extra' is needed. My point is that Meyers is merely giving us the logical conclusion of a bad analogy [5]. If it is really true that the DNA is only a string of characters, representing some complicated computer program that exists independent of media - then who wrote the program? This analogy holds that DNA is just like a language, made up of letters, and the meaning is not dependent on the physical existence of the letters, but the more abstract ideas that are associated with a given set of letters. I tell my students that, in fact, with DNA, the sequence is important because the particular order of base sequences determines the shape of the DNA helix, and it is the shape that determines function. So in this sense, I wonder whether perhaps Shapiro pushes the analogy too far in the same direction as Meyers in attempting to relate genomic evolution with information science.

In defense of the central dogma

Yes, it is technically true that the central dogma (DNA makes RNA makes protein) cannot fully explain cellular function, but there is more to the genome than merely the DNA sequence. Shapiro gives a list of genomic functions (DNA compaction, proofreading, replication, *etc.*) that cannot be explained from the central dogma. But I think these functions can be explained from the perspective of the 'sequence hypothesis': the structures of the biopolymers (DNA, RNA, and proteins) are determined by their sequences. Thus, where/how/when a piece of DNA is compacted depends on the particular sequence of nucleotides. Similarly, making sure that mRNAs are in the right place at the right time can be encoded by leader sequences (usually 5' untranslated, 5'UTR).

Having another look at the figure, the inner circles represent DNA structural properties of the reference genome sequence, with the inner-most circle showing the AT content (darker red is AT rich, turquoise is GC rich), followed by GC skew (the bias of the G's towards the leading strand - from this it is easy to see that the origin of replication is in the top right part of the circle, about -45 degrees). The next two lanes are direct and inverted repeats (blue and red, respectively), then the location of the genes is plotted, followed by a prediction of how readily the DNA sequence will be condensed by chromatin proteins - the green regions tend not to be compacted very well, and notice that they correspond to many of the gaps in other genomes. Although for some it could be a useful analogy to think of the genome from an 'informatics' point of view, on the other hand, it is also possible to build up a solid understanding of the functions from a physical/chemical perspective as well. It is possible to have different levels of explanation for the same thing.

Finally, a point where Shapiro and I would agree is that sometimes the recruitment of a single gene, in terms of the right regulator for example, can give a bacterial population the ability to adapt to an ecological niche [6]. There is evidence this has given rise to a new 'species'; I think most would agree that selection plays an obvious role here. (Shapiro makes a somewhat strange claim that "It is important to note that selection has never led to formation of a new species, as Darwin postulated... page 121, but based on what he writes in the rest of the book, I suspect he is here thinking of the need for variations to act on - so in this sense, selection is not technically 'creating' a new species.) We have found clusters of *V. cholera* specific genes that might be responsible for adaption to a particular environment, and to me it seems clear that selection is acting at the genomic level to keep these genomic islands present within a species [7]. Thus, selection is working on natural variation - randomness is still there, in the background, but there is a level of ‹jumps› that seem to defy the old adage *Natura non facit saltum*, or 'Nature does not make leaps' - sometimes it does! But this has to be seen in the larger picture of evolved complicated systems and network engineering.

Overall this book is worth the read, although I found that it progressively began to make more sense as a whole after I'd gone through it a couple of times. In my opinion, science needs theories in order to frame and interpret what we see. Shapiro is offering here a glimpse of what the framework of evolution might look like in the near future.

ACKNOWLEDGEMENTS

I would like to thank Colleen Ussery for carefully reading through the manuscript and providing grammatical and editorial assistance.

AUTHORS' ORIGINAL SUBMITTED FILES FOR IMAGES

Below are the links to the authors' original submitted files for images.

13309_2011_10_MOESM2_ESM.pdf Authors' original file for figure 1

AUTHORS' CONTRIBUTIONS

I have read the book, outlined the paper, and written this article, and I approved the final version of the manuscript.

REFERENCES

1. Shapiro J: Evolution: A View from the 21st Century. 2011, Upper Saddle River, New Jersey: FT Press Science
2. Luciano J, Agrebi R, Le Gall AV, Wartel M, Fiegna F, Ducret A, Brochier-Armanet C, Mignot T: Emergence and Modular Evolution of a Novel Motility Machinery in Bacteria. PLoS Genet. 2011, 7 (9): e1002268-10.1371/journal.pgen.1002268.
3. Crozat E, Winkworth C, Gaff J, Hallin PF, Riley MA, Lenski RE, Schneider D: Parallel genetic and phenotypic evolution of DNA superhelicity in experimental populations of *Escherichia coli*. Mol Biol Evol. 2010, 27: 2113-2128. 10.1093/molbev/msq099.
4. Meyer SC: SIGNATURE IN THE CELL: DNA and the Evidence for Intelligent Design. 2009, New York: Harper Collins
5. Ussery DW: Review of 'Signature in the cell' - An Inordinate Fondness of Bacteria. NCSE Reports. 2009, 30 (5): 39-40.

6. Mandel MJ, Wollenberg MS, Stabb EV, Visick KL, Ruby EG: A single regulatory gene is sufficient to alter bacterial host range. Nature. 2009, 458: 215-218. 10.1038/nature07660.

7. Vesth T, Wassenaar TM, Hallin PF, Snipen L, Lagesen K, Ussery DW: On the Origins of *Vibrio* Species. Microbial Ecology. 2009, 59: 1-13.

8. Hallin PF, Binnewies TT, Ussery DW: The genome BLASTatlas - a GeneWiz extension for visualization of whole-genome homology. Molecular BioSystems. 2008, 4: 363-371. 10.1039/b717118h.

CITATION

CHAPTER 1

Yamagami T, Ishino S, Kawarabayasi Y and Ishino Y (2014) Mutant Taq DNA polymerases with improved elongation ability as a useful reagent for genetic engineering. Front. Microbiol. 5:461. doi: 10.3389/fmicb.2014.00461.

CHAPTER 2

Sonja Billerbeck and Sven Panke, A genetic replacement system for selection-based engineering of essential proteins, DOI: 10.1186/1475-2859-11-110.

CHAPTER 3

Julio Perez-Marquez, SQPrimer: The Utility of Designing Homologous Primers for the Genetic Analysis Based on the PCR, doi: 10.4172/jcsb.1000162.

CHAPTER 4

Guanghua Yang, M. Gabriela Kramer, Veronica Fernandez-Ruiz, Milosz P. Kawa, Xin Huang, Zhongmin Liu, Jesus Prieto & Cheng Qian, Development of Endothelial-Specific Single Inducible Lentiviral Vectors for Genetic Engineering of Endothelial Progenitor Cells, doi:10.1038/srep17166.

CHAPTER 5

Mihaela ŠkuljEmail author, Dejan Pezdirec, Dominik Gaser, Marko Kreft and Robert Zorec, Reduction in C-terminal amidated species of recombinant monoclonal antibodies by genetic modification of CHO cells, DOI: 10.1186/1472-6750-14-76.

CHAPTER 6

Z Jin1, S Maiti, H Huls, H Singh, S Olivares, L Mátés, Z Izsvák, Z Ivics, D A Lee, R E Champlin and L J N Cooper, The hyperactive Sleeping Beauty transposase SB100X improves the genetic modification of T cells to express a chimeric antigen receptor, doi:10.1038/gt.2011.40; published online 31 March 2011.

CHAPTER 7

Béatrice Godard1, Sandy Raeburn, Marcus Pembrey, Martin Bobrow, Peter Farndon and Ségolène Aymé, Genetic information and testing in insurance and employment: technical, social and ethical issues, doi:10.1038/sj.ejhg.5201117.

CHAPTER 8

S Le Bas-Bernardet, I Anegon1 and G Blancho, Progress and prospects: genetic engineering in xenotransplantation, doi:10.1038/gt.2008.119.

CHAPTER 9

Nester EW (2015) Agrobacterium: nature's genetic engineer. Front. Plant Sci. 5:730. doi: 10.3389/fpls.2014.00730.

CHAPTER 10

Elizabeth Kelly and Stephen J Russe, History of Oncolytic Viruses: Genesis to Genetic Engineering, doi:10.1038/sj.mt.6300108.

CHAPTER 11

Joanna I Katashkina, Yoshihiko Hara, Lyubov I GolubevaEmail author, Irina G Andreeva, Tatiana M Kuvaeva and Sergey V Mashko, Use of the λ Red-recombineering method for genetic engineering of Pantoea ananatis, DOI: 10.1186/1471-2199-10-34.

CHAPTER 12

David W Ussery, Natural genetic engineering: intelligence & design in evolution? DOI: 10.1186/2042-5783-1-11.

INDEX

A

Ab-dependent cell-mediated cytotoxicity (ADCC) 174
Acidic conditions 198
Amino acid 2, 3, 12, 13, 18
antibody-dependent cell-mediated cyto-toxicity (ADCC) 173
antigen-presenting cells (aAPC3) 104
Association of British Insurers (ABI) 130, 158

B

Bacterial genomics 280
Bioinformatics tool 53
Biological systems 280

C

cation-exchange chromatography (CEX) 84
Chemotherapy blossomed 235
chimeric antigen receptor (CAR3) 103
Chinese hamster ovary (CHO) 81, 82
colony forming units (CFU) 40
compartmentalized self-replication (CSR) 3
complement-dependent cytotoxicity (CDC) 172, 173

D

During elimination 25, 27

E

electrophoretic mobility-shift assay (EMSA) 8
Endothelial progenitor cells (EPC) 55
Endothelial progenitor cells (EPCs) 56
Essential genes 23, 24
experimental therapeutic 240
Exploiting viruses 238

G

General transduction 264
genetic discrimination 124, 132, 141, 142, 149, 166, 167, 168
Genetic information 124, 127, 128, 161, 166, 167, 168, 288
Genetic tests 126, 157
green fluorescent protein (GFP) 57

H

Hepatitis viruses 238
human leukocyte antigen (HLA) 176

I

Identical primers 51, 52